Mathematical Computing

David Betounes   Mylan Redfern

Mathematical Computing

An Introduction to Programming Using Maple®

David Betounes
Mylan Redfern
Mathematics Department
University of Southern Mississippi
Hattiesburg, MS 39406-5045
david.betounes@usm.edu
m.redfern@usm.edu

Library of Congress Cataloging-in-Publication Data
Betounes, David.
    Mathematical computing : an introduction to programming using Maple / David
Betounes, Mylan Redfern.
        p. cm.
    Includes bibliographical references and index.

Additional material to this book can be downloaded from http://extras.springer.com.

    ISBN 978-1-4612-6548-1   ISBN 978-1-4613-0067-0 (eBook)
    DOI 10.1007/978-1-4613-0067-0
    1. Computer programming. 2. Maple (Computer file) I. Redfern, Mylan. II. Title.
QA76.6 .B4775   2001
005.13'3-dc21   2001042964

Printed on acid-free paper.

Maple is a registered trademark of Waterloo Maple, Inc.

Production managed by Steven Pisano; manufacturing supervised by Jacqui Ashri.
Typeset by David Betounes from the authors' LaTeX files.

9 8 7 6 5 4 3 2 1

ISBN 978-1-4612-6548-1   SPIN 10844244

Preface

The material for this book arose from a course called Mathematical Computing that we developed and taught over the years 1995-2001 with support from several NSF programs for the improvement of undergraduate education.[1] This course has recently been instituted at our university as a junior-level mathematics course for our math and science majors.

The course was designed to teach introductory computer programming using Maple as the programming language. Specific goals of the course which grew out of the original development and which guided the writing of this book are the following:

> To teach the elements of programming and construction of algorithms within the interactive user-friendly environment of a Computer Algebra System (CAS) such as Maple;

> To provide challenging programming assignments based on concepts from calculus;

> To reinforce the concepts and techniques learned in the calculus sequence;

> To develop writing and technical documentation skills.

Traditionally our majors, and science majors in general, have been required to take two or three programming courses (FORTRAN, Pascal, C, etc.) from the computer science department as part of their major requirements. Now the Mathematical Computing course can be taken as a substitute for one of these courses. Our motivation for this change was primarily to infuse more mathematically oriented programming exercises and problems into

[1]Partial support provided by NSF ILI Grants: DUE 9452213, DUE 9751156, and CCLI Grant: DUE 9950705.

their programming experience. Traditional programming courses, because of the diversity of their student audience, do not generally contain enough, if any, material requiring a solid background in mathematics through the level of calculus.

Secondarily, we were motivated to introduce the Mathematical Computing course because of the ease and facility of learning to program that a CAS such as Maple provides. All the important basic elements of computer programming can be easily learned within this interactive environment.

A third motivation for this new course was to provide another mechanism to reinforce and make relevant the concepts and techniques that students learn in the calculus sequence. Our students have found that the blend of computing and calculus makes each seem more exciting, relevant, and applied. They were frequently forced to dig out their calculus texts to seek a formula or concept associated with an assigned programming problem.

With these motivations and goals in mind, this book was written for an introductory programming course, at the sophomore or junior level, for math, science, and secondary education majors. Generally the entire calculus sequence is required as a prerequisite, but motivated students who have not had multivariable calculus can usually manage to understand and do the problems based on functions of several variables. Indeed, this can be a strong incentive for them to study the material. No prior programming experience is required for the course, and generally no prior knowledge of Maple is required. We have provided background material on Maple in the Appendix and on the CD-ROM, and this should su ce to enable the student to learn Maple as the course progresses.

It is appropriate to stress here that this book is NOT for learning how to use Maple, but rather for learning how to write and construct computer programs. We expect that the general programming principles learned in this book will greatly help students in their other programming courses (e.g., in C, Pascal, FORTRAN) particularly in their ability to analyze and write complex algorithms.

Case studies are provided in most of the chapters for more in-depth coverage of some topics and also for illustrating the methodology of analysis and design of code for more complex problems. They also are intended to provide a format for students to follow when documenting the code they write. Groups of exercises within the exercise sets are tied to these respective case studies.

For the convenience of the students, some chapters end with a section titled Maple/Calculus Notes. These sections contain material that students could find in a Maple reference or a calculus book, but we felt it best to include the information here for ease of use, uniformity of notation, and coherence in the learning process.

In teaching the material in this book, it has been our experience that Chapters 1-5, and part of Chapter 6, can be covered in a one-semester course. There is the "exibility to omit/include certain material in both the case studies and the exercises in order to fit the pace of your course to your students. The book is designed to be progressively more challenging for students from chapter to chapter. The remaining chapters in the book, especially the chapter on programming projects, we like to cover in a sequel course that is taught at the senior level under a special projects number MAT 492 and titled Computational Projects. This enables us to discuss some of the more advanced material, like graphics programming and recursion, and to assign long-term programming projects, such as those in the Programming Projects chapter.

The CD-ROM that accompanies the book is intended as an essential resource for students and provides much supplementary and complementary material. It contains the electronic versions of all the Maple code in the text along with additional discussion of examples from the text, related material, and optional exercises pertaining to those examples. The CD also contains supplementary material on the use of Maple and various aspects of calculus.

To use the CD-ROM material, open the Maple worksheet `MCtoc.mws`, which is the table of contents, and click on the hyperlinks in this worksheet to access all the other material on the CD. There are three groups of Maple worksheets, those for Maple 7, Maple 6, and Maple V, Release 5 users. Except for differences in these three versions of Maple, the worksheets in each group are the same. *Note*: All the code displayed in the book is from the Maple 7 (equivalently, Maple 6) worksheets. This differs from the Maple V code only in the use of **end do**, **end if**, and **end proc** instead of the older **od**, **fi**, and **end**.

We should mention that the figures in this book were produced using either Corel Draw, or Maple, or a combination of both. One could embellish the Maple-produced figures by using graphics programming in Maple, without resorting to Corel Draw. However, most professional illustrators would not want to do so.

To the Student: We wrote this book to help you learn how to program computers, or to be a better programmer if you already know how. We think what you learn here will be readily adapted to programming in other languages. The book starts off very simply and gradually progresses to more complex and di cult material.

We also had in mind the purpose of having you encounter once again many of the important concepts from calculus, which you may have forgotten by now. So dig out your calculus book and be prepared to consult it!

Acknowlegments: Many people contributed to bringing this book to fruition, but Ray Seyfarth from our computer science department was particularly helpful. Over the course of several years, we consulted with Ray on a regular basis and he provided invaluable advice and suggestions on computer programming and computer science in general.

The numerous students who took the mathematical computing course while it was being developed made many useful comments and criticisms. Their enthusiasm for writing algorithms and programs which require calculus made us feel that this new type of course would be truly beneficial they themselves even thought it made what they had learned in calculus more relevant!

Finally we thank the reviewers of the original manuscript for their many helpful suggestions and criticisms. In particular, the lengthy review of the material by Robert Israel of the University of British Columbia was most appreciated.

Contents

Chapter 1

Preliminaries

This initial chapter is to help you get started using and understanding Maple and to orient you to the nature of the material in the book.

Even if you already know how to use Maple, you should read this chapter. If you do not know Maple, or have only limited experience with it, you should read the tutorials on the CD-ROM and consult the reference material in the Appendix, as needed.

In essence you will learn much about Maple as you progress through the book. However, the purpose of the book is not to teach you how to use this computer algebra system (CAS), but rather to teach you how to program, that is, to write computer algorithms and code for solving specific mathematical problems which are based on concepts and ideas from calculus.

1.1 Maple as a Programming Language

Loosely speaking, a programming language is a special vocabulary that a computer understands and this vocabulary allows the computer user to communicate with the computer. Using this vocabulary, the computer user constructs a sequence of instructions, or commands, which constitute a *program* that can be submitted to a computer for execution.

Progamming languages are traditionally divided into two classes: low-level languages and high-level languages. low-level languages, which are often referred to as machine code or assembly language, are primitive and require an intimate understanding of how the computer works as a machine. Each elementary task can require numerous machine code instructions to be written for the program. For this reason, high-level languages were developed to allow the computer user to write instructions more simply without knowing the exact details of how the computer actually functions.

1

Instructions written in high-level programming languages are translated, or *compiled*, into machine code before they are submitted to the computer for execution.

You may already know some high-level programming languages such as FORTRAN, Pascal, or C. We are going to use the computer algebra system (CAS) called Maple for our programming language.[1] A CAS is different from a high-level language in that it is actually a program written in a high-level language (like C for Maple) and is continually executing in the background while waiting for input from the user. Thus, when you submit a sequence of instructions to Maple, the instructions are translated into the corresponding C code, which is then compiled and executed. For this reason, Maple programs are neither e cient nor fast and are therefore not recommended for large scale scientific, industrial, and business uses. However, Maple is very useful for learning the basics of programming and understanding how to build complex algorithms. Other recent texts which take this approach are [GKW] and [Zac].

A major feature of Maple that makes it ideal for introductory programming is its interactive nature. By means of the worksheet interface, the user can type in a line of Maple code and execute it by pressing the enter key. The output appears immediately in the worksheet. This means that the code can be analyzed and errors corrected (debugging the code) as the user proceeds to build the total program. This feature makes programming in Maple somewhat easier than programming in, say, FORTRAN or C.

Additionally, as a computer algebra system, Maple can perform symbolic computations, such as solving algebraic equations like $ax^2+5/(x+b) = c$, can perform a multitude of standard calculus operations, such as decomposing $(2x+3)/[(x-1)^3(x^2+x+1)]$ into partial fractions, and has excellent graphing capabilities. When these aspects are combined with the challenge of devising algorithms to solve problems, a truly exciting programming environment is achieved.

1.2 Analyzing Programming Tasks

When solving a technical problem of any kind it is a good idea to consider solution strategies before actually tackling the problem. Such analysis usually increases understanding of the problem and breaks it into several

---

[1]Other CASs (computer algebra systems) in current use and comparable to Maple are MatLab, Mathematica, and MathCad. We are strong proponents of Maple since we feel that it is easier for students to learn and use, and its syntax and notation most closely match the now standard mathematical notation.

simpler problems. You will be solving problems that require the concepts you learned in calculus and you will construct the solutions in the Maple worksheet environment. The introductory examples will be very simple since you might be learning the Maple language as you go. Maple/Calculus Notes are provided at the end of the chapter to help you. Also, the Appendix gives a reference list and overview of Maple commands and syntax.

Example 1.1 (Areas under Graphs) To illustrate, in an elementary way, how to approach a programming assignment, consider the following problem.

Problem: Graph the function $f(x) = 4\cos(x^2) - x/2$ on the interval $[-2, 3]$. Find the net area bounded by the curve $y = f(x)$ and the x-axis. Split the net area into its positive and negative parts.

Analysis of the Problem:

☐ Define the function in Maple and use the `plot` command to sketch the graph.

☐ Find the net area using the `int` command to evaluate (numerically) the definite integral $\int_{-2}^{3} f(x)dx$. Note that the definite integral for this function cannot be calculated exactly using formulas from calculus.

☐ To find the positive and negative areas, it is necessary to know the intervals where the function is positive and where it is negative. This amounts to finding the x-intercepts of the graph or, in other words, to solving the equation $f(x) = 0$. This equation cannot be solved exactly by hand (or otherwise), but its solutions can be approximated using the `fsolve` command.

☐ Find the positive and negative areas using the `int` command to numerically do the integrations.

An essential part of the problem is solving the equation $f(x) = 0$, approximately, using Maple's `fsolve` command. For this, *search intervals* for the solutions must be specified and these intervals can be chosen by looking at the graph of the function. Figure 1.1 shows the approximate locations of the four intercepts, or zeros, z_1, z_2, z_3, z_4 of f. From this, four intervals, say, $[-2, -1], [1, 2], [2, 2.5], [2.5, 3]$, can be chosen so that each contains only one of the intercepts.

Writing the code for each step is straightforward and the following sequence of commands solves this problem.

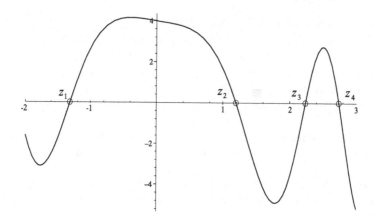

Figure 1.1: *Graph of* $f(x) = 4\cos(x^2) - x/2$ *and four of its zeros* (x-intercepts): z_1, z_2, z_3, z_4.

```
f:=x->4*cos(x^2)-x/2;
plot(f(x),x=-2..3);
netarea:=int(f(x),x=-2..3.0);
z1:=fsolve(f(x)=0,x=-2..-1);
z2:=fsolve(f(x)=0,x=1..2);
z3:=fsolve(f(x)=0,x=2..2.5);
z4:=fsolve(f(x)=0,x=2.5..3);
negarea:=int(f(x),x=-2..z1)+int(f(x),x=z2..z3)+int(f(x),x=z4..3);
posarea:=int(f(x),x=z1..z2)+int(f(x),x=z3..z4);
```

A list of instructions for performing a task is called an algorithm. An algorithm written in a computer language is called a computer program. The above sequence of commands is thus a Maple program. In languages such as Fortran and C the computer program is written, compiled, and executed before any output is given. This is not the case in Maple. In general Maple performs each command as it is given and we may use the information from a particular command to further analyze the problem on which we are working. This can be a big advantage over compiled programming languages like C. For instance, we saw in the example above that the x-ranges in the four `fsolve` commands can be chosen by looking at the graph which is produced immediately after the `plot` command is executed (see Figure 1.1). Later we will study a programming construction in Maple called a procedure, which is not so interactive and behaves like the code in a compiled language.

Another elementary programming task that can be easily accomplished in Maple is the display of secant lines described below.

Example 1.2 (Secant Lines) A primary motivation for the derivative of a function in calculus is the tangent line problem: For a function f the slope m of the tangent line to its graph at the point $(a, f(a))$ is found as the limit of a sequence of slopes of secant lines

$$m = \lim_{h \to 0} \frac{f(a+h) - f(a)}{h}.$$

This example illustrates how to produce some graphics that will geometrically enhance this definition.

Problem: For the function $f(x) = x^5 - 2x^3 - 4x^2 + x + 4$ and $a = -0.4$, plot a sequence of five secant lines that approach the tangent line at the point $(a, f(a))$. Also include plots of the function on the interval $[-1, 2]$ and the tangent line. Display all plots in the same figure.

Analysis of the Problem:

- [ ] Do an initial plot of f in order to decide on an appropriate sequence of secant lines.

- [ ] Calculate the slopes $m(h) = (f(a+h) - f(a))/h$ for the five selected values of h. Calculate the true slope $m = f'(a)$ of the tangent line.

- [ ] Plot the five secant lines with equations $y = f(a) + m(h)(x - a)$. For convenience, the result of each plot is stored in variables named p1,p2,p2,p4,p5.[2] Likewise, the plots of the graph of f and the tangent line at $(a, f(a))$ are stored in separate named variables.

After the analysis of the problem the following code is easily written. The values $h = 1.2, 0.8, 0.4, 0.2, 0.1$ are chosen for determining the five secant lines. It is convenient to name the endpoints of the interval $[-1, 2]$ by $c = -1, b = 2$. Notice that some of the commands end with a colon instead of a semicolon. The use of the colon suppresses the output from the command from being printed to the screen.

---

[2] The output from a plot command is usually displayed directly in the worksheet, either inline or in a separate plot window. When the output is stored in a variable name it is called a *plot structure* and can be manipulated in various ways.

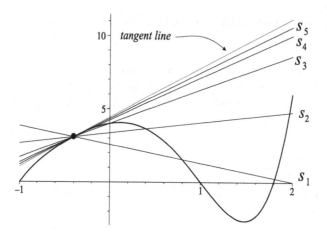

Figure 1.2: *Plots of five secant lines approaching the tangent line to the graph of the function $f(x) = x^5 - 2x^3 - 4x^2 + x + 4$ at the point $(a, f(a))$, where $a = -0.4$.*

```
with(plots,display);
f:=x->x^5-2*x^3-4*x^2+x+4;
a:=-0.4;b:=2;c:=-1;
m1:=(f(a+1.2)-f(a))/1.2:
m2:=(f(a+0.8)-f(a))/0.8:
m3:=(f(a+0.4)-f(a))/0.4:
m4:=(f(a+0.2)-f(a))/0.2:
m5:=(f(a+0.1)-f(a))/0.1:
m:=D(f)(a):
p1:=plot(f(a)+m1*(x-a),x=c..b,color=black):
p2:=plot(f(a)+m2*(x-a),x=c..b,color=black):
p3:=plot(f(a)+m3*(x-a),x=c..b,color=black):
p4:=plot(f(a)+m4*(x-a),x=c..b,color=black):
p5:=plot(f(a)+m5*(x-a),x=c..b,color=black):
tp:=plot(f(a)+m*(x-a),x=c..b,thickness=2):
p:=plot(f(x),x=c..b,color=black,thickness=3):
display({p,p1,p2,p3,p4,p5,tp});
```

Figure 1.2 shows the resulting display of all plots. Maple's `display` command, which is part of the `plots` package, must be included in the code to produce the figure. Note that the plots of the graph of f and the tangent line were done with larger line thickness than the default `thickness=1`. Also the tangent line is plotted with the default color **red**, which of course does not show in Figure 1.2. The labels in the figure are not provided by the Maple code above but were added by the typesetter.

1.3 Documentation and Coding

Analyzing a programming task before writing the program is an important skill to develop and leads to better and more e cient programming. It is also important to include comments and explanations (documentation) as part of the program. You should think of a programming assignment not only as a technical problem (writing the code to perform the task) but also as a writing assignment with the goal of communicating, to anyone who reads the program, what the program is about and how the different parts of it work.

Documentation of the code you write to solve each exercise and assigned project should include the following.

> **Description of Code:** A brief discription of what the code is supposed to do and how the algorithm and data are structured to accomplish the task. As a general rule: the longer the code needed to solve a problem, the longer and more in depth should be the documentation.

> **Variable Names:** A description of the names used in the code and what they stand for or represent mathematically. When possible the names should be chosen to suggest the quantity they represent.

Example 1.1 above, on finding net, positive, and negative areas, is really *too simple* to document, but for the sake of illustration we offer the following documentation of it.

Description of Code: The code finds the net, positive, and negative areas under the graph of the function $f(x) = 4\cos(x^2) - x/2$ on the interval $[-2, 3]$. The code is *specific* to this function and generally will have to be modified to work for other functions. The code consists of a command to define the function and a command to plot f on the interval $[-2, 3]$. The plot is needed to determine the points where the function changes from being positive to negative and vice versa. These points, the x-intercepts, are called $z1, z2, z3, z4$, and are found by using the **fsolve** command. The search intervals used in the **fsolve** command were chosen by inspection from the graph. The net area under the graph is $\int_a^b f(x)dx$, where $a = -2$ and $b = 3$ are the endpoints of the given interval. The negative area under the graph is the sum of all the definite integrals of f over those intervals where its graph is below the x-axis. Similarly the positive area involves the intervals where f is positive. Each definite integral is computed using the command int(f(x),x=a..b), for the appropriate values of a,b.

Variable Names: f is the function, while z1,z2,z3,z4 are the x intercepts (or zeros) of the function on the given interval. The net area, negative area, and positive area are named **netarea**, **negarea**, and **posarea**, respectively.

The documentation of the code and programs that you write should be done on the Maple worksheet which contains the code. Maple is also a good word processor for writing mathematical and scientific documents. Over time you will learn more and more of its word processing features (such as how to put in math symbols). For now, however, you should read the CD-ROM material on how to create and use Maple worksheets.

You will inevitably find that there are several ways to write the code to solve a particular programming task, and, while you will usually be happy to have discovered at least one way, you should nevertheless always re"ect on whether the way you chose is the clearest and most e cient. The words clearest and e cient are di cult to define and there is no universal agreement about how to achieve clarity and e ciency. Here are some rough guidelines:

> **Clarity:** Your code should not have unnecessary routines. This does not mean that you should look for the absolute minimum number of lines of code to accomplish the task, since sometimes a longer version is more natural to write and easier to understand by someone reading it. Likewise for variables. Using redundant ones, like a,b,c in a:=2;b:=a;c:=a*sin(b), can be confusing unless there is some compelling reason for having them.

> **E ciency:** Try to accomplish the programming task in a simple and direct way using some of the standard programming constructs (which you will learn in this book). For example, in the next chapter you will see that using do loops is more e cient than the code in Example 1.2 for calculating and storing the secant line data. At a broader level, e cient programs are important for conserving resources. Most, if not all, the programs in this book together with the ones you write, are not complex or long enough for us to have to worry about the length of time they take to execute or the amount of memory they consume.

In your future jobs, this may become an issue in the programming you do, and so you should begin to consider e ciency now, as you progress through this book.

Once you have your code clearly and e ciently written (and working), you might want to consider the names that you used for the variables. There are (at least) two schools of thought about how you should name your variables.

Keep It Long: Some people think that it adds clarity to the code if you name variables for what they stand for, and *not* abbreviate the name too much. This also may help you (or others) know what the variables mean a year from now if you cannot find (or worse, never wrote) the documentation for the code. For example, this school of thought might advocate that you use the name `tangentline` or `tangent_line` for a variable that holds the data for the plot of a tangent line. Using the names like `tanline`, `tline`, or TL for this variable would not be recommended. Of course, judgment must always be used, since using `least_common_denominator` as a name would be rather cumbersome and, besides, there is a generally accepted abbreviation for this, namely, `lcd` or LCD.

Keep It Short: Some people think that code is more readable if the variable names are very short, maybe even one or two characters long. After all, much of the advancement in postrenaissance mathematics has come from the clever use of notation and symbols to stand for things.[3] This school of thought recognizes that good documentation will explain exactly for what all the variables stand. Of course, when a descriptive name for a variable is also short, judgment says to use it. So, for instance, use `slope` for the slope, but maybe `maxv` or `mv` for maximum value.

Our recommendation is that you choose whatever naming style that you (or your instructor) prefer, or maybe try a combination of each. *Regardless of what style you use, you must document your code as suggested above.*

We do have a stricter recommendation for naming variables connected with calculus objects: *Try to have the variable names match, as closely as*

---

[3] Descartes and others introduced much helpful notation that has now become standard. For example the exponent notation, the use of letters a, b, c near the beginning of the alphabet to stand for known quantities, and letters x, y near the end of the alphabet for unknowns. Without this notation, renaissance mathematicians were stuck with writing the cube of the thing plus four times the thing is equal to eight, instead of $x^3 + 4x = 8$.

possible, the mathematical notation. Thus, if a function f is to be evaluated at points x_1, x_2, x_3, giving y values $y_i = f(x_i)$, $i = 1, 2, 3$, then maybe use x1,x2,x3 and y1:=f(x1);y2:=f(x2);y3:=f(x3) for the Maple names. This will make it easier to see how the notation in the statement of the calculus programming problem matches the notation in the code used to solve it.

We have one last suggestion about clarity in your code and documentation. Many of the longer programs that you will write later can be made more readable if you (a) indent the code appropriately, and (b) include comments within the code. We will discuss this in later chapters.

Ground Rules: Use only the basic Maple constructs in writing your code (unless your instructor or the problem's instructions permit otherwise). This means you should *not* use most of Maple's built-in procedures (such as max to find the maximum number and sum to sum a sequence of numbers, etc).

Exercise Set 1

When you submit your answers to the assigned exercises, be sure to follow the guidelines for documentation and good writing suggested in the reading.

1. Use Maple to define the following functions. For each function plot its graph on the specified intervals, find the net area under the graph, and split the net area into its respective positive and negative parts.

    (a) $f(x) = e^{-x} \cos 4x$, on the interval $[0,3]$.

    (b) $g(x) = x \sin 4x \cos x$, on the interval $[0,5]$.

    (c) $f(x) = e^{-x^2}(x^4 - x^2)$, on the interval $[-3, 3]$. Use symmetry to cut the work in half.

2. Let f and g be the functions

$$f(x) = x^{\sin x}, \ g(x) = x^{\cos x}$$

    defined for $x > 0$. Plot f and g in the same picture on the interval $[1,15]$. Let R be the region bounded by the curves $y = f(x), y = g(x), x = 1$, and $x = 15$. Find the area of the region R.

3. Modify the code in Example 1.2 to produce plots of five secant lines approaching the tangent line to each of the following functions f at $(a, f(a))$. Choose the secant lines appropriately. Annotate your figures (by hand after printing out) to identify which secant line corresponds to which value of h. You can use different colors in the Maple session to help with this.

    (a) $f(x) = x^5 - 2x^3 - 4x^2 + x + 4$, with $a = 0.6, c = -1, b = 2$.

    (b) $f(x) = e^{-x} \cos 4x$, with $a = 1.6, c = 0, b = 3$.

(c) $f(x) = e^{-x^4}(x^4 - x^2)$, with $a = 1.2, c = 0, b = 3$.

4. Let $f(x) = 4\sin(x^3) - x^2$. Plot the graph of f on the interval $[-2, 2]$ and use the mouse to find the approximate values of the critical points $c_i, i = 1, \ldots, 5$ of f and the corresponding local maximum and local minimum values $f(c_i), i = 1, \ldots, 5$. Record these and write them on the graph after you print out your worksheet. Use the **fsolve** command to find better approximations for the values of the critical points and max/min values. Mark these on a different plot (after printing out).

5. For a point $P = (a, f(a))$ on a the graph of f, the *normal line* at P is the line through P which is perpendicular to the tangent line at P. See Figure 1.3.

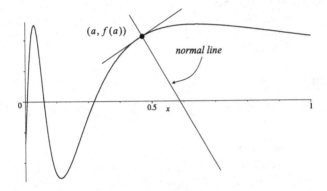

Figure 1.3: *The normal line to the graph of f at the point $P = (a, f(a))$.*

Recall that two lines are perpendicular if and only if their slopes are the negative reciprocals of each other. Since the slope of the tangent line is $m = f'(a)$, the slope of the normal line is $-1/m = -1/f'(a)$. Thus the equation for the normal line at $(a, f(a))$ is

$$y = f(a) - \frac{1}{f'(a)}(x - a).$$

In the exceptional case when $f'(a) = 0$ (i.e., the tangent line is horizontal), the normal line is vertical and has equation $x = a$.

In analogy with the sequence of secant lines, write some code to plot a sequence of five normal lines N_i to the graph of f at points $(a+h_i, f(a+h_i))$, $i = 1, \ldots, 5$, with $h_1 > \cdots > h_5 > 0$ and h_5 small. Also plot the normal line N at $(a, f(a))$. By limiting the domains/ranges in the **plot** commands, choose suitable lengths for the normal lines N_i so that (i) they all intersect the normal line N, (ii) they start at the point $(a + h_i, f(a + h_i))$ on the curve, and (iii) they are not so long as to make the graph of f too small. Plot the graph of f (on the given interval $[c, b]$) and all six normal lines in the same picture and clearly label which normal line corresponds to which value of h_i.

(a) $f(x) = \frac{1}{4}(x^5 - 2x^3 - 4x^2 + x + 4)$, with $a = 0.3, c = -1, b = 2$.

(b) $f(x) = x^2 e^{-x}$, with $a = -0.3, c = -1, b = 2$.

(c) $f(x) = \sin x - \frac{1}{2} \cos 3x$, with $a = 1, c = 0.2, b = 2$.

In theory, as the sequence of normal lines N_i approaches the normal line N at $(a, f(a))$, their points of intersection with N will approach a definite point Q in the plane and this point is the center for the circle of curvature at $(a, f(a))$.

1.4   Maple/Calculus Notes

The following notes are provided as review and reference for some of the aspects of Maple needed in this chapter.

1.4.1   Basic Syntax

The arithmetic operations in Maple are performed using the symbols in Table 1.1.

symbol	operation	example
+	addition	a+b
−	subtraction	a-b
*	multiplication	a*b
/	division	a/b
^	exponentiation	a^b

Table 1.1: *Symbols for the arithmetical operators.*

Grouping is done using parentheses. Parentheses are also used in the function notation. For example, $\sin 2x$ is written in Maple as `sin(2*x)`, while the natural exponential function e^x is written as `exp(x)`. Note that while you can write `e^x` in Maple, this will not define the natural exponential function unless `e` is assigned the exact value $e = \lim_{n \to} (1 + 1/n)^n$ 2.718.

To *define* a function in Maple we use the arrow notation: `->` in conjunction with an assignment statement. For example, consider the function

$$g(x) = \frac{2 + \cos(x)}{1 + x^2}.$$

The Maple code to define this function is

```
g:=x->(2+cos(x))/(1+x^2);
```

To plot this function on, say, the interval $[-8, 8]$, we use the basic `plot` command:

```
plot(g(x),x=-8..8);
```

1.4.2   Derivatives in Maple

To differentiate a function in Maple, which has previously been defined in your worksheet, use the `diff` command:

```
fprime:=diff(f(x),x)
```

The command `diff(f(x),x)` differentiates the expression `f(x)` with respect to `x`. It is important to note that the result, which we have called `fprime` here, is an *expression* in x not a *function* of x. While Maple will allow you to plot an expression in x, it will not allow you to use it like a function of x. Thus, for instance, evaluating `fprime` at 2 by writing `fprime(2)`, makes no sense. If you want to turn an expression in x into a function of x, use the `unapply` command. For example,

```
f1:=unapply(fprime,x)
```

will convert `fprime` to a function `f1` of x. The process of computing the derivative and converting it to a function can be done all at once by combining the two commands:

```
f1:=unapply(diff(f(x),x),x)
```

Important Tip: The easier way to differentiate a function and have the result be a function is to use the `D` operator. In essence, it combines the two commands `diff` and `unapply` into one. So you can simply write

```
f1:=D(f)
```

1.4.3   Integrals in Maple

To compute the indefinite integral $\int f(x)\,dx$ of a function f that has previously been defined in your worksheet, use the `int` command:

```
F:=int(f(x),x)
```

As with the `diff` command this makes F an *expression*, not a *function* of x. If you wish for F to be a function of x, use the `unapply` command as before:

> `F:=unapply(int(f(x),x),x)`

The definite integral $\int_a^b f(x)\,dx$ of a function f over an interval $[a, b]$ can be computed, either exactly or approximately, by using the following form of the `int` command:

> `m:=int(f(x),x=a..b)`

When computing the definite integral, if a and b are integers Maple tries all of its routines to first compute the indefinite integral exactly and if this is successful, Maple then uses this result to compute the exact value of the definite integral. If, however, the indefinite integral cannot be computed in closed form in terms of other basic functions, Maple returns the symbol for the integral (indicating that there is no exact answer). To evaluate such an integral numerically (approximately), you could use the command: `evalf(int(f(x),x=a..b))`. However, you should realize that for some functions f, the process of first trying to compute $\int f(x)\,dx$ can take a while, so if all you want is the numerical approximate value of $\int_a^b f(x)\,dx$, then it is quicker to use the following combination of commands:

> `m:=evalf(Int(f(x),x=a..b))`

Note that this uses `Int` not `int`. The command `Int` is known as the *inert* form of the `int` command, since it does not cause the integral to be evaluated.

Also, the numerical routine will be invoked if either one of the limits of integration is a "oating-point number; for instance

> `m:=int(sin(x^2),x=1..3.0);`

would yield the approximation $m := 0.4632942252$.

1.4.4 Combining Plots

To plot several functions f, g, h on the interval $[a, b]$ in the same picture you can use the `plot` command in either of the following forms:

> `plot({f(x),g(x),h(x)},x=a..b);`
> `plot([f(x),g(x),h(x)],x=a..b);`

The only difference between these is the use of set braces { } as opposed to square brackets [ ]. Each command produces the same picture   the three plots with three colors assigned to them by Maple. If you wish to assign colors yourself, say, black, red, blue to f, g, h, respectively, use the command:

```
plot([f(x),g(x),h(x)],x=a..b,color=[black,red,blue]);
```

Note the use of brackets [ ] here. This enforces an order on the functions and their respective colors. If you use set braces, { }, you cannot be sure that Maple will color f black, g red, and h blue.

On occasion the functions you want to plot in the same picture are defined on different intervals. For example, suppose T is a function defined on $[c, d]$. To plot f and T in the same picture, plot each separately and then use the **display** command to display the plots in the same picture. The **display** command is part of the plots package and must be loaded into your Maple session before you can use it.

```
with(plots):
ps1:=plot(f(x),x=a..b,color=black):
ps2:=plot(T(x),x=c..d,color=blue):
display({ps1,ps2});
```

Observe that the **display** command has a set of *plot structures* as its argument. These plot structures, in this example, were created by assigning the output of the plot commands to two variables **ps1,ps2**. *Note*: The colon is used instead of the semicolon in these assignment statements. This suppresses the output from being written into the worksheet.

Chapter 2

Basic Aspects of Maple

When learning a programming language it is necessary to first master a few basic elements of the language before even the simplest program can be written. This chapter is designed to introduce you to some key elements in Maple's programming language. Many of these will be discussed in later chapters, but for now, these elements will su ce to get you started with your initial programs.

2.1 Variables and Constants

Any computer program works with (operates on) basic information, whether this information represents numbers, names of people, or complex mathematical structures. The process of manipulating this information is facilitated by storing the information in certain physical locations in memory and, for convenience, any high-level programming language allows you to give a *name* to each piece of information. The piece of information itself is known as the *value* associated with the name (or the value of the name). For example, you may wish to store the number 2.71828 in memory and refer to this value by the name E. Pictorially you can think of a storage location in memory as a box labeled by its name and storing an assigned value for the name, for example,

$$E \quad \boxed{\qquad 2.71828 \qquad}$$

A program can then manipulate the value stored in E by making various operations on the name itself, e.g., to add the number to its square and store the resulting value in a memory location named A, might be accomplished by using $A = E + E^2$, depending on the syntax of the particular programming language.

$$E \quad \boxed{ 2.71828 }$$

$$A \quad \boxed{ 10.10732616 }$$

Similarly, when programming in Maple we need to be able to assign names to things so that we can keep track of them and perform operations on them. A **name** in Maple is usually a symbol formed from letters, digits, and underscores with uppercase and lowercase letters distinct. For example,

```
area, Area, area2, x, X, x5, net_area
```

are all legitimate names and result in seven different storage locations being allocated in memory. Naming is case sensitive in Maple.

A **name** may be assigned some fixed value throughout a Maple session or it may take on different values in the course of the session. For this reason we also refer to names as variables. As in the general discussion above, it is helpful to think of a Maple name as a label for a storage location in the computer. Some fixed value may be stored at this location or different values may be stored there at different times. Because Maple is also a symbolic manipulation language, names that have no values associated with them are treated as formal symbols. For example, if the name x has no value, then the statement y:=x+x^3 will store the formal expression $x + x^3$ in the memory location named y.

$$x \quad \boxed{ x }$$

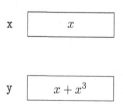

On the other hand if x is assigned the value two, x:=2, then the statement y:=x+x^3 results in y having the value 10:

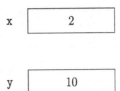

It is important to note that once a name has been assigned a value, you can not use that name as if it were a variable (i.e., as if it had no value associated with it). In the above example x has value 2 associated with it, so if later in your session you forget this and try to use x as a variable in a command such as `plot(x^2,x=0..1)`, you will get an error message: `Error, (in plot) invalid arguments`. If you wish to use x as a variable, you can get rid of the value associated with it by using the statement

`x:=evaln(x);`

The command `evaln` means evaluate to a name.

Some names always have the same value and these are called *constants*. There are different types of constants in Maple numeric constants and symbolic constants. Integers, fractions, and "oating-point numbers are numeric constants while `Pi`, which stands for , and `I`, which stands for $\overline{-1}$, are symbolic constants. We will encounter other symbolic constants as we go.

There are certain names and words, called *protected names* and *reserved words*, which cannot be used for variables in your worksheets. These are words that Maple already uses for something. For instance, `Pi` is a protected name and `if,for,do` are reserved words. There are others and if you try to assign values to these reserved words and protected names you will get error messages.

2.2 Expressions and Assignments

When we think of the word *expression* in math, we probably think of an arithmetic expression, such as

$$4^2 - 6(2+8) - 12/7,$$

or an algebraic expression such as

$$2x^2y + 3yz - 2z(x^2 + y^2).$$

What was typed	What was intended
`ans:= 4((x+7)/(6*x+7))^2`	`ans:= 4*((x+7)/(6*x+7))^2`
`A:=3*[(x+3)^2+y^2]^(1/2)`	`A:=3*((x+3)^2+y^2)^(1/2)`
`g(x):= 3*x-6*e^5`	`g:=x->3*x-6*exp(5)`
`f:=x->3*(cos x)^2`	`f:=x->3*(cos(x))^2`

Table 2.1: *Some hard to find syntax errors.*

These are also called expressions in Maple, but so is the set $\{a, b, c\}$, as well as the equation $x = 4$. In Maple, constants, variables, and objects that can be created using Maple operations are generally referred to as expressions. It is usually convenient to assign names to expressions and this is done with an assignment statement. For example,

```
p:= (x^3+1)^4;
```

assigns the polynomial $(x^3 + 1)^4$ to the variable named p. Notice that := is used for the assignment statement, *not* the symbol = . Some other assignment statements are

```
i:= 2;
R:= {2,5,6} union {-3.6,8,10};
eqn:= (x^3-9)=(x^5+2*x-6);
```

Once an expression has been assigned a name, that name can be used in place of the expression in further operations. For example, with the above assignments, the following commands can be executed:

```
S:=R intersect {5,10,20};
solve(eqn,x);
```

2.3   Notation in Mathematics and in Maple

As we see in the examples in the previous section, the notation used to communicate with Maple is very similar to mathematical notation. Because of this similarity, learning how to use Maple is made easier, but we must be careful to appreciate the differences. Often Maple will refuse to execute our commands because we have written them in correct mathematical notation but not in correct Maple notation. Observe the commands shown in the left-hand column of Table 2.1. Maple will *not* give error messages for the first three commands even though they will not produce the desired output. In fact the first one will result in $ans = 16$, which indicates that Maple interprets the command as assigning 4^2 to ans. Here 4 is acting as the constant function whose value is 4. Such errors are sometimes hard to find since they are not errors in mathematical notation, but rather errors in the Maple notation. Often debugging a program may be a matter of locating such mistakes.

However, for the most part, the notation in Maple is the same, or close to the same, as that in mathematics. This is especially important with regard to functions, since in calculus you have been warned many times to distinguish carefully between the name of a function and its value at a particular element in its domain. Thus, the mapping notation f:=x->x^2-x^4 defines a function named f in Maple (just as in mathematics) and the expression f(x) designates the value of this function at x. Similarly for functions of several variables. The arrow notation g:=(x,y)->x^2-y^2 defines a function named g and g(x,y) refers to the value of this function at the point (x,y).

Care must be taken when grouping expressions in Maple. *Only parentheses are used for grouping in Maple*, not brackets: [ ] or braces: { }. This is what the second incorrect command listed Table 2.1 illustrates. Brackets are generally used for selection of particular elements from arrays, lists, sets, and other complicated expressions. For instance, if S is a list with five elements then S[3] refers to the third element of S. Similarly if a 3×4 matrix is represented by a two-dimensional array A in Maple, then A[2,3] refers to the element in the second row and third column of A. Mathematical notation generally uses subscripts for this: A_{23}. So one use of brackets [ ] in Maple is for referring to subscripted objects. On the other hand, braces { } are used exclusively in Maple for sets. For example, S:={3,a,b,10,7/6} defines a set with five elements.

Another case where Maple notation and math notation differ is in the notation for vectors. Mathematically a vector-valued function is often defined by angled brackets, for example, $\mathbf{r}(t) = \ t, t^2, 1 - t$, or in the arrow notation $\mathbf{r} = t \cdot \ t, t^2, 1 - t$. However in Maple there is no boldface and square brackets are used: r:=t->[t,t^2,1-t].

2.4 Sequences, Lists, Sets, and Arrays

In mathematics we often have need to work with sequences and sets. A *sequence* is an ordered collection of objects, while a *set* is a collection of objects in no particular order (and no repeated objects). Sequences and sets also exist in Maple and are handled in pretty much the same way that they are in mathematics.

For example, the sequence of numbers $1, \frac{1}{3}, \frac{1}{9}, \frac{1}{27}, \frac{1}{81}, \frac{1}{243}$, can be defined and stored in Maple by using the following command:

```
b:= 1,1/3,1/9,1/27,1/81,1/243;
```

Then we can refer to this sequence by the name **b**. Maple not only stores the sequence in the name **b** but does it in such a way that we can access each element in the sequence. Explicitly, `b[1],b[2],b[3],b[4],b[5],b[6]` are the names given to the first through sixth elements in the sequence. The elements of a *sequence* in Maple may be any expressions (not just numbers), are in a specified order, and are separated by commas. Here is an example, which defines **C** as a sequence of four expressions in **x**:

```
C:=x*x,x^2,x^3,1,sqrt(x+1);
```

Note: There is a special command called **seq** in Maple that sometimes can be used to define sequences. For example the sequence that we would mathematically denote by $\{1/3^n\}_{n=0}^5$, can be defined in Maple by

```
b:=seq(1/3^n,n=0..5)
```

This will give the same sequence **b** discussed above. The general form for the **seq** command is

$$\boxed{\text{seq}(\textit{expression},I=I_1..I_2)}$$

Here *expression* is usually a Maple expression that depends on the index I and I_1, I_2 are the beginning and end values for the index.

Sequences can be used to create other data structures in Maple. Specifically, a set in Maple is a sequence enclosed in braces { }, while a list in Maple is a sequence enclosed in brackets [ ].

For example, the sequence `C:=x*x,x^2,x^3,1,sqrt(x+1)` can be turned into a set by enclosing it in braces:

```
S:={x*x,x^2,x^3,1,sqrt(x+1)};
```

Note, however, that this results in the following output

$$S := \{1, x^2, \ \overline{x+1}, x^3\}.$$

This shows that Maple recognizes x*x and x^2 as the same expression in x and since the elements in a set are not duplicated, Maple eliminates the redundant expression. Additionally, you should note that the order in which the elements are stored is not determined by the user, as the above output demonstrates. The set S could also be defined using the previously defined sequence C and the **seq** command:

```
S:={seq(C[i],i=1..5)};
```

or just

```
S:={C[1..5]};
```

Enclosing a sequence in square brackets creates a *list*. The order of the elements in a list is specified by the user and the same expression can be repeated several times. For example, from the sequence C discussed above, we can create a list:

```
L:=[seq(C[i],1=1...5)];
```

with the resulting output

$$L := [x^2, x^2, x^3, 1, \overline{x+1}],$$

showing that the order of the elements and the repetition of elements is maintained. As with sequences and sets the ith element in the list L is referred to by L[i].

The *array* data structure is also available in Maple. A one-dimensional array is *conceptually* the same as a list, however, the internal structure by which each is represented in Maple is different. This difference need not concern us now. A two dimensional array can be thought of as a table with a given number of rows and a given number of columns. A simple example is the 3 by 4 array (3 rows and 4 columns):

$$
\begin{array}{cccc}
1 & 2 & 3 & 4 \\
2 & 4 & 8 & 16 \\
3 & 9 & 27 & 81
\end{array}
$$

To store this array, named A, in Maple we could use the following code:

```
A:=array(1..3,1..4);
A[1,1] := 1;  A[1,2] := 2;  A[1,3] := 3;  A[1,4]:=4;
A[2,1] := 2;  A[2,2] := 4;  A[2,3] := 8;  A[2,4]:=16;
A[3,1] := 3;  A[3,2] := 9;  A[3,3] := 27;  A[3,4]:=81;
```

The following single command can also be used to create the array A and include its entries at the same time.

```
A:=array([[1,2,3,4],[2,4,8,16],[3,9,27,81]]);
```

Arrays can be especially useful when keeping track of several groups of information. In particular, the elements in an array need not be numbers and need not be all of the same type. The following example demonstrates this for a typical problem in calculus.

Example 2.1 (Tangent Lines) Suppose we want to graph the function $f(x) = 4e^{-x/3}\sin 3x$ on the interval $[0,6]$ and the tangent lines to the graph for the x-values $.5, 1.5, 2.5, 3.5, 4.5, 5.5$. Let us label these $x_1, x_2, x_3, x_4, x_5, x_6$. Since we need the equations for the 6 tangent lines for each x_i, we will calculate the y coordinate, $f(x_i)$, and the slope of the tangent line at a point $(x_i, f(x_i))$, namely $f'(x_i)$. Then the equation of the tangent line at the point $(x_i, f(x_i))$ is

$$y = f(x_i) + f'(x_i)(x - x_i).$$

Now we have 6 values of x and for each of these we need two pieces of information. Visualizing this data in a table with 6 rows and 3 columns, as shown in Table 2.2, suggests that we use a two-dimensional array, with dimensions 6 by 3, to store the data in computer memory.

x_i	$f(x_i)$	tangent line
0.5	3.38	$y = 3.38 - .41(x - 0.5)$
1.5	-2.37	$y = -2.37 - .74(x - 1.5)$
2.5	1.63	$y = 1.63 + 1.26(x - 2.5)$
3.5	-1.10	$y = -1.10 - 1.41(x - 3.5)$
4.5	.72	$y = .72 + 1.35(x - 4.5)$
5.5	$-.46$	$y = -.46 - 1.20(x - 5.5)$

Table 2.2: *Equations for the six tangent lines.*

To set up this array, which we will name G, we first declare it with the command

```
G:=array(1..6,1..3);
```

Note that the ranges `1..6,1..3` indicate that the array is to consist of 6 rows and 3 columns. The entry in the ith row and jth column of the table is referred to by `G[i,j]`. To assign all the entries their respective values we use the following code.

```
f:=x->4*exp(-x/3)*sin(3*x);
fprime:=unapply(diff(f(x),x),x);
G[1,1]:=0.5;
G[1,2]:=f(0.5);
G[1,3]:=f(0.5)+fprime(0.5)*(x-0.5);
                .

                .

                .
G[6,1]:=5.5;
G[6,2]:=f(5.5);
G[6,3]:= f(5.5)+fprime(5.5)*(x-5.5);
```

This inputs the information from Table 2.2 into the array G. In particular, the name G[6,3] is assigned the y value in the tangent line equation at the sixth point $(5.5, f(5.5))$. Note that the ellipsis $\cdots$ in the above code indicates to repeat the pattern. You cannot use an ellipsis in Maple, but rather, you will have to type in the 12 lines of code that we have glossed over by using the ellipsis above. In the next section we will learn a method for assigning the information from Table 2.2 to the array other than laboriously typing it in piece by piece as indicated above.

2.5 The Do Loop

The do loop in any programming language is a basic feature which allows one to repeat a certain action a number of times, say, N times, while an index, or counter, i varies from 1 to N. In Chapter 3 we will discuss this feature in general, but here we present a basic form of the do loop in Maple, so that you can begin using this indispensable element in your programming tasks.

A simple example illustrating the need for a repetition statement is the following.

Example 2.2 Recall that the second derivative of a function f is the derivative of the first derivative of f:

$$f'' = (f')' = D(Df) = \frac{d}{dx}\ \frac{df}{dx}\ \ .$$

Similarly the third derivative of f is the derivative of the second derivative of f.

$$f''' = (f'')' = D(D(D(f))) = \frac{d}{dx}\frac{d}{dx}\ \frac{df}{dx}\ \ .$$

Generally the nth derivative of f, denoted by $f^{(n)}$ or $D^n f$, comes from repeatedly differentiating f a total of n times.

If we assume that f has been previously defined in a Maple worksheet, then the following code will calculate the first, second, and third derivatives of f.

```
F1:=D(f); F2:=D(F1); F3:=D(F2);
```

But clearly this type of thing would be impractical if you wanted to calculate a large number of the higher derivatives of f. Hence, we need a do loop to automate the general task.

To do the calculation of $Df, D^2 f, D^3 f, \ldots, D^n f$, in Maple and store the results in an array named F, we can use the following commands:

```
F:=array(0..n);
F[0]:=f;
for i from 1 to n do
F[i]:=D(F[i-1])
end do;
```

The code assumes that f has been previously defined (as a function, not an expression in x) and that n has been assigned an integer value larger than 0. Note that arrays in Maple can have indices other than positive integers (see Chapter 6).

The repetition statement we used above has the form:

```
for I from I₁ to I₂ do statements end do
```

There is a more general form of this which is discussed in Chapter 3. Here I stands for the name of the varying index, I_1 is the starting value for the index, and I_2 is the end value for the index. The designation *statements* stands for any sequence of valid Maple statements, separated by semicolons. Note that these statements must be bracketed by the reserved words do and end do.

In the next example, we return to Example 2.1 in the last section, which illustrates the need for a do loop to simplify repetitive tasks.

Example 2.3 (Tangent Lines Revisited) The problem of graphing the function $f(x) = 4e^{-x/3} \sin 3x$ along with the tangent lines at six given points was discussed in Example 2.1 and Table 2.2 shows the resulting data. Using

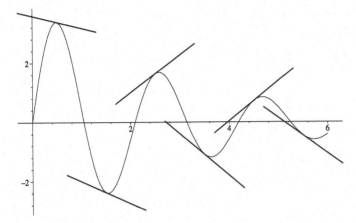

Figure 2.1: *Graph of $f(x) = 4e^{-x/3}\sin 3x$ with tangent lines.*

a do loop to store the information from the table in a 2-dimensional array G saves us the time of tediously typing it in. Also, storing the graphs in a 1-dimensional array P facilitates using the `display` command to plot all the graphs in the same figure.

```
f:=x->4*exp(-x/3)*sin(3*x);
G:=array(1..6,1..3);P:=array(0..6);
fprime:=D(f);
for i from 1 to 6 do
  G[i,1]:= 0.5+(i-1);
  G[i,2]:= f(G[i,1]);
  G[i,3]:= G[i,2]+fprime(G[i,1])*(x-G[i,1]);
  P[i]:=plot(G[i,3],x=G[i,1]-0.8..G[i,1]+0.8,thickness=2):
end do;
P[0]:=plot(f(x),x=0..6,color=black):
with(plots):
display({seq(P[i],i=0..6)});
```

When the above code is executed we get the graph shown in Figure 2.1.

Thus, you can readily see how the do loop would be extremely useful if you wanted to plot the tangent lines at 100 points.

The use of arrays to store results often simplifies and clarifies a programming task, as the array G does here. It also enables the results to be used for other purposes. Thus, we can print a table of the data stored in G by using the following do loop

```
for i from 1 to 6 do  G[i,1],G[i,2],G[i,3] end do;
```

which produces output like

.5, 3.377445107, 3.581086134 - .4072820542 x

1.5, -2.371607949, -1.256037847 - .7437134011 x

2.5, 1.630612438, -1.529956507 + 1.264227578 x

3.5, -1.095760382, 3.845369266 - 1.411751328 x

4.5, .7173941912, -5.374730636 + 1.353805517 x

5.5, -.4552002392, 6.122004159 - 1.195855345 x

Alternatively, since G is a two-dimensional array, we can simply evaluate it

```
eval(G);
```

to produce the table:

$$
\begin{array}{lll}
.5 & 3.377445107 & 3.581086134 - .4072820542x \\
1.5 & -2.371607949 & -1.256037847 - .7437134011x \\
2.5 & 1.630612438 & -1.529956507 + 1.264227578x \\
3.5 & -1.095760382 & 3.845369266 - 1.411751328x \\
4.5 & .7173941912 & -5.374730636 + 1.353805517x \\
5.5 & -.4552002392 & 6.122004159 - 1.195855345x
\end{array}
$$

Note that just using the command G; will not produce the above output. This is because G is an array data structure and arrays are not evaluated fully by the command G;. See Section 2.7 on evaluation rules.

If only the graphical display in Figure 2.1 is required, then it is possible to accomplish the task without the use of the array G (exercise). This is often desirable in programming tasks where the use of arrays would consume quite a bit of memory and so avoiding their use will make the program more e cient. However, the elimination of the use of arrays should always be weighed against the sacrifice in clarity that might result from this.

2.6 Procedures: A First Glance

A fundamental feature of any programming language is the ability to use a given algorithm for doing a certain task when writing the code to do something else. One could just repeat the code for the given algorithm within the new code being written, but that would be time consuming and could make the new code hard to understand, especially if the given algorithm is long or if it is needed in several places in the new code. To enable the code for a given algorithm to be used over and over again in new programming tasks, all programming languages have some facility that allows the code for the algorithm to be stored as a subprogram, subroutine, or procedure and be called, or referenced, by any new program you write.

This fundamental facility in Maple is the *procedure*. You can make a procedure out of any sequence of Maple commands by enclosing them between a *procedure declaration* and an *end procedure* command. Here is an elementary example.

Example 2.4 (Products) While Maple has a built-in procedure, called product, for finding products of collections of numbers, it is instructive to write the code for such an algorithm from scratch. So here is a sample procedure to find the product of all the numbers in the array A=array(1..N) of numbers. The strategy for calculating the product is to start by letting $p = 1$, and then successively compute a new value of p from the previous one.

$$
\begin{aligned}
p &= pA_1 = A_1 \\
p &= pA_2 = A_1 A_2 \\
p &= pA_3 = A_1 A_2 A_3 \\
&\vdots \\
p &= pA_{N-1} = A_1 A_2 A_3 \cdots A_N.
\end{aligned}
$$

This is easily accomplished by the following procedure:

```
PR:=proc(A,N)
   p:=1;
   for i from 1 to N do p:=p*A[i] end do:
end proc;
```

The procedure is named PR and the code above only *defines* the procedure. No products are taken until you *use* the procedure. The reserved words proc and end proc are used to indicate the beginning and end of the statements comprising the procedure. The input to the procedure is the array A and the number N indicating the number of elements in the array. Once this procedure is defined in the Maple session, it can be used any time thereafter (but not before). For example, we could now use the commands

```
B:=array([2,-5,7,16]);
c:=PR(B,4);
```

to compute the product $c = 2(-5)(7)(16) = -1120$. Notice that the first command creates the array B by listing its entries. The procedure will work for lists as well as one-dimensional arrays. Thus,

```
L:=[2,-5,7,16];
PR(L,3);
```

will define a list L and compute the product of the first three elements in L, giving -70 as the result.

The general form for the procedure statement in Maple is the following:

---

proc(*parameters*) *statements* end proc

Here *parameters* is a sequence $p1, p2, \ldots$, of names. These are often called *formal parameters* since these names are used in writing the procedure. When the procedure is actually used (or invoked) the user will enter names of actual variables for each of the parameters, so the parameters become *actual parameters*. Some of the parameters are used to pass information to the procedure, while other parameters can be used to return information from the procedure, like the value of a calculation. The latter type of parameter must evaluate to a name, i.e., have no value assigned to it, at the time you invoke the procedure. This only makes sense because the procedure will assign values to these parameters. *Note:* Generally the result of the last command in the procedure will be returned as output from the procedure. In the product procedure PR discussed above there are no output parameters and the last command executed in the procedure is `p:=p*A[N]`, which results in the total product being returned.

The body of the procedure consists of a sequence of Maple commands separated by semicolons or colons. Calling or invoking the procedure will cause this sequence of commands to be executed.

Additional examples will help clarify what procedures are and how to write elementary ones. See Chapter 5 for more details on procedures.

Example 2.5 (Binomial Expansion) Consider the problem of finding the expansion of the binomial $(a + b)^n$, where n is a positive integer. The Binomial Theorem gives us the formula for the expansion:

$$(a+b)^n = \sum_{k=0}^{n} \frac{n!}{k!(n-k)!} a^{n-k} b^k. \qquad (2.1)$$

It is easy to write a procedure which has as input values for a, b, n and computes the expansion of $(a + b)^n$ shown on the right side of the above equation. The following code accomplishes this.

```
binom:=proc(a,b,n,result::evaln)
  s:=0;
  for k from 0 to n do
      s:=s+(n!/(k!*(n-k)!))*a^(n-k)*b^k;
  end do;
  result:=s;
  RETURN()
end proc:
```

The name of the above procedure is **binom** and there are three formal input parameters **a,b,n**. The output parameter is **result**, and since it must not have any value associated with it when the procedure is used, we have included the declaration **::evaln** in the parameter list. This forces Maple to evaluate it as a name (i.e., no value is assigned to it).

Suppose now we wish to use the procedure to expand $(x - 2)^5$ and store the expansion in a variable named **answer**. Then we replace the formal parameters **a,b,n,result** with the actual parameters **x,-2,5,answer** and execute the command **binom(x,-2,5,answer)**. There is no output written to the screen when this is done. This is because the last statement in the procedure is a **RETURN** statement (actually an empty return). Without this statement, the last command in the procedure would be **result:=s** and so when invoked, the procedure would not only assign the binomial expansion to the actual parameter **answer**, but also would write the expansion on the screen. Since we did not want to see the latter, we used the empty return. To see the value stored in **ans** we could (as usual) simply use the command

```
answer;
```

to get the following output written on the screen

$$x^5 - 10x^4 + 40x^3 - 80x^2 + 80x - 32.$$

To embellish the output we could make the command in the form

```
(x-2)^5=answer;
```

to get the following result:

$$(x - 2)^5 = x^5 - 10x^4 + 40x^3 - 80x^2 + 80x - 32.$$

Note the use of = (as opposed to :=) in the above command. The *equal* sign is used in Maple for equations. An *equation* is a Maple expression, not an assignment statement. Other calls to this procedure

```
binom(1,(cos(x))^2,3,answer);
(1+(cos(x))^2)^3 = answer;
binom(2*x,-1,4,answer);
(2*x-1)^4 = answer;
```

give us the following output written to the screen:

$$(1 + \cos^2 x)^3 = 1 + 3\cos^2 x + 3\cos^4 x + \cos^6 x$$

$$(2x - 1)^4 = 16x^4 - 32x^3 + 24x^2 - 8x + 1$$

2.7 Evaluation Rules

The rules whereby Maple determines the values of variables during an interactive session are very simple *with the exception of the* array, table, *and* matrix *data structures, all expressions are evaluated fully, down to their lowest levels*. What this means is explained as follows.

If we make the assignments a:=b; b:=c; c:=d; d:=16;, then four storage locations are created for the variables a,b,c,d and the assigned values are stored in the respective variables. The variable a has value b at the first level. However, evaluated to the second level a has value c, and to the third level it has value d. When fully evaluated (to the fourth level), a has value 16. Thus, now when you make a command like e:=a+4, Maple will trace through all the previous assignments to determine the value of a at the lowest level, which is 16, and then assign e the value of 20.

Here's another example to help you understand full evaluation. Suppose we make the following assignment statements (in the indicated order).

```
a:=x+y;
x:=5;
y:=6;
z:=a^2;
```

Then the following diagram shows schematically the values stored in the four variables.

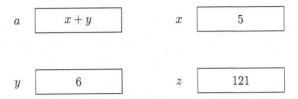

This diagram represents the values stored at the *first level*. Thus, when a is initially assigned the value which is the expression x+y, this is its value at the first level. When x and y are next given the values of 5 and 6, Maple does not re-evaluate a to have value 11. No action is taken until a is referred to. When z is assigned the value of a^2, the expression a^2 is evaluated fully, meaning that a is evaluated to 11 down at its second level, and this is squared, resulting in the value of 121 being assigned to z.

During an interactive session with Maple, you will generally not have to worry about evaluation and levels of evaluation, since Maple evaluates everything fully to the lowest level. This is what you would want and expect. There are, however, several exceptions to this. Arrays, tables, and matrices are *not* fully evaluated (for technical reasons) and the way variables are evaluated when they are part of a procedure is different as well. These aspects of evaluation will be discussed in later chapters.

Exercise Set 2

1. In each of the following write a do loop to store the sequence in a 1-dimensional array.

   (a) $1, 3, 9, 27, 81, 243$.
   (b) $1, -1, 1, -1, 1, -1, 1, -1, 1, -1, 1, -1$.
   (c) $1, 4, 7, 10, 13, 16, 19$.

2. Write a pair of nested do loops to store the array

$$\begin{array}{cccc} 1 & 1 & 1 & 1 \\ 2 & 4 & 8 & 16 \\ 3 & 9 & 27 & 81 \end{array} \ .$$

3. Consider the function
$$f(x) = x^{\sin x},$$

   which is defined for all $x > 0$. Use the **fsolve** command to find the approximate values for the in"ection points $Q_i = (p_i, f(p_i)), i = 1, \ldots, 6$ of f on the interval $[0.1, 16]$. Plot, in the same picture, the function f and the tangent lines $T_i, i = 1, \ldots, 6$ to f at the six in"ection points. The ith tangent line should be plotted just on the interval $[p_i - 1, p_i + 1]$. In the picture, after it is printed out, mark the coordinates of each in"ection point.

4. For each of the following functions use an array and a do loop to compute the first five derivatives $F_i = f^{(i)}$, $i = 0, 1, \ldots, 5$. Graph all the resulting functions in the same figure, using an array to assign colors to each function. In a second plot, graph f, f , and f in the same picture and interpret the geometric information this provides about the graph of the original function f. Be sure to restrict the y ranges in the plots, if necessary, so that the graphs show suitable information.

   (a) $f(x) = e^{-x} \sin x$ on $[0, 3]$.

   (b) $f(x) = \sin x - x \cos x$ on $[0, 5]$.

   (c) $f(x) = \tan(x)$ on $[-1, 1]$.

   (d) $f(x) = 1/(x^2 + 1)$ on $[-2, 2]$.

5. Write procedures to do the following tasks

   (a) Sum all the numbers in a list L of numbers.

   (b) Calculate $n!$, where n is a positive integer.

6. Make the code in Example 2.3 into a general procedure which has for input a function f, an interval $[a, b]$ on which the function is defined, and two points $x_0 < x_5$ in the interval. Have your procedure calculate the entries in the array G and plot the tangent lines (of suitable length) as in Example 2.3. Here, however, the points at which the tangent lines occur are $(x_{i-1}, f(x_{i-1}))$, for $i = 1, \ldots, 6$, and the x's are given by $x_{i-1} = x_0 + (i - 1)\Delta x$, with Δx $(x_5 - x_0)/5$.

7. **(Plane Curves: The Implicit Description)** This exercise and the next deal with curves in the plane, i.e., $\mathbb{R}^2$. There are essentially two ways to represent such curves: by a single equation involving two variables, say, x and y (this is the implicit description) and parametrically by using a parameter to describe all the points (x, y) on the curve (the parametric description). We will use the implicit description in this exercise. The parametric description is discussed in the next exercise.

   If H is a function of two variables then the equation

   $$H(x, y) = 0$$

   is the implicit description of the curve. Technically, the curve is the set of points $C = \{ (x, y) \mid H(x, y) = 0 \}$, i.e., the curve is the solution set of the above equation. It is customary to just write the equation and refer to it as the equation for the curve. Standard examples of curves you have encountered are the parabola $y - x^2 = 0$, the hyperbola $x^2 - y^2 - 1 = 0$, and the ellipse $x^2/4 + y^2 - 1 = 0$.

You will be assigned one of the following curves to study. For your assigned curve

(a) $H(x, y) = x^3 - xy^2 + 1$

(b) $H(x, y) = x^3 - 2y^2 + \frac{3}{2}$

(c) $H(x, y) = x^3 + y^2 - 2xy - 1$

(d) $H(x, y) = x^3 - x - y^2 + 1$

(e) $H(x, y) = x^3 y - x - y^2 + 1$

(f) $H(x, y) = x^3 y - x - xy^2 + 1$

(g) $H(x, y) = x^3 - y - x^2 y^2 + 1$

(h) $H(x, y) = x^3 y - x - x^2 y^2 + 1$

(i) $H(x, y) = x^3 y - x - x^2 y^2 - y^2 + 1$

(j) $H(x, y) = x^2 y + x^3 + xy^3 + y + 1$

do the following

(i) Use the `implicitplot` command to plot the portion of the curve that lies in the rectangle $[-4, 2] \times [-4, 5]$ (see Maple/Calculus Notes). Use a fine enough grid so that the curve appears smooth. In the same picture plot the circle $x^2 + y^2 = 1$ and use the `fsolve` command to find all points of intersection of this circle with the assigned curve. Annotate your picture, identifying the respective curves.

(ii) For $h = -.5, 0, .5$, use the `fsolve` command to find all solutions to $H(1 + h, y) = 0$. This will give one or more points $(1 + h, y)$ on the curve. Plot the tangent lines at these points (with suitable length) and the curve itself (for x $[0, 2]$) in the same picture. Use colors to distinguish the respective plots. Print out and annotate your picture.

(iii) Define two 1-dimensional arrays `p[i]`,`c[i]`,`i=1..10`. Store your favorite 10 colors in the array `c` (such as `c[2]:=magenta`). Then use a do loop to store the plots of the curves $H(x, y) - i/10 = 0$, for $i = 1, \ldots, 10$, in the array `p`, with `p[i]` in color `c[i]`. Display all the plots in the same picture. Annotate this picture to clearly exhibit which curve goes with which value of i.

8. **(Plane Curves: The Parametric Description)** If two functions $f, g :$ $[a, b]$ $\mathbb{R}$ are given on some interval $[a, b]$, then a curve is given parametrically by the two equations:

$$\begin{aligned} x &= f(t) \\ y &= g(t). \end{aligned}$$

Technically the curve is the set of points $C = \{(f(t), g(t)) \,|\, t \quad [a, b]\}$. The functions f and g are called the x and y coordinate functions, respectively. The independent variable t is called the parameter and often represents time. For each $t \quad [a, b]$ there is a corresponding point in the plane with coordinates $(f(t), g(t))$. Plotting these points as t varies over $[a, b]$ gives a collection of points in the plane which constitutes the curve. When t represents the time, the point $(f(t), g(t))$ represents the position of a particle moving in the plane and as t varies from a to b, the particle traces out the curve, which is also known as the trajectory of the particle. Note that in this sense, a parameterized curve has a direction associated to it, i.e., the direction in which the particle moves on its trajectory.

Also recall from calculus that a parameterized curve can alternatively be thought of as a vector-valued function $\mathbf{r} : [a, b] \quad \mathbb{R}^2$, expressed by the formula

$$\mathbf{r}(t) = [f(t), g(t)],$$

for $t \quad [a, b]$. Different calculus texts will use different notation for the component expressions for vectors. Here we use square brackets: $\mathbf{v} = [v_1, v_2]$, since this is how Maple represents vectors, namely, as lists, or 1-dimensional arrays. From a physical point of view, $\mathbf{r}(t)$ is the position vector of the particle at time t. While a vector can be plotted anywhere in the plane, the position vector is usually plotted with its initial point at the origin. Then its terminal point is $(f(t), g(t))$.

The first and second derivatives of the vector-valued function $\mathbf{r}$ are also vector-valued functions and in physics are called the velocity and acceleration function for the moving particle:

$$\mathbf{v}(t) \;=\; \mathbf{r}'(t) = [f'(t), g'(t)]$$
$$\mathbf{a}(t) \;=\; \mathbf{r}''(t) = [f''(t), g''(t)].$$

The vector $\mathbf{v}(t)$, when plotted with its initial point at $(f(t), g(t))$ is tangent to the curve. The vector $\mathbf{a}(t)$, when plotted with its initial point at $(f(t), g(t))$ should point in the direction in which the curve is curving.

You will be assigned one of the following curves to study:

(a) $\mathbf{r}(t) = [t^4 - t, 2t^3 - t^5]$, for $t \quad [-1.55, 1.55]$.

(b) $\mathbf{r}(t) = [t^5 - t^2 + 1, 2t^3 - t^2 - t + 1]$, for $t \quad [-0.8, 1.2]$.

(c) $\mathbf{r}(t) = [t^6 - t^2 + 1, t^3 - t^2 - t + 1]$, for $t \quad [-1.1, 1.25]$.

(d) $\mathbf{r}(t) = [t^4 - t^2 + 1, 2t^2 - t + 1]$, for $t \quad [-0.8..1.2]$.

(e) $\mathbf{r}(t) = [t^4 - t^3 + 1, 2t^3 - t^2 - t + 1]$, for $t \quad [-0.8, 1.2]$.

(f) $\mathbf{r}(t) = [t^4 - t^3 + 1, 2t^4 - t^2 - t + 1]$, for $t \quad [-0.8, 1.2]$.

(g) $\mathbf{r}(t) = [t^5 - t^2 + 1, 2t^4 - t^2 - t + 1]$, for $t \quad [-0.8, 1.2]$.

(h) $\mathbf{r}(t) = [t^5 - t^2 + 1, 2t^5 - t^2 - t + 1]$, for $t \quad [-0.8, 1.2]$.

(i) $\mathbf{r}(t) = [t^5 - t + 1, 2t^6 - t^2 - t + 1]$, for t $[-1, 1.2]$.

For your particular curve do the following:

(i) Study the way the velocity and acceleration change as the curve is swept out. Do this by choosing a point $Q = \mathbf{r}(t_0)$ on the curve (your choice, but choose an interesting one) and then plotting the tangent (or velocity) lines and acceleration lines at the points $Q_i = \mathbf{r}(t_0 + i\Delta t)$, for $i = 0, 1, 2, 3$. For this choose a suitably small Δt so that the points Q_i are not too far apart, yet far enough apart to be distinguished from each other. Store the plots of the velocity and acceleration lines in two arrays vel and accel. For the velocity lines, you will have to experiment in order to choose a suitable length, but plot each as a line segment starting at its respective Q_i and extending in the forward direction. Use an array of colors c[i],i=0..3 to render each velocity line in a different color. Do a similar thing for the acceleration lines. Plot all of these and the curve itself in the same picture. Mark the directions on the lines and curve and annotate the figure (by hand after printing out).

(ii) Use fsolve to find all points of intersection of your curve with itself. Print out and annotate a figure with this information.

(iii) Use fsolve to find all points of intersection of your curve with the curve:

$$\mathbf{r}(t) = [t^3 - t, t^4 - t^2],$$

for t $[a, b]$. You will have to select a suitable interval $[a, b]$ by experimentation. Plot both curves in the same picture and mark your answer on the printout of the picture.

2.8   Maple/Calculus Notes

We review here a number of topics from both calculus and Maple for dealing with plane curves. Recall that a plane curve can be represented in two ways: by an equation $H(x, y) = 0$ (the implicit description) and parametrically by a pair of equations involving a single parameter (the parametric description).

2.8.1   Planes Curves Given Implicitly

(1) To define a function $H(x, y) = xy^2 - y^3 + 1$ of two variables in Maple use the arrow notation much as for a function of one variable:

```
H:=(x,y)->x*y^2-y^3+1;
```

To plot the curve with equation $H(x, y) = 0$, use the implicitplot command. This command is part of the plots package so you have to load this package into memory first.

```
with(plots):
implicitplot(H(x,y)=0,x=a..b,y=c..d,color=blue);
```

If $G(x, y) = 0$ is another curve, you can plot both curves together with the command

```
implicitplot({H(x,y)=0,G(x,y)=0},x=a..b,y=c..d,color=blue);
```

If you want to use different ranges for x and y or to use different options (such as color, grid size, etc.) for the curves, you will have to plot each curve separately and then display them in the same picture:

```
p:=implicitplot(H(x,y)=0,x=a..b,y=c..d,color=blue):
q:=implicitplot(G(x,y)=0,x=e..f,y=r..s,color=black,
               grid=[n,k]):
display({p,q});
```

The grid=[n,k] option causes Maple to apply its plotting routine with a grid of points obtained by subdividing $[e, f]$ into $n - 1$ equal subintervals and $[r, s]$ into $k - 1$ subintervals to produce a grid of nk points. The default is a 25×25 grid of 625 points.

(2) You can use the fsolve command to solve systems of equations, much as you have used it to solve a single equation. There must be as many equations as there are unknowns. For example, suppose there are two equations with two unknowns:

$$H(x, y) = 0$$
$$G(x, y) = 0$$

Assuming the functions H and G have been previously been defined in your Maple session, you can find a solution in the rectangle $[a, b] \times [c, d]$ with the command

```
fsolve({H(x,y)=0,G(x,y)=0},{x,y},x=a..b,y=c..d);
```

Of course there may not be a solution in this rectangle, in which case the command is just echoed back as output. Geometrically, the solutions (x, y) of the system give the points of intersection of the two curves with equations $H(x, y) = 0$ and $G(x, y) = 0$.

(3) To calculate the slope of the tangent line to the curve $H(x, y) = 0$ at a point $P = (x_0, y_0)$, you need to use implicit differentiation and find the derivative of y with respect to x. In Maple use the `implicitdiff` command,

```
implicitdiff(H(x,y),y,x);
```

to get the formula for dy/dx. Here as in the `diff` command the result is an expression, not a function of x and y. To convert it to a function use the `unapply` command.

```
M:=unapply(implicitdiff(H(x,y),y,x),(x,y));
```

Now you can evaluate M at the point P to get the number $M(x_0, y_0)$, which is the slope of the tangent line at P.

2.8.2 Plane Curves Given Parametrically

(1) If a plane curve has parametric representation $x = f(t), y = g(t)$, for t $[a, b]$, then it can easily be plotted using the following version of Maple's `plot` command:

```
plot([f(t),g(t),t=a..b],color=black, numpoints=200);
```

Note the special form of `[f(t),g(t),t=a..b]`, which is somewhat different from what you would expect. Mathematically a parameterized curve also arises from a vector-valued function $\mathbf{r} : [a, b]$ $\mathbb{R}^2$, with $\mathbf{r}(t) = [f(t), g(t)]$. See the comments in Exercise 7. You could define such a function in Maple by `r:=t->[f(t),g(t)]` (which makes it list-valued). However, doing so does *not* allow you to plot it by using the command `plot(r(t),t=a..b)`, even though this would seem the logical way to do it.[1]

If $\mathbf{q}(s) = [h(s), k(s)]$, for s $[c, d]$ is another parameterized curve, then both curves can be plotted in the same picture using

---

[1]However, for a curve in space, given by vector-valued function $\mathbf{r} : [a, b] \to \mathbb{R}^3$, de-"ned in Maple by `r:=t->[f(t),g(t),h(t)]`, it *is* possible to plot the curve by using the command `spacecurve(r(t),t=a..b)`.

```
plot({[f(t),g(t),t=a..b],[h(s),k(s),s=c..d]});
```

Of course you can use the same parameter name t instead of s in doing
the second plot if you wish. To have different colors (or other options)
for each curve, you will have to create separate plot structures and
then use the `display` command.

(2) If $\mathbf{r}(t) = [f(t), g(t)]$ and $\mathbf{q}(s) = [h(s), k(s)]$, for t $[a, b]$ and s $[c, d]$
are two parameterized plane curves, then the points where these curves
intersect can be found by solving a system of equations. To see this,
note that if the curves intersect at a point $P = (x_0, y_0)$, then there
are values t, s of the respective parameters for which $\mathbf{r}(t) = P$ and
$\mathbf{q}(s) = P$. Thus, we get the vector equation

$$\mathbf{r}(t) = \mathbf{q}(s),$$

which, when written out fully, gives a pair of two equations

$$\begin{aligned} f(t) &= h(s) \\ g(t) &= k(s) \end{aligned}$$

for the two unknowns t, s. If we can find a solution of this system, i.e.,
a t and s (which generally will be different numbers), then we can use
either t or s to compute P.

Chapter 3

Looping and Repetition

A computer is an ideal instrument for doing some things repeatedly, while many other types of machines (often controlled by computers) routinely carry out complex but repetitive tasks in manufacturing and industry. By nature human beings do not like being repetitive and, indeed, often think of ourselves as being clever when we devise an algorithm to have a machine do the work for us. This chapter discusses this essential programming technique: writing algorithms to repeat an action, or set of actions, a specified number of times.

3.1 The Basic Loop

One primitive way to have a statement S repeatedly executed by a computer is just to write the statement, for the required number of times N, in the code

$$S, S, \ldots, S$$

Of course, if $N = 1000$ this is hardly practical and more importantly is forcing you the programmer to do the repetition. Thus, nearly all programming languages have a shorthand way of writing this in the code   some way to say   do statement S a total of N times.   Since a computer executes the statements in a program sequentially, in order, the repetition of the single statement S a number of times in succession is thought of as a *loop* because the progress through the program is interrupted while S is repeatedly executed. Similarly, the process of repeatedly executing a whole block of contiguous statements, say,   do $S1, S2, \ldots, SP$ a total of N times,   is considered as a loop in the program. It is equivalent to writing out the block of statements in the code:

$$S1, S2, \ldots, SP, S1, S2, \ldots, SP, \ldots, S1, S2, \ldots, SP.$$

a total of N times.

Each programming language has its own particular way of writing such repetitive processes, or loops, and since many of them, like FORTRAN, C, Pascal, Algol, etc., use the reserved word do in their syntax, these processes are commonly referred to as do loops. Maple's do loop is identical to that of Algol[1] and is similar to those in many of the other languages. It has a number of different optional features which make it very useful. However, the most basic version of it is quite simply this. To do the statement S a total of N times in succession, use the Maple command

```
for i from 1 to N do S end do;
```

When the command is executed, the variable i is created and given the initial value 1. Then the relation i N is tested. If the relation is true, the statement S is executed, the index i is incremented by 1 and the relation i N is tested again. The process is repeated until $i > N$, in which case the execution is complete. Note that the index variable i (also called a counter) for the loop varies in value during the execution and so if the statement S involves expressions that depend on i, they too will vary from one execution of the statement to the next. Also, the final value of i when the program exits the loop is $N + 1$.

For example, if a variable x has value 5, then the do loop

```
for i from 1 to 20 do x:=x+x end do;
```

will, after execution, have assigned x the value $5 \cdot 2^{20}$ and the index i will have value 21. On the other hand, the do loop

```
for n from 1 to 40 do b[n]:=evalf(n^(1/n)) end do:
```

will calculate the "oating-point approximations to $n^{1/n}$, for the first 40 natural numbers n and store the results in the 1-dimensional array called **b**. The value of n at the end of the loop is 41. The results are shown graphically in Figure 3.1, and can be used as *experimental* evidence for the mathematical result

$$\lim_{n \to} n^{1/n} = 1,$$

which follows from L' Hopital's rule. Note that Maple has a built-in proce-

---

[1]Algol is an old programming language which is no longer in use. Its last version, "nalized in 1968, was called Algol 68 and it in"uenced the structuring of many present-day languages. After you learn Maple s programming construct, browse through a book on Algol 68 (say [BW]) and compare. For a history of programming languages see [BG], [Wex].

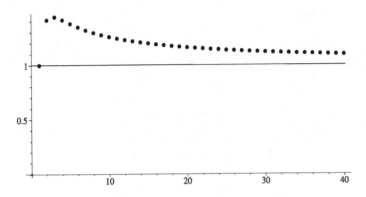

Figure 3.1: *Plot of the points $(n, n^{1/n})$, for $n = 1, \ldots, 40$.*

dure lim for experimentally calculating limits, however, in keeping with the philosophy of this book, you should not use it, but rather write your own code (exercise).

Another common example of a single statement do loop involves summing a finite sequence of numbers:

$$a_1, a_2, \ldots, a_N.$$

Again Maple has a procedure called sum which will easily do this. However, that procedure consists of several pages of code and is designed to do all sorts of summing (like finding closed form formulas, when they exist). So it seems rather silly to use sum when you can simply use the do loop

```
u:=0;
for i to N do u:=u+a[i] end do;
```

Here we are assuming that the numbers forming the sequence have been stored in an array called a and that N has been assigned a particular value. Also u is a temporary variable used to hold the partial sums that accrue during the summation.

Multiple statement do loops are constructed in the obvious fashion. If S, T, U are three statements which you want executed N times in Maple, then use the code

```
for i from 1 to N do S;T;U end do;
```

This is equivalent to writing each of the three statements in the code N times in the order

```
S;T;U;S;T;U; ... S;T;U;
```

Note that semicolons must separate the statements in the do loop and that it is not necessary (but is OK) to put a semicolon after the last statement (right before the **end do**).

The basic form of the do loop that we have been using is

> **for** I **from** I_1 **to** I_2 **by** K **do** *statements* **end do**

Here I is the index (or counter), I_1, I_2 are integers that delimit the range of values of I, and K is a nonzero integer (an increment or decrement) that is added to I each time through the loop. The designation: *statements* stands for a sequence of Maple statements separated by semicolons (or colons).

If $I_1 \le I_2$, then K must be positive (otherwise the loop does not execute) and the loop works as follows. I is assigned the initial value $I = I_1$ and if $I \le I_2$, then the statements in the loop are executed. Then I is assigned the value $I = I_1 + K$ and if $I \le I_2$, then the statements in the loop are executed. This process continues until I is assigned a value $I = I_1 + nK$ for which $I > I_2$. Then the loop terminates and control passes to the next statement after the loop (the statement after **end do**).

The situation just described has a variant whereby the index I decrements from I_1 down to I_2. Thus, if $I_1 \ge I_2$, then K must be negative (otherwise the loop does not execute) and the loop works as follows. I is assigned the initial value $I = I_1$ and if $I \ge I_2$, then the statements in the loop are executed. Then I is assigned the value $I = I_1 + K$ and if $I \ge I_2$, then the statements in the loop are executed. This process continues until I is assigned a value $I = I_1 + nK$ for which $I < I_2$. Then the loop terminates and control passes to the next statement after the loop.

There are default settings for I_1 and K. As you have seen, if the **by** K part is omitted from the loop command then by default $K = 1$. On the other hand, you may omit the **from** I_1 part in the loop command and then by default $I_1 = 1$.

Any do loop D can contain another do loop as one of its statements S, in which case the do loops are called *nested* with D as the *outer loop* and S as the *inner loop*. This type of situation occurs frequently in many programming algorithms. For example, consider the process of calculating the sequence of partial sums for a series $\sum_{n=1} a_n$. By definition the kth partial sum s_k is the sum of the first k terms of the series:

$$s_k = \sum_{n=1}^{k} a_n.$$

We can calculate the sequence $s_1, s_2, \ldots, s_N$ of partial sums for any $N \geq 1$ by using a nested pair of do loops as follows. The design of this code is based on viewing the computation as consisting of the following sequence of steps

$$
\begin{aligned}
s_1 &= a_1 \\
s_2 &= a_1 + a_2 \\
&\vdots \\
s_N &= a_1 + a_2 + \cdots + a_N.
\end{aligned}
$$

So we use an outer loop to create the sequence and an inner loop to sum the numbers as we did above.

```
for k from 1 to N do
    psum:=0;
        for n from 1 to k do
        psum:=psum+a(n);
    end do;
    s[k]:=psum;
end do;
```

Note that now we are assuming that the sequence a is a function (as is usual in mathematics) and that it has been previously defined in the Maple session. We have also indented the code and spread it over several lines so that the inner and outer do loops are easily discerned.[2] Immediately after calculating the kth partial sum s_k we use the command s[k]:=psum; in order to store the result in s[k] and also to have the result displayed on the screen.

There is a way to accomplish the same calculation by using a *single* do loop. This is based on the observation that the s_k comes from s_{k-1} by

---

[2] The key words do and end do mark the beginning and end of the statements in a loop. After do, you can put the ensuing statements on the same line or successively hit enter after each statement. Maple will keep remainding you that the do loop is not complete until you "nally enter end do. Our general practice is to put the header for the do loop (e.g., for k from 1 to N do) on one line, the statements comprising the loop on successive lines, and the ender, end do, on a separate last line. However, if there is only one statement in the loop, and it is short, we will put everything on one line.

```
1.000000000 1.250000000 1.361111111 1.423611111
1.463611111 1.491388889 1.511797052 1.527422052
1.539767731 1.549767731 1.558032194 1.564976638
1.570893798 1.575995839 1.580440283 1.584346533
1.587806741 1.590893161 1.593663244 1.596163244
1.598430818 1.600496934 1.602387293 1.604123404
1.605723404 1.607202694 1.608574436 1.609849946
1.611039007 1.612150118 1.613190701 1.614167264
1.615085538 1.615950590 1.616766917 1.617538522
1.618268982 1.618961503 1.619618965 1.620243965
```

Table 3.1: *A table of the partial sums* $s_k = \sum_{n=1}^{k} \frac{1}{n^2}$, *for* $k = 1, \ldots, 40$.

adding in one more term, i.e., $s_k = s_{k-1} + a_k$, for $k = 2, \ldots, N$. Thus, the following do loop suffices for the calculation

```
s[1]:=a(1);
for k from 2 to N do s[k]:=s[k-1]+a(k) end do;
```

We can now use this code to study the convergence of the famous $p = 2$ series

$$\sum_{n=1}^{\infty} \frac{1}{n^2} = 1 + \frac{1}{2^2} + \frac{1}{3^2} + \cdots$$

The first 40 partial sums of this series are shown in Table 3.1. This table was generated using a do loop that increments the counter k by 4 each time through the loop:

```
for k from 1 to 40 by 4 do s[k],s[k+1],s[k+2],s[k+3] end do
```

As a result the output appears with 4 values to the line.

As we have mentioned repeatedly, there are many ways to solve even simple programming problems. Some ways are longer (like the two loop version above) but perhaps clearer conceptually (easier to understand a year after you wrote it). However, short code, especially code that does not create a lot of memory allocation for unnecessary arrays, often is preferable if one is concerned with execution time. These examples are too simple to make much of a difference.

Another good example of the use of looping is the following:

Example 3.1 (Riemann Sums) This example illustrates how to calculate Riemann sum approximations of areas using the left-hand rule. In particular, we will approximate the area below a curve $y = f(x)$ which is

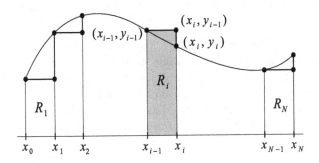

Figure 3.2: *The rectangles R_i constructed by using the heights $y_{i-1} = f(x_{i-1})$ at the left-hand endpoints of the subintervals $[x_{i-1}, x_i]$, $i = 1, \ldots, N$.*

the graph of a nonnegative function f on an interval $[a, b]$. The actual area is given by

$$area = \int_a^b f(x)dx.$$

One way to approximate the true area is to divide the interval $[a, b]$ into N subintervals all of which have the same length: $\Delta x = \frac{b-a}{N}$. The sequence of division points in the partition is then $x_i = a + i\Delta x$, for $i = 0, \ldots, N$. The y values corresponding to these division points are $y_i = f(x_i)$, for $i = 0, \ldots, N$. Over each of the N subintervals $[x_{i-1}, x_i]$, $i = 1, \ldots, N$, we construct a rectangle by evaluating f at the *left-hand* endpoint, to get a height $y_{i-1} = f(x_{i-1})$ for the ith rectangle R_i. Then the area of R_i is $f(x_{i-1})\Delta x$. This is shown in Figure 3.2. Thus, the area (and definite integral) is approximated by the sum

$$area = \int_a^b f(x)dx \approx \sum_{i=1}^N f(x_{i-1})\Delta x.$$

This is known as a Riemann sum constructed by the left-hand rule and the calculation of this sum can be easily automated in a number of different ways as follows. We assume that $f, a, b,$ and N have been previously defined in the session. A basic version of the code is

```
dX:=(b-a)/N;
X:=a; Y:=f(X);
s:=0;
for i from 1 to N do
    s:= s+Y*dX;
```

```
    X:= X+dX; Y:=f(X)
  end do;
```

The first statement calculates $\Delta x = (b-a)/N$, while the next two assign initial values (corresponding to $i = 0$) to x, y, and s. The variable s will store the value of the sum $s = \sum_{i=1}^{N} y_{i-1}\Delta x$. The do loop has index i running from 1 to N, but note that the statements in the do loop do not involve i. The loop computes the sum by successive additions of a new value of $y\Delta x$ to the previous value of s. You should convince yourself that this will give the correct result, i.e., the 1st, 2nd, $\ldots, N$th time through the loop give

$$s \; = \; s + y_0\Delta x = y_0\Delta x$$
$$s \; = \; s + y_1\Delta x = y_0\Delta x + y_1\Delta x$$
$$\vdots$$
$$s \; = \; s + y_{N-1}\Delta x = y_0\Delta x + y_1\Delta x + \cdots + y_{N-1}\Delta x.$$

Note that we have used capital letters X, Y in the code for the variables. This is because, in all likelihood, we will want to use the lower-case letters x, y elsewhere in our Maple session.

Often the code for doing a given task can be made clearer by using arrays even though the task can be done without them. Thus, for instance, in the last example we could have set up two arrays X,Y to store the values of x_i, y_i, $i = 0, \ldots, N$, and then have written the code as follows.

```
  X:=array(0..N); Y:=array(0..N);
  dX:=(b-a)/N;
  X[0]:=a; Y[0]:=f(X[0]);
  s:=0;
  for i from 1 to N do
      s:= s+Y[i-1]*dX;
      X[i]:= X[i-1]+dX; Y[i]:=f(X[i])
  end do;
```

Of course clarity comes at the expense of using storage space in memory for the values of $x_i, y_i, i = 0, \ldots, N$. In this example this waste of storage is not critical. However, for large arrays the choice between clarity and conserving memory can be a crucial decision. On the other hand, the next example shows that it is often necessary to sacrifice this storage space by creating arrays because the values of $x_i, y_i, i = 0, \ldots, N$, are needed not only in the calculation of the sums, but also at other places in the code.

Example 3.2 (Riemann Sums with Graphics) Suppose that in addition to calculating the Riemann sum approximation to the area, we also produce a picture showing an outline of the approximating area.[3] This outline is to consist of the top parts of all the rectangles (a stairstep figure), the two vertical bounding line segments, and the line segment $[a, b]$.

We assume the function f on the interval $[a, b]$ has been defined and that the number N of equal subintervals has been assigned. As above we set up two arrays to hold the computed points (x_i, y_i), $i = 0, \ldots, N$, on the graph:

```
X:=array(0..N):Y:=array(0..N):
```

We can then use the do loop discussed above to compute the Riemann sum approximation. However, while in the loop doing this computation, we want to also produce the graphical data for the desired outline. The main part of this is the stairstep figure running across the tops of the rectangles. As shown on Figure 3.2, on the ith rectangle, a tread, joining (x_{i-1}, y_{i-1}) to (x_i, y_{i-1}) and a riser joining (x_i, y_{i-1}) to (x_i, y_i) contribute to the overall stairstep figure. So we draw these two line segments each time through the do loop and store the results in a tread-riser array: `tr:=array(1..N)`. You should convince yourself that the choice of points here does not depend on whether the function is increasing or decreasing on the ith subinterval. The following code calculates the Riemann sum and produces the graphical data.

```
dX:=evalf((b-a)/N);X[0]:=a;Y[0]:=f(X[0]);s:=0;
vert0:=plot([[X[0],0],[X[0],Y[0]]]):
for i from 1 to N do
  s:=s+Y[i-1]*dX;
  X[i]:=X[i-1]+dX; Y[i]:=f(X[i]);
  tr[i]:=plot([[X[i-1],Y[i-1]],[X[i],Y[i-1]],[X[i],Y[i]]]):
end do:
vertN:=plot([[X[N],0],[X[N],Y[N-1]]]):
horiz:=plot(0,x=a..b):
p:=plot(f(x),x=a..b,color=black):
```

The plot structures `vert0` and `vertN` contain the beginning and ending vertical line segments, which are not contained in the tread-riser plots constructed in the do loop. The name `horiz` contains the plot structure for the line segment $[a, b]$.

---

[3]Maple has a command for drawing all the rectangles that comprise the approximating area, but here we are concerned with writing our own code for a slightly different figure.

We can now see the end value found for **s** and display the graphical information by

```
s;
display({seq(tr[i],i=1..N),vert0,vertN,horiz,p},
        scaling=constrained,axes=none);
```

Figure 3.3 shows the results from using this code twice with the function $f(x) = x^2 - x^4 + 1$ on the interval $[0, 1]$, and with approximations of $N = 10$ and $N = 30$ rectangles.

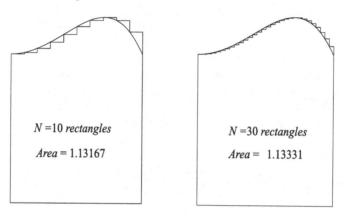

N =10 rectangles

Area = 1.13167

N =30 rectangles

Area = 1.13331

Figure 3.3: *Plot of the tread-riser outlines for the left-hand rule approximations $N = 10$ and $N = 30$.*

The actual area under the curve is easily calculated exactly by using the Fundamental Theorem of Calculus:

$$\int_0^1 (x^2 - x^4 + 1)\, dx = \frac{1}{3} - \frac{1}{5} + 1 = \frac{17}{15} = 1.133\overline{3}.$$

3.2   The Do Loop with All Its Features

While the basic do loop in any language, combined with other programming elements, can accomplish any repetitive task, most languages provide some extra features in the loop syntax that automatically perform frequently needed tasks. For example, as you have seen, in some tasks it would be convenient to have the index, or counter, for the loop be incremented by 2 instead of 1 after each step of the repetition. Or there might be a side condition that does not depend on the index, which you might want to check before entering the loop at each step. Or alternatively, you might want to have the loop controlled by this condition with no index used at all. Thus, most languages contain extended versions of their basic do loops for your convenience.

Maple has two full-featured versions of its do loop. One is called the `for-from` loop (the short version of which is what we used above) and the other is called the `for-in`. They are both equivalent in the sense that either can be used to accomplish any repetitive task. However, one version is sometimes more convenient and natural than the other.

3.2.1 The for-from Loop

The complete structure of Maple's `for-from` do loop (which is identical to that of Algol 68 and is similar to many other languages) is expressed by

```
for I from I₁ to I₂ by K while C do statements end do
```

Here I_1 is the initial value for the index I and I_2 is the value against which I is tested each time before executing the loop. K is the increment (or decrement) which is added to I after all the *statements* in the body of the loop are executed.

The `while` part of the do loop is optional (as all the examples so far have demonstrated), however, it is useful to include when you need additional control over execution of the loop. Generally you will want the loop executed the specified number of times, but often when a side condition C occurs during the looping you will want to terminate execution of the loop.

A useful version of the `for-from` loop is the one where `for` and `from` are omitted! This version, called a `while` loop has the form:

```
while C do statements end do
```

This form is invariably used when the number of repetitions needed to accomplish the task is not known in advance. A good example occurs in numerical approximations of the value of a convergent infinite series.

Example 3.3 (Alternating Series) Suppose we want to estimate the value of an alternating series $\sum_{n=1} (-1)^{n+1} a_n$, whose terms satisfy (1) $a_n >$

$a_{n+1} > 0$, for all n, and (2) $\lim_{n \to \infty} a_n = 0$. These two conditions are
the hypotheses of a theorem, called the *Alternating Series Test* that guarantees the convergence of the alternating series (not all alternating series
converge). Most calculus books also mention (and maybe prove) that if S
is the sum of the series

$$S = \sum_{n=1}^{\infty} (-1)^{n+1} a_n,$$

and if $s_k = \sum_{n=1}^{k} (-1)^{n+1} a_n$ is the kth partial sum, then

$$|S - s_k| < a_{k+1},$$

for all k. This inequality enables us to determine how large k has to be
in order to achieve a desired degree of accuracy, say, $\epsilon = 10^{-5}$, in the
approximation s_k to S. We simply choose k so that $a_{k+1} < \epsilon$. Then $|S - s_k| < \epsilon$, or

$$s_k - \epsilon < S < s_k + \epsilon.$$

So s_k is the desired estimate of the value of S.

Writing the code to compute a partial sum $s_k = \sum_{n=1}^{k} (-1)^{n+1} a_n$ that is
within a given ϵ of the total sum S can easily be done with a `while` loop.
We do not need to know in advance what k is, but we keep adding terms
until $|S - s_k|$ is less that ϵ. By the theory this will happen when $a_{k+1} \leq \epsilon$.
Thus, we keep summing while the condition $a_{k+1} > \epsilon$ is true and stop
once it is not.

The following code will implement this. The terms a_n are taken to be

$$a_n = \frac{1}{n(n+1)},$$

which we define as a function in Maple. Note that the kth partial sum is
given by $s_k = s_{k-1} + (-1)^{k+1} a_k$. We store the partial sums in an array
called s and initialize with `s[0]:=0`. The error is taken to be $\epsilon = 0.001$.

```
a:=n->1/(n*(n+1));
epsilon:=0.001;
k:=0; s[0]:=0;
while a(k+1)>epsilon do
    k:=k+1:
    s[k]:=s[k-1]+(-1)^(k+1)*a(k);
end do:
```

Note that the table structure s for the partial sums is defined *implicitly* by the code.[4] This is because we do not know how many elements s will have and so we cannot define it *explicitly* by s:=array(0..N).

When the loop terminates, k has the first value for which a(k+1) is less than or equal to 0.001. For this example, checking the value of k will give $k = 31$. This means, according to the theory, that s_{31} is within 0.001 of the true value S for the sum of the series. We compute one more partial sum (specifically s_{32}) and then display the values with the following code:

```
s[k+1]:=s[k]+(-1)^(k+1)*a(k+1):
for i from 1 to k-2 by 4 do s[i],s[i+1],s[i+2],s[i+3] end do;
```

This gives the following output:

```
.5000000000, .3333333333, .4166666666, .3666666666
.3999999999, .3761904761, .3940476190, .3801587301
.3912698412, .3821789321, .3897546897, .3833444333
.3888389388, .3840770340, .3882437007, .3845672301
.3878352040, .3849112274, .3875428063, .3851618539
.3873263561, .3853500715, .3871616657, .3854949990
.3870334605, .3856089591, .3869317104, .3857001833
.3868496086, .3857743398, .3867824043, .3877293740
```

Figure 3.4 illustrates these results and how they oscillate above and below the exact value S for the sum of the series. The calculations and the theory give us the following approximation of the value of S. Since $s_{31} = .3867824043$ and $|S - s_{31}| < .001$, we have that $.3857824043 < S < .3877824043$. One can show (exercise) that the exact value is $S = 2\ln 2 - 1$.

You should bear in mind that the above code for the while loop assumes that the terms a_n in the series satisfy conditions (1) and (2) of the Alternating Series Test. When this is not the case the code could become an infinite loop, i.e., one that never terminates. To see this note that if you used the code on the alternating series of plus and minus ones:

$$\sum_{n=1} (-1)^{n+1} = 1 - 1 + 1 - 1 + 1 - 1 + \cdots,$$

then, since $a_n = 1$ for every n, the loop condition a(k+1)>e will always be true and so the loop never terminates. To avoid this, you can always include the condition x<N, with N a large integer, in the control logic of the loop. For example, to make sure the loop does not continue past 100,000 repetitions, you could do the following.

---

[4] A table data structure is similar to an array data structure in its use. See Chapter 6.

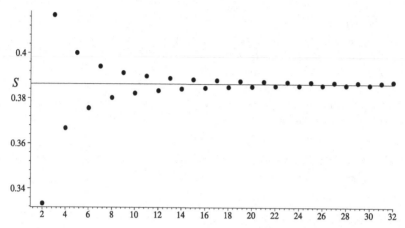

Figure 3.4: *Oscillation of the partial sum approximations s_k, $k = 2, \ldots, 32$ about the exact value $S = \sum_{n=1} \frac{(-1)^{n+1}}{n(n+1)}$ for the sum of the series.*

```
epsilon:=0.001; N:=100000;
k:=0; s[0]:=0;
while a(k+1)>epsilon and k<=N do
   k:=k+1:
   s[k]:=s[k-1]+(-1)^(k+1)*a(k);
end do:
```

Of course, you could also use the following variation to accomplish the same thing.

```
epsilon:=0.001; N:=100000;
s[0]:=0;
for k from 1 to N+1 while a(k)>epsilon do
   s[k]:=s[k-1]+(-1)^(k+1)*a(k);
end do:
```

3.2.2 The for-in Loop

The structure of Maple's **for-in** loop is very similar to its **for-from** loop:

---

for x in S while C do *statements* end do

Here S is a set (sequence or list and possibly empty) and C is a condition that can be true or false. The *statements* in the loop are executed once for each element x in the set S. However, before each execution, the condition C is checked and if it is false the looping terminates. No action (looping) is taken when S is the empty set. The `while` clause is optional just as it is in the `for-from` do loop.

As an elementary example illustrating this type of loop, consider calculating the product of all the numbers in a finite set S of numbers:

```
S:={-2,4,6,-10};
p:=1;
for x in S do
   p:=p*x
end do;
```

This results in p being assigned the value 480. Of course, the above loop could have been written in an equivalent way using a `for-from` loop:

```
S:={-2,4,6,-10};
p:=1;
for i from 1 to 4 do
  p:=p*S[i]
end do;
```

In general any `for-in` loop can be converted into an equivalent `for-from` loop. This requires knowing the number of elements in the set S, which can be found by using the **nops** command (shorthand for number of operands). In this case **nops(S)** returns the number of elements in the set S.[5]

On the other hand any `for-from` loop can be converted to a `for-in` loop, and thus these two constructs are equivalent. Your personal preferences will dictate using one of these constructs over the other. Often clarity of the code and convenience in writing the algorithm will suggest one construct as being more natural than the other.

3.3 Case Study: Iterated Maps

This is the first case study in the book and so a word of explanation may be helpful. A case study is like a long example but is intended to show you how to take a statement of a programming problem, analyze it, and design the code that solves the problem.

---

[5]See Chapter 6 for descriptions and uses of the **nops**,**ops**, and **op** commands in working with sets, lists, etc.

This case study introduces you to the topic of iterated maps, which may be new to you even though it appears in many aspects of mathematics. Newton's method for approximating solutions x of an equation $f(x) = 0$, is one example of an iterated map that you may have studied in your calculus sequence and this will be the main focus of this case study.

The general topic of an iterated map involves a map $T : U \quad U$, from a set U into itself. For a point $x \quad U$, the image $T(x)$ is also a point in U and so we can apply T again to get another point $T(T(x))$ in U. Applying T once again gives yet a third point $T(T(T(x)))$ in U. These are known as *iterates* of x under the map T, specifically the first, second, and third iterates. They are denoted by the notation $T^1(x) = T(x)$, $T^2(x) = T(T(x))$, and $T^3(x) = T(T(T(x)))$. Generally $T^n(x)$ denotes the nth iterate of x and is the result of applying T a total of n times in succession.

An important example of an iterated map T arises in Newton's method for approximating solutions of an equation $f(x) = 0$.

Problem: Write some Maple code to implement Newton's method and produce graphics illustrating the convergence of the approximations to a true root of the equation $f(x) = 0$, where f is a function of your choice. Specifically, for a given initial approximation x_0 to the a true root r and a given number > 0, have your code produce a sequence $\{x_n\}_{n=0}^{N}$ of approximations to r, stopping when $|f(x_N)| \quad$ (which indicates that x_N is within of being a true root of $f(x) = 0$). Include a control variable to avoid an infinite loop. Use a two dimensional array A to store the calculated values: $A_{n1} = x_n$, $A_{n2} = f(x_n)$. For the graphics, store the necessary information to produce a picture showing the sawtooth polygon of the iterates (like that in Figure 3.6 below) along with the graph of the function f on an appropriate interval.

Analysis of the Problem: This method states that if x_0 is a fairly good approximation to a root of the equation, then we can find a better approximation by using the tangent line at the point $(x_0, f(x_0))$. Figure 3.5 illustrates what is meant by this.

The better approximation is x_1 and from elementary geometry of similar triangles we see that

$$\frac{f(x_0)}{x_0 - x_1} = \frac{f'(x_0)}{1}.$$

Rearranging this equation, we get the following formula for x_1:

$$x_1 = x_0 - \frac{f(x_0)}{f'(x_0)}.$$

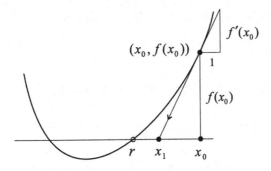

Figure 3.5: *Newton s method produces, from x_0, a better approximation x_1 to a root r of the equation $f(x) = 0$.*

Thus, starting with x_0, the formula allows us to compute a better approximation x_1 to the desired root. If we now replace x_0 by x_1 in the formula we get yet a better approximation x_2:

$$x_2 = x_1 - \frac{f(x_1)}{f'(x_1)}.$$

Continuing in this fashion, we get x_3, x_4, etc., indeed an infinite sequence $\{x_n\}_{n=0}$ of approximations which, according to Figure 3.6, should converge

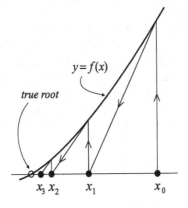

Figure 3.6: *Newton s method, applied repetitively, gives a sequence of approximations which converges, in this example, to a root of the equation $f(x) = 0$.*

to a true root r of the equation $f(x) = 0$. It is important to note that the approximations $x_1, x_2, x_3, \ldots$, arise from a map T applied successively to x_0. Namely, if we define

$$T(x) \quad x - \frac{f(x)}{f'(x)}, \tag{3.1}$$

then $x_1 = T(x_0)$ and $x_2 = T(x_1) = T(T(x_0)) = T^2(x_0)$. Generally the nth approximation is $x_n = T^n(x_0)$.

Design of the Program: The code begins with the definition of the function f and the endpoints a, b of an interval on which f has a zero, i.e., on which $f(x) = 0$ has a solution in $[a, b]$. Next the definition of the iteration function T given in Formula (3.1) is made. The repetition (or iteration) of T, applied to an initial approximate root, for an unspecified number of times, is easily accomplished by a `while` loop. An initial value of an approximate root, called `x0`, is assigned before entering the loop. Within the loop the next approximate root is simply computed by `x1:=evalf(T(x0));` and the closeness of this to being a true root is measured by the value of `E:=abs(f(x1))`. To repeat this process the next time (and all the ensuing times) through the loop, we make the assignment `x0:=x1` before looping back. The looping continues until the condition `E>e` is false, where `e` is the error accuracy (i.e., the   ) that we specifiy before entering the loop. Using `x1:=evalf(T(x0))` instead of `x1:=T(x0)` forces the calculations to be done using "oating-point instead of exact arithmetic. This prevents the use of increasingly complicated exact expressions.

To keep the do loop from becoming infinite, we use a control variable `n`, which counts the number of times through the loop (the number of iterations). We stop the looping when n is greater than or equal to an assigned value `N`, which we take to be 100 iterations. The variable `n` will also be convenient to use as an index in assigning values to a matrix (table) `A` whose entries are the approximate roots and the values of f at these approximate roots. This two dimensional table is created implicitly by the code in the loop.

The sawtooth polygon in Figure 3.6 is easily produced in pieces, i.e., each time through the loop we draw the   tooth   consisting of the two sides with vertices

$$[\texttt{x0}, 0], [\texttt{x0}, \texttt{f(x0)}], [\texttt{x1}, 0].$$

(See Figure 3.5.) This plot is saved in a plot structure `ps` which is added to a total plot structure `tps`. Upon completion of the loop we can then display `tps` to get the sawtooth picture required. We treat `tps` as a set and use Maple's `union` command to build it from the individual teeth. We initialize tps by assigning it the empty set: `tps:={}`. For the sake of this example, we use the function $f(x) = x^2 - 1$, which clearly has the two roots $x = \pm 1$.[6]. We choose $x_0 = 2.0$ as an initial approximation to the root $x = 1$.

| Code for Newton s Method |

```
f:=x->x^2-1: a:=0.8; b:=2;
T:=x->x-f(x)/D(f)(x):
e:=0.0001; N:=100;
n:=0; tps:={};
x0:=2.0; E:=evalf(abs(f(x0)));
while ( E>e and n<N ) do
  A[n,1]:=x0;
  A[n,2]:=E;
  x1:=evalf(T(x0));
  E:=abs(f(x1));
  ps:={plot([[x0,0],[x0,f(x0)],[x1,0]],color=black,
            thickness=2)}:
  tps:=tps union ps:
  n:=n+1:
  x0:=x1:
end do:
```

Note that each time through the loop, `ps` is a set consisting of a single element, namely the plot structure for the tooth calculated at that iteration.

After executing this code, we can display the calculated values of the approximate roots and the errors E by using the following loop

```
for j from 0 to n-1 do x[j]=A[j,1], E =A[j,2] end do;
```

This produces the following output.

$$x_0 = 2., E = 3.$$
$$x_1 = 1.250000000, E = .562500000$$
$$x_2 = 1.025000000, E = .050625000$$
$$x_3 = 1.000304878, E = .000609849$$

The quotes on E prevent evaluation of the variable E and cause only its name to be printed as output. You can check that the value of E is .000000092 when the loop finishes.

---

[6]It is always recommended to initially test your code with a simple example for which the resulting output is known

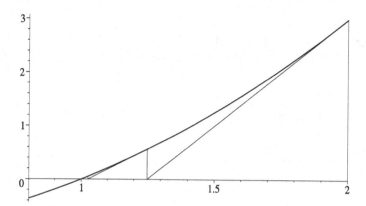

Figure 3.7: *Convergence of the approximate roots to the root $x = 1$ of the equation $x^2 - 1 = 0$.*

Next we can display the picture showing convergence of the approximating roots to the true roots by first combining a plot of the function f with the sawtooth polygon and then displaying the result. The following code,

```
pf:={plot(f(x),x=a..b)}:
pic:=pf union tps:
display(pic,tickmarks=[3,3]);
```

gives the picture shown in Figure 3.7

To test the code on another example, consider $f(x) = \sin x - x \cos x$ on the interval $[-1, 8]$. Figure 3.8 indicates the equation $f(x) = \sin x - x \cos x$ has three roots in the interval $[-1, 8]$.

One root is clearly $x = 0$, but the other two cannot be found exactly by algebraic methods. So we try our code to approximate the root r that lies in the interval $[4, 6]$. Figure 3.9 gives the graphics produced by the code with $x_0 = 5.5$ as the initial approximation to r.

For the choice of   in the code, the iteration stops at $n = 2$, giving the approximate value $x_2 = 4.503123428$ for the true root r. A finer estimate with $= .000001$, gives $x_3 = 4.493430093$. Thus, for this example, the algorithm converges very quickly (i.e., just a few iterations to achieve the given accuracy).

However, depending on the choice for the initial approximate root x_0, you will get results that may surprise you. Generally Newton's method will work better when the choice of x_0 is near the actual root that you are trying to approximate. This only makes sense based on the geometry behind the

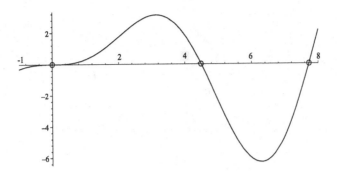

Figure 3.8: *Graph of the function* $f(x) = \sin x - x \cos x$ *on the interval* $[-1, 8]$, *showing the three roots of the equation* $\sin x - x \cos x = 0$ *in that interval.*

method. However, to illustrate what can go wrong, consider using $x_0 = 6$ in the present example. The algorithm produces a sequence of approximations that converges to the root $x = 0$, rather than the root $x = r$ 4.4934. This may have surprised you! The plot in Figure 3.10 graphically illustrates this.

Exercise Set 3

1. The following are variations of the code used in Examples 3.1 3.2 to compute Riemann sums. Analyze what each does and give the mathematical formula for what it is actually computing.

   (a) ```
        dX:=(b-a)/N; X:=a; Y:=f(a); s:=0;
        for i from 1 to N do
            X:= X+dX; Y:=f(X);
            s:= s+Y*dX;
        end do;
        ```

   (b)  ```
        X:=array(0..N); Y:=array(0..N);
        dX:=(b-a)/N; X[0]:=a; Y[0]:=f(X[0]); s:=0;
        for i from 1 to N do
            X[i]:= X[i-1]+dX; Y[i]:=f(X[i]);
            s:= s+Y[i-1]*dX;
        end do;
        ```

2. **(Riemann Sums)** As in Example 3.2, write some code to compute the Riemann sum approximations and draw the outline (boundary) of the approximating areas for the following situations.

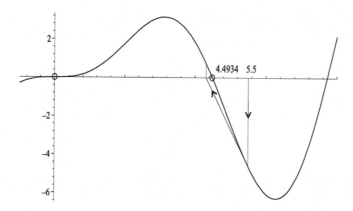

Figure 3.9: *For initial approximation $x_0 = 5.5$, Newton s method converges to the root r   4.4934.*

(i) (Right-Hand Rule) For the ith rectangle on the interval $[x_{i-1}, x_i]$, use the value $f(x_i)$ at the right-hand endpoint for the height of the rectangle.

(ii) (Midpoint Rule) For the ith rectangle on the interval $[x_{i-1}, x_i]$, use the value $f(m_i)$ at the midpoint $m_i = (x_{i-1} + x_i)/2$, for the height of the rectangle.

Apply your code to the functions that you are assigned from the following.

(a) $f(x) = x^2 - x^4 + 1$ on the interval $[0, 1]$ (this is from Example 3.2).

(b) $f(x) = 1 + x^{\sin x}$ on the interval $[1, 5]$.

(c) $f(x) = 1 + e^{-x} \sin 10x$ on the interval $[0, 2]$.

(d) $f(x) = 1 + \sin 10x \sin x$ on the interval $[0, 2]$.

In each case, do the Riemann sum approximations and graphics for $N = 10, 30, 50, 100$. Annotate your graphics with the value of N and the corresponding value of the approximating Riemann sums. Use Maple to compute a good "oating-point approximation to the true value of the definite integral. If you are assigned (c) or (d), compute the exact value by hand. Show your work!

3. **(Alternating Series)** At the end of the discussion in Example 3.3, we made the claim that the alternating series being studied numerically had an exact value for its sum:

$$\sum_{n=1}^{\infty} \frac{(-1)^{n+1}}{n(n+1)} = 2\ln 2 - 1.$$

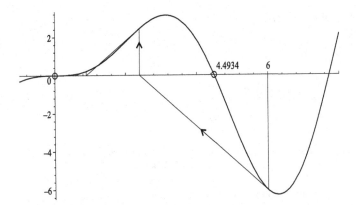

Figure 3.10: *For initial approximation $x_0 = 6$, Newton s method converges to the root $r = 0$.*

Determine how we got this. *Hint*: First show that

$$\ln(1+x)\,dx = (1+x)\ln(1+1) - x.$$

Then expand the integrand in a Maclaurin series. In addition, choose one (or more, if you wish) of the following alternating series to study as in Example 3.3. Determine the exact sum of the series by methods as suggested in the above hint.

(a) The alternating series

$$\sum_{n=1} \frac{(-1)^{n+1}}{2n-1} = 1 - \frac{1}{3} + \frac{1}{5} - \frac{1}{7} + \cdots,$$

is a famous alternating series. Use the code in Example 3.3 to compute the partial sums s_k of this series until k is the first value for which a_{k+1} , with $= .001$. What is the value of $k = k_0$ for which this occurs? Can you predict this value k_0 in advance? Produce a table of the last 100 values of the partial sums s_k, $k = k_0 - 100, \ldots, k_0$, with five values per line. Does this series converge very slowly? Why?

Show that the series converges to $/4$,

$$\sum_{n=1} \frac{(-1)^{n+1}}{2n-1} = \frac{}{4}.$$

Hint: Use a method like that in the preceding exercise, but now with

$$\frac{1}{1+x^2}\,dx = \tan^{-1}(x).$$

(b) The following alternating series converges to the value shown (the sum of the series).

$$\sum_{n=0}^{\infty} \frac{(-1)^n}{(n+2)n!} = 1 - \frac{2}{e}.$$

Show this by considering the integral

$$\int xe^{-x}\,dx = -xe^{-x} - x.$$

Also study this series numerically using the code from Example 3.3. Compute the partial sums s_k of this series until k is the first value for which $a_{k+1} < \epsilon$, with $\epsilon = 10^{-9}$. What is the value of $k = k_0$ for which this occurs? Can you predict this value k_0 in advance? Print out a table of the values of the s_k's, four values to a line. Does this series converge very rapidly? Why?

4. (**Approximating Lengths of Curves**) Write a Maple procedure to compute the approximate length for a curve that is the graph of a function $f : I \to \mathbb{R}$ defined on a closed interval $I = [a, b]$. Recall that the graph of f is the set of points

$$C = \{\, (x, f(x)) \mid x \in R \,\},$$

in $\mathbb{R}^2$. Here are some instructions for designing your procedure:

Approximate the curve C by a polygon for which you can calculate the lengths of its individual sides. Construct this polygon as follows. Partition I into subintervals by using the standard partitions of $[a, b]$ into N subintervals of equal lengths. Then the endpoints of the subintervals in the partition of I are $x_i = a + (b-a)i/N$, for $i = 0, \ldots, N$. The ith subinterval is $I_i = [x_{i-1}, x_i]$, for $i = 1, \ldots, N$. Corresponding to this subdivision is a polygonal curve P_N in $\mathbb{R}^2$ with vertices $Q_i = (x_i, y_i)$, $i = 1, \ldots, N$, where $y_i = f(x_i)$. Connecting these points with straight line segments produces a polygonal approximation to the curve C. See Figure 3.11. Your procedure should compute the approximating length to be the sum of the lengths of all the sides in this polygon, i.e.,

$$AL(f) = \sum_{i=1}^{N} d(Q_{i-1}, Q_i), \qquad (3.2)$$

where $d(Q_{i-1}, Q_i)$ denotes the length of the ith side of the polygon. Here are some further directions and parts to the exercise.

(a) The distance formula gives

$$d(Q_{i-1}, Q_i) = \sqrt{(\Delta x_i)^2 + (\Delta y_i)^2}, \qquad (3.3)$$

where $\Delta y_i = y_i - y_{i-1}$. This is the length of the ith side of the approximating polygon. Use this in your procedure.

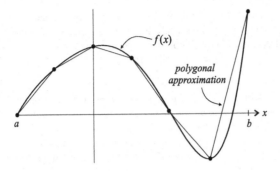

Figure 3.11: *A polygonal approximation to the curve C which is the graph of f on the interval* $[a, b]$.

(b) Have f, a, b, and N as input to your procedure and the approximating length $AL(f)$ as output. Also have your procedure plot the approximating polygon and the graph of f in the same picture.

(c) Test your procedure with $f(x) = 4x/3$ on $[0, 3]$, for which you know the exact length. Does your procedure give the exact answer for any choice of N? Why?

(d) Use your procedure to approximate the length of the curve that is the graph of $f(x) = x^{\cos x}$ for $x \quad [1, 15]$. Use $N = 10, 20, 50, 100$ and any other values you feel are appropriate. How large does N have to be before the graphs of f and the approximating polygon with N sides are indiscernible?

(e) Do (d) again but now with $f(x) = x^3/6 + 1/(2x)$ on the interval $[0.1, 2]$.

(f) Show that Equation (3.3) can be written as

$$d(Q_{i-1}, Q_i) = \quad \overline{1 + \quad \frac{\Delta y_i}{\Delta x_i}^{\;2}} \; \Delta x_i \qquad (3.4)$$

and use this to argue that the exact lengths of the curve which is the graph of f on $[a, b]$ is:

$$L(f) = \quad \int_a^b \sqrt{1 + (f\ (x))^2}\, dx. \qquad (3.5)$$

(g) Use Formula (3.5) to find the exact lengths of the curves that are the graphs of the functions in parts (d) and (e). For (d) you will have to use Maple (so the answer is only a good approximation), but for (e) do the work by hand (turn in your work).

5. **(Infinite Series and Products):** This exercise deals with elementary programming constructs involving do loops, Maple's **seq** command, and some graphical display. The mathematical topics covered are infinite series and infinite products (the latter might be new to you).

Infinite Series: For a series $\sum_{n=1} a_n$, the convergence or divergence of the series, i.e., the question of whether we can sum the infinitely many terms in the series, is, by definition, dependent upon whether the sequence $\{s_n\}_{n=1}$ of partial sums converges or diverges (i.e., $\lim_{n\to} s_n$ exists or not). This is a theoretical question and *cannot* be determined by a computer. However, if we can show, using one of the many tests for convergence, that the series converges, then the computer can be used to estimate the value for the sum of the series. To explore this do the following.

(a) The standard p-series is the infinite series

$$\sum_{n=1} \frac{1}{n^p},$$

with $p > 0$. Do an experimental study of this type of series for $p = 1, 2, 3$. Specifically:

(i) Define the terms a_i of the series (which also depend on p) as a function of two variables `a:=(n,p)-> 1/n^p`. Use a two-dimensional array `s` and do loops to calculate the partial sums s_n for $n = 1, \ldots, 180$ and $p = 1, 2, 3$. *Note*: use an **evalf** to force "oating-point evaluation (otherwise Maple will give you fractions).

(ii) For $p = 1, 2, 3$, display in your worksheet the values of `s[n,p]` for $n = 100, \ldots, 180$, with five values per line. You can do this with a do loop and the **seq** command.

(iii) For $p = 1, 2, 3$, build a list
    `L[p]:=[[1,s[1,p]], . . . , [180,s[180,p]]]`
of points `[n,s[n,p]]` for plotting. Build a plot structure `ps[p]` using the plot command
    `ps[p]:=plot(L[p],style=point,color=c[p]):`
Here `c` is an array of your favorite three colors. Display all three plots in the same picture. Annotate this picture with the values of p and your best guess of the values $\lim_{n\to} s_n$. For this, study the values displayed in part (b). How many decimal places of accuracy are certain?

(b) Do a study, similar to that in part (a), for the alternating series

$$\sum_{n=1} \frac{(-1)^{n+1}}{n^p}.$$

(c) Do a study, similar to that in part (a), for the logarithmic series

$$\sum_{n=1}^{\infty} \frac{1}{(n+1)(\ln(n+1))^p}.$$

Here, to save space, omit the tables of values, and just use graphical displays as in Part (a)(iii). Click on the graphs to estimate the values of the sums for the series. Also use the integral test to determine, in general, for what values of $p > 0$ the series converges (hand in this work too).

(d) Do a study, similar to that in part (a), for the series

$$\sum_{n=1}^{\infty} \frac{n^{p+1}}{(p+1)^n}.$$

Here, to save space, omit the tables of values, and just use graphical displays as in part (a)(iii). Click on the graphs to estimate the values of the sums for the series. Also use the root test to determine, in general, for what values of $p > 0$ the series converges (hand in this work too).

Infinite Products: The concept of taking the product of infinitely many numbers is a natural analog of forming the sum of infinitely many numbers. Both are ideal concepts and involve using limits to formulate precisely. The notation for an infinite product is, naturally enough,

$$\prod_{n=1}^{\infty} b_n = b_1 b_2 b_3 \cdots,$$

where $\{b_n\}_{n=1}^{\infty}$ is a sequence of real numbers. Just as with infinite sums, we have to agree upon (or define) what we mean by such a product, since calculating manually, or by computer, infinitely many products is impossible in practice. Of course we can always take the product of *finitely* many terms, and thus the *sequence of partial products*

$$
\begin{aligned}
pr_1 &= b_1 \\
pr_2 &= b_1 b_2 \\
&\vdots \\
pr_n &= b_1 b_2 \cdots b_n \\
&\vdots
\end{aligned}
$$

is a well-defined sequence (and we can compute as many of the pr_n's as we wish with a computer). Thus, we get a sequence $\{pr_n\}_{n=1}^{\infty}$, which is called the *sequence of partial products* and, by definition, the infinite product is said to *converge* if the sequence of partial products converges.

In this part of the exercise, you are to study the infinite products of the following two types:

$$(1) \quad \prod_{n=1} (1 - a_n), \qquad (2) \quad \prod_{n=1} (1 + a_n),$$

where $a_n = 1/(n+1)^p$, for values $p = 1, 2, 3$. Specifically:

Set up two dimensional arrays `prod1` and `prod2` and use them to store the values of the partial products for $n = 1, \ldots, 180$ and $p = 1, 2, 3$. Then display the information graphically (as you did for series) with five sequences of partial products in the same pictures (five colors will help you keep things straight). The five sequences should be the type (1) products for $p = 1, 2, 3$, and the type (2) products for $p = 2, 3$. From your picture, try to determine if the infinite products converge and to what values. Annotate your picture with the results. What happens in the type (2) product for $p = 1$?

6. **(Newton s Method)** Use the code from the case study (cf. the CD-ROM) to find, approximately, all roots of the equation $f(x) = 0$ in the interval I, where f and I are from one of the items below. Also: (i) produce the graphics showing convergence of Newton's method to the root (or roots), (ii) compare the results found by our code with the results found by using Maple's `fsolve` command, and (iii) annotate your graphics appropriately.

   (a) $f(x) = x^2 - x^4 - 0.1$, for x in $I = [-1.2, 1.2]$.

   (b) $f(x) = x^3 - x - 0.2$, for x in $I = [-1.5, 1.5]$.

   (c) $f(x) = \sin x - \cos 2x + 0.2$, for x in $I = [0, 6]$.

7. **(Iterated Maps)** Work the exercises on the CD-ROM in CD Chapter 3 that pertain to iterated maps of an interval into itself.

3.4 Maple/Calculus Notes

3.4.1 Riemann Sums and the Definite Integral

Most calculus books motivate and define the definite integral $\int_a^b f(x)\,dx$ in pretty much the same way. However, there are slight differences in the definitions and the notation, and so we summarize a common approach here. You should consult your calculus book for how this is done as well as reading this.

We will assume that f is a continuous function on the interval $[a, b]$. Even though discontinuous functions are important and prevalent in the physical sciences, the assumption of continuity simplifies the discussion quite a bit.[7]

A *partition* of $[a, b]$ is a set of numbers $P = \{x_0, x_1, x_2, \ldots, x_N\}$, with $x_0 = a, x_N = b$, and $x_{i-1} < x_i$, for $i = 0, 1, \ldots, N$. The numbers x_i are called the partition points of P and they divide the interval $[a, b]$ into N different subintervals $[x_{i-1}, x_i]$ of lengths $\Delta x_i \quad x_i - x_{i-1} \ (i = 1, \ldots, N)$. Note that in the text we used partitions where the partition points are equally spaced, and thus there was a formula for them[8] and the length of each subinterval was the same, namely, $\Delta x = (b - a)/N$. Such partitions are called *regular* partitions. Now we want to consider the more general situation, allowing for arbitrary partitions, not just regular ones.

In the more general case, the measure of fineness for the partition P is the maximum of all the lengths of the subintervals:

$$ P \quad \max\{\Delta x_1, \Delta x_2, \ldots, \Delta x_n\}. $$

When P is small, the partition divides $[a, b]$ into many small subintervals, and P is thought of as a fine partition (as opposed to a coarse one).

For a partition $P = \{x_i\}_{i=0}^N$, suppose that $C = \{c_1, c_2, \ldots, c_N\}$ is a *selection* (*based on* P) of points $c_i \quad [x_{i-1}, x_i]$, for $i = 1, \ldots, N$. Then the sum

$$ S_{P,C}(f) \quad = \quad \sum_{i=1}^{N} f(c_i) \, \Delta x_i \tag{3.6} $$

$$ = \quad f(c_1) \, \Delta x_1 + f(c_2) \, \Delta x_2 + \cdots + f(c_N) \, \Delta x_N, \tag{3.7} $$

is called a *Riemann sum* (or the Riemann sum corresponding to partition P and selection C).

Now consider the set of all possible Riemann sums

$$ \mathcal{R}(f) = \{ S_{P,C}(f) \,|\, P \text{ is a partition, } C \text{ is a selection based on } P \}. $$

The set contains Riemann sums corresponding to partitions that are as fine as we wish, and as the fineness goes to (but never reaches) zero, the Riemann sums will differ little from each other. Thus, there should be a common number that all these Riemann sums are approaching. [9] There is

---

[7]Consult an advanced calculus book (e.g., [MH, p. 205, 448] and [Hof, p. 130]) for how to de"ne the Riemann integral for functions that are not continuous.

[8]In the equally spaced case, the formula is $x_i = a + i(b - a)/N$.

[9]The modern notion is that $\mathcal{R}(f)$ is a *lter base* or *net* and there is a de"nition for the limit L of a "lter base, $\mathcal{R}(f) \to L$. See [Dug].

a major theorem that asserts the existence of a unique number L with the following property. For each > 0, there is a > 0 such that

$$|S_{P,C}(f) - L| < \, , \tag{3.8}$$

for *all* partitions P with $P <$ and *all* selections C, based on the partition P. Roughly speaking this says that all the Riemann sums with fine enough partitions are within of L, and this is so regardless of the selection C. The number L is called the definite integral of f over $[a, b]$:

$$\int_a^b f(x)\, dx \quad L.$$

It is this theorem (property (3.8)) that allows us to numerically approximate the definite integral in so many ways the arbitrariness in the selection C and type of partition P (as long as $P <$) gives us plenty of options. A standard option (used in this text and many calculus books) is to use only partitions with equally spaced partition points, $\Delta x_i = \Delta x = (b - a)/N$, for every i. The standard choices for the selections C are left-hand points, right-hand points, and midpoints. The fact that we can use other choices than the standard ones is helpful in certain numerical work. For instance, if f varies more rapidly on one region of $[a, b]$, you will get better approximations if the partitions are finer in that region.

Chapter 4

Conditionals - Flow of Control

Many elementary computer programs consist of only a simple list of statements to be executed sequentially, one after another from first to last. The "ow from one statement to the next can be visualized as a straight line, or linear, "ow. Examples in Chapter 1 illustrate this type of program.

More complex computer programs require *loops* and *branches* to encode their complicated algorithms. In the last chapter we saw how loops, or repetition statements, alter the linear "ow, causing a specific section of code to be executed a number of times. In this chapter we will study *conditional statements* which allow the linear "ow to branch into one of two sections of code based upon the truth or falsity of a control expression. When nested together, conditional statements can create an elaborate path of multiple branching through which the execution can "ow and thus, model any complex decision process.

4.1 Logic in Mathematics

In mathematics one is generally concerned with determining if a given statement is true or false. Many mathematical statements are very complicated and determining whether they are true or false can be a challenge. Complicated statements are often composed of simpler statements and the logical operators: and, or, not. If p and q are statements then

p and q, p or q, not p

are also statements and their truth values are determined according to the truth values of p, q and the natural interpretation of and, or, not. For example the statement that *Isaac Newton was an Englishman and Isaac Newton*

71

p	q	p and q	p or q	not p
true	true	true	true	false
true	false	false	true	false
false	true	false	true	true
false	false	false	false	true

Table 4.1: *Truth table.*

was a 17th century mathematician is true because both of the statements connected by the **and** are true. On the other hand the statement *Gottfried Leibniz was an Englishman and Gottfried Leibniz was a 17th century mathematician* is false because even though it is true that Leibniz was a 17th century mathematician, he was German not English. Based on this reasoning, if we change the **and** to an **or**, we see that the compound statement: *Gottfried Leibniz was an Englishman or Gottfried Leibniz was a 17th century mathematician* is true.

Table 4.1 summarizes how to analyze the truth value of compound statements from the truth values of their basic parts. Thus, we can determine the truth value of a complicated statement if it can be broken down into a sequence of basic statements connected by logical operators and if we know the truth values of the basic statements. The truth table automates this in a rote way.

Some expressions, such as x>4 and A=B, are neither true nor false as they stand, but rather have truth values that depend on the values of x,A,B. These are called *conditional expressions* and in programming they are known as *Boolean expressions*. Whether in mathematics or programming, x,A, and B must be known before the truth table can be applied.

Our challenge in programming is to design programs which are mathematically sound (logical), that is, to create computer algorithms which function as we intend. This may not always be easy, especially if several cases must be analyzed and different paths must be taken depending on which one of several conditions is true. The "ow of control along different paths, or branches, in a computer program is known as branching. The decision about which branch to take is based on the Boolean value (true or false) of a mathematical statement.

4.2   Relational and Logical Operators

As discussed above **and, or,** and **not** are logical operators. Operators such as **=, >,** and **<** are called relational operators since they indicate a relationship between two quantities. Other relational operators in Maple notation are : **>=** (greater than or equal to), **<=** (less than or equal to), and **<>** (not equal to), see Table 4.2.

=	equal to
<	less than
>	greater than
<=	less than or equal to
>=	greater than or equal to
<>	not equal to

Table 4.2: *Relational operators.*

In certain contexts these operators are used to determine which of two possible routes is to be taken in order to proceed. A simple example to illustrate the need for this routing is the familiar definition of a piecewise-defined function.

Example 4.1 Consider the function defined by

$$h(x) = \begin{cases} x^2 & \text{if } x \leq 1 \\ 2 - x & \text{if } 1 < x \leq 2 \\ \sin x & \text{if } x > 2 \end{cases}.$$

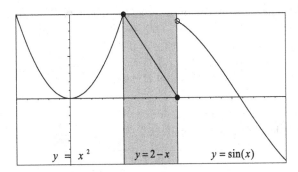

Figure 4.1: *Graph of the piecewise defined function h in Example 4.1.*

The graph of h on the interval $[-1, 4)$ is shown in Figure 4.1. For a particular value of x the value of $h(x)$ depends on the interval in which x is located. In this example, the value of $h(x)$ is determined from one of the three formulas x^2, $2 - x$, or $\sin x$ based upon which of the three conditional statements $x \leq 1$, $1 < x \leq 2$, or $x > 2$ is true. Note that $1 < x \leq 2$ is a compound statement standing for $1 < x$ and $x \leq 2$. This example illustrates the need to be able to evaluate the truth of an expression in the programming language. This situation comes up often when a decision must be made at a particular juncture before continuing with the algorithm.

4.3   Boolean Expressions

In Example 4.1 the expressions $x \leq 1$, $1 < x \leq 2$, and $x > 2$ are evaluated as true or false for a particular value of x instead of being evaluated to a number and for this reason they are called *Boolean expressions*. The possible values of Boolean expressions are known as *Boolean constants*, and thus `true` and `false` are Boolean constants. In Maple there is another Boolean constant: `FAIL`. This result appears if we ask Maple to decide whether a Boolean expression is true or false and it cannot do so. For example,

$$\texttt{is(c*2,positive)}$$

where `c` has not previously been assigned a numerical value will result in `FAIL`. Hence there are three possible answers when evaluating a Boolean expression: `true,false,FAIL`.

4.4   The if-then-else Statement

The basic method of branching in many computer programming languages is to use a construct of the form: `if-then-else`. The "ow of control in the program goes one way (in the *then* direction) when the *if* clause is true and in a different direction (the *else* direction) when it is false.

To do this branching in Maple we would use the `if-then-else-end if` statement, which has the following syntax:

> if C then S_1 else S_2 end if

The condition C is a Boolean expression which evaluates to either true or false and S_1, S_2 are sequences of Maple statements separated by semicolons (or colons). If the C evaluates to true the statements S_1 are executed and control is then transferred to the statement following the words **end if**. If condition C is false, then the statements S_2 are executed and control is transferred to the statement following the **end if**.

The **else** S_2 is optional in this command and the simpler case would be

$$\text{if } C \text{ then } S \text{ end if}$$

Here if C evaluates to true, the statements S are executed and then control goes to the statement following the **end if**. If C is false, then statements S are not executed and control is transferred to the statement after the words **end if**.

Note: Each **if** statement must be terminated by **end if**.

Example 4.2 Suppose we have Maple draw a picture of the prime staircase from 1 to 30. This is a step function where the step risers are of height one and occur only at prime integers. The treads for the steps are then of variable width and connect the risers. See Figure 4.2.

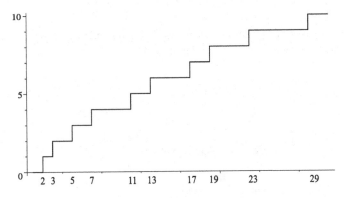

Figure 4.2: *Prime staircase.*

For each integer from 1 to 30 we need code to do one of two things: (1) draw a riser and part of the next tread or (2) draw a part of a tread. If the integer is prime we want to draw a riser (of one unit) and a one unit part of the next tread. If the integer is not prime we want to draw a unit part of a tread. Letting n be one of the integers, the condition that determines the "ow of control is is n prime. If this condition is true we want the draw a riser-tread code executed. If this condition is false we want the draw a tread code executed. We can easily do this using an `if-then-else-end if` construction. The Maple statement for is n prime is `is(n,prime)`. This evaluates to *true* if n is prime and to *false* if n is not prime. Our code then will consist of a loop for n running from 1 to 30. We will store the riser-treads and treads as plot structures in an array called `ps` and use the `display` command to draw the prime staircase shown in Figure 4.2.

```
ps:=array(1..30):
y:=0:
for n from 1 to 30 do
  if is(n,prime)
    then ps[n]:=plot([[n,y],[n,y+1],[n+1,y+1]]):
    y:=y+1;
    else ps[n]:=plot([[n,y],[n+1,y]]):
  end if;
end do;
with(plots):
display({seq(ps[n],n=1..30)});
```

Example 4.3 Consider the piecewise-defined function in Example 4.1. For each value of x one of the three formulas x^2, $2-x$, or $\sin x$ must be selected to define $h(x)$. There are several ways to write the code for defining this function in Maple. Using the logic illustrated in Figure 4.3,

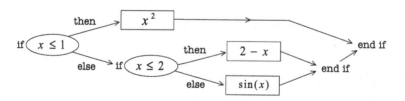

Figure 4.3: *Flow of control for Example* 4.1.

and the form

if C_1 then S_1 else if C_2 then S_2 else S_3 end if end if,

for two nested `if` statements, we can define h using a procedure in Maple as follows.

```
h:=proc(x)
```

```
    if x<=1 then x^2;
    else if x<=2 then 2-x;
        else sin(x);
        end if;
    end if;
end proc:
```

Notice that in constructing this procedure we used the fact that the real numbers are linearly ordered: if the statement $x \quad 1$ is false then the statement $x > 1$ is true. Similarly, if the statement $x \quad 2$ is false then the statement $x > 2$ must be true. This code consists of two nested `if-then-else` statements and the logic is a bit complicated. When several nested `if-then-else` statements are used, it is important to make the code readable (using spacing and indentations) so that we can easily trace through the logic if the procedure does not work properly.

In the next section we will see an equivalent method for defining the function h in this example. For a function like h, this method is not simpler than the one in the example, but you will find it simpler and clearer for functions involving many nested logic statements in their definitions.

4.5 The if-then-elif-then Statement

In Example 4.3 we used two nested `if-then-else-end if` statements. If there are many cases to consider the logic may become very complicated when using nested statements. The `if-then-elif-then` statement in Maple provides an alternative, perhaps clearer, way to express the logic. This statement has the following general form.

```
    if C₁ then S₁ elif C₂ then S₂ end if
```

The word elif is a combination of the words else and if . If we have several cases to check more than one `elif-then` statement can be used. We can rewrite the procedure in Example 4.3 using the `elif-then` construction:

```
h:=proc(x)
  if x<=1 then x^2
    elif x<=2 then 2-x
    else  sin(x)
  end if;
```

```
end proc:
```

Notice that the logic is much easier to follow. It might be clearer, though less e cient, if we do not use the fact that the real numbers are linearly ordered and use two `elif`'s:

```
h:=proc(x)
  if x<=1 then x^2
    elif x>1 and x<=2 then 2-x
    elif x>2 then sin(x)
  end if;
end proc:
```

This type of construction may include as many nested `elif`'s as needed to check all the cases.

```
    if C₁ then S₁
  elif C₂ then S₂
  elif C₃ then S₃
        ⋮
  elif Cₙ then Sₙ
    end if;
```

Figure 4.4 shows the diagram for the "ow of control for this collection of nested `elif` statements.

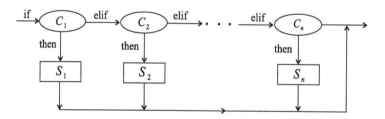

Figure 4.4: *Flow of control for nested* `elif` *statements.*

Some languages have a specific structure, often called a *case statement*, that controls the branching to a number of different cases based on the value of a variable k. The case statement tests if the value of k is in $\{1, 2, 3, \ldots, n\}$, and if it is, then the "ow of control passes to the block of statements S_k.

While Maple does not have such a case statement, we can always construct one out of nested `elif`'s as shown above. The following example is one situation where you might want to use such a case statement.

Example 4.4 (Step Functions) A type of piecewise-defined function that occurs frequently in calculus is the following. A function $f : [a, b] \longrightarrow \mathbb{R}$ is called a *step function* if the interval $[a, b]$ can be divided into subintervals using a partition $x_0 = a < x_1 < \cdots < x_{n-1} < x_N = b$, such that f is constant on each subinterval $[x_{i-1}, x_{i+1})$, $i = 1, \ldots, N - 1$. Thus, a step function is just a particular type of piecewise-defined function, i.e., one where the formula for f on each subinterval is of the form $y = c$. An example of a step function is

$$f(x) = \begin{cases} 3 & \text{if } 0 \le x < 1 \\ -1 & \text{if } 1 \le x < 2 \\ 2 & \text{if } 2 \le x < 3 \\ 1 & \text{if } 3 \le x < 4 \\ -2 & \text{if } 4 \le x < 5 \end{cases},$$

which is defined on the interval $[0, 5]$. This step function can be defined in Maple using nested `elif` statements as indicated above (exercise).

Example 4.5 (Step Functions of Two Variables) The idea of a step function can easily be generalized to two variables. A function f of two variables is a *step function* if its domain is a rectangle $R = [a, b] \times [c, d]$ and R can be subdivided into subrectangles: $R_{ij} = [x_{i-1}, x_i) \times [y_{j-1}, y_j)$, by partitions: $a = x_0 < \cdots < x_N = b$, of $[a, b]$, and $c = y_0 < \cdots < y_M = d$, of $[c, d]$, such that f is constant on each subrectangle.

A simple example of such a function, which involves a subdivision into only four subrectangles, is:

$$f(x, y) = \begin{cases} 1 & \text{if } (x, y) \in [0, 1) \times [0, 1) \\ 2 & \text{if } (x, y) \in [0, 1) \times [1, 2) \\ 1.5 & \text{if } (x, y) \in [1, 2) \times [0, 1) \\ 1 & \text{if } (x, y) \in [1, 2) \times [1, 2) \end{cases}$$

The graph of this function is a surface consisting of four square sections as shown in Figure 4.5.

The definition of f here is divided into four cases and we use nested `elif` statements to define this function in Maple as suggested in the discussion above.

```
f:=proc(x,y)
```

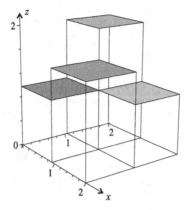

Figure 4.5: *The step function f in Example* 4.5.

```
if (0<=x and x<1)and (0<=y and y<1) then 1;
   elif (0<=x and x<1) and (1<=y and y<2) then 2;
   elif (1<=x and x<2) and (0<=y and y<1) then 1.5;
   elif (1<=x and x<2) and (1<=y and y<2) then 1
end if;
end proc:
```

Having defined the function f by the above procedure, we can use Maple to graph it. However, there are two things you should be aware of in doing this. First, since the procedure involves Boolean logic to decide which sub-rectangle the point (x, y) lies in, you must delay evaluation of the Boolean statements (otherwise you will get an error message).[1] To get around this, you can include the procedure call and its parameters in single quotes when doing the plot:

```
plot3d( f(x,y) ,x=0..2,y=0..2);
```

You can also plot this function using the command

```
plot3d(f,0..2,0..2);
```

where the function f instead of the expression $f(x, y)$ is plotted. If you execute either of the above plot commands, you will see the second thing that you should be aware of in this example. Since the function f is discontinuous,

---

[1]This has to do with the way parameters, such as x and y, get evaluated when passed to a procedure. See Chapter 5 for details of this.

it is di cult for Maple (or any CAS) to render it correctly. Maple does the best it can, but includes nearly vertical planes in making the steps among the four values of f. Figure 4.6 shows this.

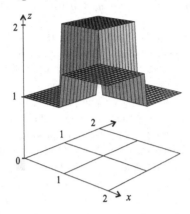

Figure 4.6: *The step function f plotted using the* `plot3d` *command.*

The plot command used to produce this figure is

```
plot3d({0, f(x,y) },x=0..2,y=0..2,axes=normal,
        scaling=constrained,orientation=[-37,75],
        projection=0.8,tickmarks=[3,0,2],
        shading=XYZ,lightmodel= light3 );
```

For step functions, it is possible to write your own plotting procedure so that this problem with the discontinuities of f does not occur (see the Case Study below). We did this to produce Figure 4.5, which accurately represents the graph of f.

In the last example, the domain $[0, 2] \times [0, 2]$ was divided into four equal subintervals using the partition $0 = x_0 < x_1 = 1 < x_2 = 2$ of $[0, 2]$ and thus it was easy to write a procedure with four nested `if`'s to define the step function. However, a general step function may have a large number, say K, of subrectangles in its definition, and thus a procedure to define it as above would require K nested `if`'s. The next example shows how to handle this with do loops and also provides for plotting of arbitrary step functions.

Example 4.6 Let us write a procedure to define an arbitrary step function f of two variables, whose domain is the rectangle $R = [a, b] \times [c, d]$. Suppose the $a = x_0 < x_1 < \cdots < x_N = b$ is a partition of $[a, b]$ and $c = y_0 < y_1 < \cdots < y_M = d$ is a partition of $[c, d]$ and f has the following form

$$f(x, y) = t_{ij} \quad \text{if } (x, y) \quad [x_{i-1}, x_i) \times [y_{j-1}, y_j),$$

where the t_{ij}'s are constants. The input parameters are to be the partitions, entered as two lists, and the constants t_{ij}, entered as an $N \times M$ array, as well as arguments x and y. This problem is similar to Example 4.5 in that the procedure takes a point (x, y), checks to see which subrectangle it lies in, and then assigns the value $f(x, y)$ accordingly. The partition points of the intervals $[a, b]$ and $[c, d]$ as well as the constants t_{ij} giving the values of the function must be passed to the procedure through input parameters.

We will let A be the list whose elements contain the partition points of the interval $[a, b]$ and B the list of partition points for $[c, d]$. Thus, we do not assume that the subintervals are all the same length. To set up the loops that assign the function values we will need the number of partition points for each interval. We can use the **nops** command to determine the number of elements in the lists A and B. The array T will contain the constant values; $T[i, j]$ is the value of the function on the rectangle

$$[A[i], A[i + 1]) \times [B[j], B[j + 1]).$$

We will use the local variable **val** to store the function value $f(x, y)$.

```
f:=proc(A,B,T,x,y)
  local n,m,i,j,val;
  n:=nops(A);m:=nops(B);
    for i from 1 to n-1 do
      for j from 1 to m-1 do
      if (A[i]<=x and x<A[i+1]) and
         (B[j]<=y and y< B[j+1]) then val:=T[i,j];
      end if;
      end do;
    end do;
  val;
end proc:
```

By using a compound statement as the condition to be checked we need only use a simple **if-then** statement.

Now, for given (or fixed) A,B,T, this procedure defines a function of (x, y). For instance, the following code:

```
A:=[0,1,2]; B:=[0,1,2];
T:=array([[1,2],[1.5,1]]);
g:=(x,y)->f(A,B,T,x,y);
```

defines a function g which coincides with the step function in the last example above.

Example 4.7 (Conic Sections) As another example of branching, we will look at the classification of conic sections given by the equation

$$Ax^2 + By^2 + Cx + Dy + E = 0. \qquad (4.1)$$

Since there is no xy term, the conic sections will have horizontal and/or vertical axes of symmetry. We will organize our procedure around the coefficients of the square terms A and B. We have the following cases:

(1) ($A \neq 0$ and $B \neq 0$) Completing the square on both of the squared terms in equation (4.1) yields

$$A\left(x + \frac{C}{2A}\right)^2 + B\left(y + \frac{D}{2B}\right)^2 = M,$$

where $M = \frac{BC^2 + AD^2 - 4EAB}{4AB}$.

 (i) If $M = 0$ and if

 (a) A and B have the same sign, then the graph is just a point $(-C/2A, -D/2B)$.

 (b) A and B have opposite signs, then the graph is two lines $x + \frac{C}{2A} = \pm \left|\frac{B}{A}\right|(y + \frac{D}{2B})$.

 (ii) If $M \neq 0$, then the equation can be rewritten as

$$\frac{(x + C/2A)^2}{M/A} + \frac{(y + D/2B)^2}{M/B} = 1.$$

 (a) If M/A and M/B are both positive, the graph is an ellipse.

 (b) If M/A and M/B have opposite signs, the graph is a hyperbola.

 (c) If M/A and M/B are both negative, there is no graph.

(2) ($A = 0$ and $B \neq 0$) Equation (4.1) in this case is equivalent to

$$B\left(y + \frac{D}{2B}\right)^2 = -Cx - E + \frac{D^2}{4B}.$$

(i) If $C \neq 0$, then the equation becomes

$$B\left(y + \frac{D}{2B}\right)^2 = -C\left(x - \frac{E}{C} + \frac{D^2}{4BC}\right),$$

which is a parabola.

(ii) If $C = 0$, then the equation is

$$\left(y + \frac{D}{2B}\right)^2 = N,$$

where $N = \frac{-E}{B} + \frac{D^2}{4B^2}$.

(a) If N is positive, the graph is two lines.

(b) If N is zero, the graph is a point.

(c) If N is negative, there is no graph.

(3) $(A \neq 0 \text{ and } B = 0)$ This is similar to case (2) and we leave the analysis as an exercise.

(4) $(A = 0 \text{ and } B = 0)$ Equation (4.1) in this case reduces to

$$Cx + Dy + E = 0,$$

which is a line or all of $\mathbb{R}^2$.

The code for this procedure then would be the following:

Code for Classifying Conic Sections

```
conics:=proc(A,B,C,D,E)
  local N,M,ans;
  if A<>0 and B<>0 then
      M:=evalf((B*C^2+A*D^2-4*E*A*B)/(4*A*B));
        if M=0 then
              if A*B>0 then ans:= "point"
              else ans:="two lines"
              end if;
        elif M<>0 then
          if (M/A)>0 and (M/B)>0 then  ans:="ellipse"
          elif (M/A)<0 and (M/B)<0 then ans:="no graph"
          else ans:="hyperbola"
```

```
          end if;
        end if;
   elif A=0 and B<>0
        then N:=-E/B+D^2/(4*B^2);
          if C<>0 then ans:="parabola"
          elif C=0 then
            if N>0 then ans:="two lines"
            elif N=0 then ans:="point"
            else  ans:="no graph"
            end if;
          end if;
   elif A<>0 and B=0   then ans:="Exercise"
   elif A=0 and B=0 then
        if C<>0 or D<>0 then ans:="line"
        elif E=0 then ans:="the whole plane"
        else ans:="no graph"
        end if;
   end if;
   ans;
 end proc:
```

This example should illustrate how the `elif` construction helps clarify the branching into the various cases given in the prior analysis. The four main cases (1) (4) in the above analysis correspond to the `if` and three `elif`'s which are the least indented in the code. The subcases, like (i) (ii) and (a) (c), in the analysis correspond to the respectively more indented `if`'s and `elif`'s in the code. Notice that the value `ans` to be printed is, not a number, but rather a phrase consisting of one or more words. Enclosing words in double quotes, as shown in the code, produces a *string* data type. This data type can be assigned to a name and can be output from a procedure. For more information about strings see Chapter 6.

4.6 Case Study: Riemann Sums for a Double Integral

Consider a rectangle R in the plane, say, $R = [a,b] \times [c,d]$. The double integral, over R, of a function f of two variables is denoted by

$$\int_R f(x,y)\,dA,$$

and its definition is a direct extension of the definition of the single integral for a function of a single variable. See the Maple/Calculus Notes at the end of this chapter for the precise definition. The definition in essence amounts to taking limits of Riemann sums as in the one-variable case, but now these sums are actually *double sums*.

This case study develops ways of approximating $\int_R f(x,y)\,dA$ by Riemann sums and is a direct analog of the one-variable case in Examples 3.1-3.2. This is an easy task, but the extension of the code to allow for double integrals over more complicated regions D than rectangles R is more challenging and involves logic statements. Producing graphics for the approximating parallelepipeds is also more involved than the one-variable case.

Problem 1: Approximate the double integral $\int_R f(x,y)\,dA$, where R is a rectangle. Do this by using a subdivision of R into $NM = N \times M$ congruent rectangles and *lower left-hand corner*, double Riemann sums, i.e., use evaluations of f at the lower left-hand corners of the subrectangles in the subdivision of R. Produce a picture of the approximating parallelepipeds.

Analysis of Problem 1: We subdivide the rectangle R into NM subrectangles by choosing a partition $a = x_0 < x_1 < \cdots < x_{N-1} < x_N = b$ of $[a,b]$ and a partition $c = y_0 < y_1 < \cdots < y_{M-1} < y_M = d$ of $[c,d]$. Then the i-jth subrectangle is

$$R_{ij} = [x_{i-1}, x_i] \times [y_{j-1}, y_j],$$

and has area $\Delta x_i \Delta y_j$, where

$$\Delta x_i = x_i - x_{i-1}, \qquad \Delta y_j = y_j - y_{j-1}.$$

To motivate how the double Riemann sum approximations arise, consider the case where f is nonnegative on the rectangle R. Then we can form parallelepipeds (boxes) over each subrectangle R_{ij} by using the subrectangle as the base and taking the height to be the number

$$z_{i-1,j-1} = f(x_{i-1}, y_{j-1}),$$

which is the value of f at the lower, left-hand corner: (x_{i-1}, y_{j-1}) of R_{ij}. See Figure 4.7. Then $V_{ij} = f(x_{i-1}, y_{j-1})\Delta x_i \Delta y_j$ represents the volume of this parallelepiped. Hence, the volume bounded by the surface $z = f(x,y)$ and the xy plane would be approximated by summing these volumes over all i and j:

$$V = \int_R f(x,y)\,dA \approx \sum_{i=1}^{N} \sum_{j=1}^{M} f(x_{i-1}, y_{j-1})\Delta x_i \Delta y_j.$$

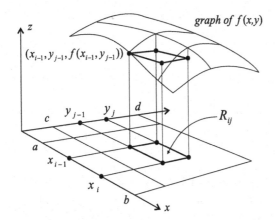

Figure 4.7: *A parallelepiped formed using $f(x_{i-1}, y_{j-1})$ as the height, where (x_{i-1}, y_{j-1}) is the lower, left-hand vertex of the i-jth subrectangle.*

Computing the double Riemann sum is quite simple. We just use two do loops to sum up the values for $f(x_{i-1}, y_{i-1})\Delta x_i \Delta y_j$. The challenging part of this problem will be producing the graphics. The surface forming the top faces of the parallelepipeds can be defined as a step function and the code for doing this is similar to that in Example 4.6. Plotting this function using the plot command does not give a very good picture. Instead we can create a nice picture by actually drawing in all the edges of each parallelepiped and the top surface as a polygon. This will involve

(a) writing the code to generate all the vertices of all the polygons, and

(b) creating plot structures for all the edges and the top surfaces of the parallelepipeds.

Design of the Program: We will assume f, a, b, c, d, N, M have been previously defined in the session. We will assume all the subintervals of $[a, b]$ have length $\Delta x = \frac{b-a}{N}$ and those of $[c, d]$ have length $\Delta y = \frac{d-c}{M}$. Thus, the partition points of $[a, b]$ are $x_i = a + i\Delta x$, for $i = 0, \dots, N$, and those for $[c, d]$ are $y_j = c + j\Delta y$, for $j = 0, \dots, M$.

The following code then computes the double Riemann sum.

```
dx:=(b-a)/N;dy:=(d-c)/M;
s:=0;
for i from 0 to N-1 do
```

```
  X:=a+i*dx;
 for j from 0 to M-1 do
   Y:=c+j*dy;
   s:=s+f(X,Y)*dx*dy;
 end do;
end do;
```

Producing the picture is a bit more involved. We want to produce a plot of all the parallelepipeds which have volumes given by the summands. We will draw the top surface of each parallelepiped as a polygon and the edges of the remaining surfaces as spacecurves.[2] We will store all of the pictures in plot structures and use the `display` command at the very last to graph the entire picture. First we will store the coordinates of the partition points in arrays X and Y. Then we will store the function values $z_{ij} = f(x_i, y_j)$ in a 2-dimensional array Z.

```
vert:=array(0..N,0..M);basex:=array(0..N);basey:=array(0..M);
top:=array(0..N,0..M);
X:=array(0..N);Y:=array(0..M);
Z:=array(0..N,0..M);
for i from 0 to N do
  for j from 0 to M do
    X[i]:=a+i*dx;Y[j]:=c+j*dy;Z[i,j]:=f(X[i],Y[j]);
  end do;
end do;
```

To draw the bottom faces of the parallelepipeds we need to draw the grid on the rectangle R. The line segments parallel to the y-axis in this grid will be stored in the array `basex` and the ones parallel to the x-axis in the array basey.

```
with(plots):
for i from 0 to N do
  basex[i]:=spacecurve([[X[i],Y[0],0],[X[i],Y[M],0]],
                                          color=black);
end do:
for j from 0 to M do
  basey[j]:=spacecurve([[X[0],Y[j],0],[X[N],Y[j],0]],
```

---

[2]Polygons can be drawn in two and three diemensions by using Maple s `ploygonplot` and `polygonplot3d` commands.

```
                                        color=black);
   end do:
```

Now we draw the top faces, stored in array **top**, and the vertical edges, stored in array **vert**.

```
for i from 0 to N-1 do
   for j from 0 to M-1 do
      top[i,j]:=polygonplot3d([[X[i],Y[j],Z[i,j]],
                 [X[i+1],Y[j],Z[i,j]],[X[i+1],Y[j+1],Z[i,j]],
                 [X[i],Y[j+1],Z[i,j]], [X[i],Y[j],Z[i,j]]]);
      vert[i,j]:=spacecurve({[[X[i],Y[j],0],
                 [X[i],Y[j],Z[i,j]]],[[X[i+1],Y[j],0],
                 [X[i+1],Y[j],Z[i,j]]],[[X[i+1],Y[j+1],0],
                 [X[i+1],Y[j+1],Z[i,j]]],[[X[i],Y[j+1],0],
                 [X[i],Y[j+1],Z[i,j]]]},color=black);
   end do;
end do;
```

To display all of the plot structures, we combine them in a single set. We do this using several **seq** commands.

```
display({seq(seq(top[i,j],i=0..N-1),j=0..M-1),
         seq(seq(vert[i,j],i=0..N-1),j=0..M-1),
         seq(basex[i],i=0..N),seq(basey[j],j=0..M)});
```

Figure 4.8 shows the parallelepipeds used in the Riemann sum approximation $N = M = 15$, for $f(x,y) = x^2 + y^2$ and $R = [-1,1] \times [-1,1]$.

***Problem* 2:** Devise a scheme to approximate $\int_D f(x,y)dA$, where D is the triangular region determined by the points $(1,2), (4,5)$, and $(8,4)$.

***Analysis of Problem* 2:** For suitable regions D in the plane, we can devise a double Riemann sum approximation as follows. We first enclose the entire region in a rectangle R, preferably one as small as possible. Then for each subdivision of R into subrectangles R_{ij}, we produce a grid that covers D by retaining all those subrectangles that have at least one vertex in D and eliminating those that do not. Then the approximating Riemann sum for the double integral of f over the region D is taken to be a sum over all the subrectangles in the grid:

$$\sum_{i,j \text{ with } R_{ij} \text{ grid}} f(x_{i-1}, y_{j-1}) \, \Delta x_i \Delta y_j.$$

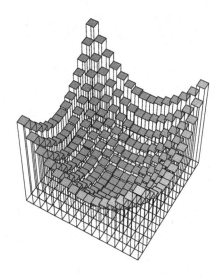

Figure 4.8: *The 225 parallelepipeds used in the $N = M = 15$, Riemann sum approximation for* $\int_R(x^2 + y^2)\, dA$*, where* $R = [-1, 1] \times [-1, 1]$.

For the triangular region D in this problem (see Figure 4.9), we can use the rectangle $R = [1, 8] \times [2, 5]$ for this type of double Riemann sum. Thus, dividing [1,8] into N subintervals $[x_{i-1}, x_i], i = 1, \ldots, N-1$, each of length 7/N and dividing [2,5] into M subintervals $[y_{j-1}, y_j], j = 1, \ldots, M)$, each of length 3/M we get NM subrectangles R_{ij}. Now to decide which of these is in the grid, notice that D is bounded by the graphs of three lines

$$-x + y = 1, \quad x + 4y = 24, \quad -2x + 7y = 12.$$

With a little further work you can convince yourself that the region D consists of precisely those points (x, y) that satisfy the three inequalities:

$$-x + y \quad 1, \quad x + 4y \quad 24, \quad -2x + 7y \quad 12.$$

These inequalities will be our test conditions to determine if any of the four vertices of R_{ij} lie in D and thus if R_{ij} is to be included in the grid.

Design of the Program: We will use the function $f(x, y) = 1$ and $N = M = 10$ to be specific. First we need to define the boundaries of D as functions.

```
N:=10; M:=10;
```

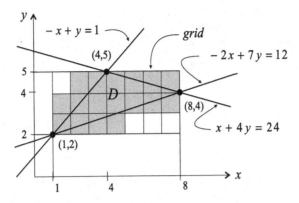

Figure 4.9: *The triangular region D determined by the points $(1,2), (4,5)$, and $(8,4)$. An approximating grid of rectangles (shown in gray).*

```
a:=1; b:=8; c:=2; d:=5;
f:=(x,y)->1;
F:=(x,y)->y-x-1;
G:=(x,y)->x+4*y-24;
H:=(x,y)->-2*x+7*y-12;
```

We can compute the double Riemann sum using a pair of do loops, much like we did in *Problem 1*. The difference here is that we do not include in the Riemann sum those terms corresponding to *all* the rectangles R_{ij}, but only those corresponding to rectangles in the grid. So for each R_{ij} we must put in a check (an **if-then** construction) to see if $f(P)\Delta x\Delta y$, where P is the lower left-hand corner of R_{ij}, will be included in our sum. This check involves the four vertices of R_{ij} as shown in Figure 4.10.

```
dx:=(b-a)/N; dy:=(d-c)/M;
Rsum:=0;
for i from 0 to N-1 do
   for j from 0 to M-1 do
     X:=a+i*dx;Y:=c+j*dy;
     P:=X,Y;
     Q:=X+dx,Y;
     R:=X+dx,Y+dy;
     S:=X,Y+dy;
     if  (F(P)<=0 and G(P)<=0
       and H(P)>=0 )or
```

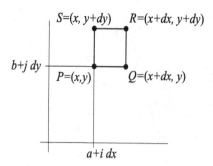

Figure 4.10: *The four vertices to be tested as to whether at least one of them lies in D.*

```
        (F(Q)<=0 and G(Q)<=0
        and H(Q)>=0 )or
        (F(R)<=0 and G(R)<=0
        and H(R)>=0 )or
        (F(S)<=0 and G(S)<=0
        and H(S)>=0 )
     then
        Rsum:=Rsum+evalf(f(P)*dx*dy);
     end if;
   end do;
end do;
```

For the function $f(x,y) = 1$, the Riemann sum above will give the approximate area, $\int_D 1\, dA$, of the triangular region D. We chose this particular f so that we could check our code by directly calculating, by hand, the area of the triangle. This is always a good practice to use before using the code on other functions f.

Exercise Set 4

1. Write a procedure to define each of the following functions.

(a)

$$f(x) = \begin{array}{ll} \sin(3x) & \text{if } x \quad 0 \\ \sin(x/3) & \text{otherwise} \end{array} .$$

(b)

$$g(x) = \begin{cases} 1 + \sin(x) & \text{if } x \quad 0 \\ 9 - x^2 & \text{if } 0 < x \quad 3 \\ (x-3)^2 & \text{if } x > 3 \end{cases}.$$

2. Plot the functions in Exercise 1. Since these are procedures having Boolean expressions involving the input parameter x, you must use the plot command with your function expression enclosed in single quotes `plot( h(x) ,x=a..b)` to delay evaluation, or plot the function `plot(h,a..b)`.

3. (a) Using nested `elif` statements, write a procedure to define the step function given in Example 4.4. Have Maple plot this function.

   (b) Write a procedure to define an arbitrary step function of one-variable whose domain is the interval $[a, b]$. This is a one variable version of Example 4.6.

   (c) Let f be a function which is continuous on the interval $[a, b]$ and $a = x_0 < x_1 < \cdots < x_N = b$ be a partition of $[a, b]$ with $\Delta x = x_i - x_{i-1} = (b-a)/N$. Adjust the procedure in part (b) to define the following step function f_N in Maple.

   $$f_N(x) = f(x_{i-1}), \qquad \text{if } x \quad [x_{i-1}, x_i).$$

   (d) For the function $f(x) = 12e^{-x} \sin x$, on the interval $[0, 3]$ define the functions f_{10}, f_{15}, f_{20}, and f_{30} using your procedure in part (c). Plot these step functions and f, all in different colors, in the same picture.

   (e) What is the relationship between $\int_a^b f_n(x)\, dx$ and $\int_a^b f(x)\, dx$?

4. (a) Let f be a function which is continuous on the rectangle $R = [a, b] \times [c, d]$ and consider the standard partition of R into subrectangles $R_{ij} = [x_{i-1}, x_i) \times [y_{j-1}, y_j)$, $i = 1, \ldots, N$, $j = 1, \ldots, M$. Adjust the procedure in Example 4.6 to define the step function $f_{N,M}$ given as follows:

   $$f_{N,M}(x, y) = f(x_{i-1}, y_{j-1}), \qquad \text{if } (x, y) \quad R_{ij}.$$

   (b) Let $f(x, y) = 1 - x^2 - y^2$ on the rectangle $R = [-4, 4] \times [-4, 4]$. Partition R into 9 congruent subrectangles $R_{ij} = [x_{i-1}, x_i) \times [y_{j-1}, y_j)$, where $i = 1, 2, 3$ and $j = 1, 2, 3$. Write a procedure to plot the graph of $f_{3,3}$, similar to Figure 4.5, so that the problems with the discontinuities do not occur.

   (c) For the general step function $f_{N,M}$ as in part (a), what is the relationship between $\int_R f_{NM}(x, y)\, dA$ and $\int_R f(x, y)\, dA$?

5. Complete Example 4.7 (Classifying Conic Sections) by including the case where $A \neq 0$ and $B = 0$.

6. Calculate, by hand, the area of the triangle in Figure 4.9. Use the code
   in problem 2 of the case study to calculate the approximations to this area
   with N, M having values $10, 50, 100, 150, 200$. How do these compare with
   the true area? Calculate, by hand, the double integral $V = \int_L f(x,y)\, dA$,
   where $f(x,y) = 16 - x - y$. Show your work. Interpret this as the volume
   of a certain solid and draw this solid. Use the code in problem 2 of the
   case study to calculate the approximations to this volume with N, M having
   values $10, 50, 100, 150, 200$.

7. Write a procedure to calculate the approximate value of the double integral

$$\int_L f(x,y)\, dA,$$

where D is the region bounded by $y = g(x)$, $y = h(x)$, $x = a$, $x = b$. Here
it is assumed that $g(x) < h(x)$, for all $x \in [a, b]$. See Figure 4.11.

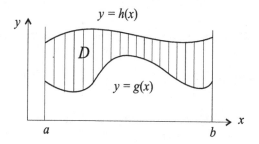

Figure 4.11: *The region D bounded by the graphs of two functions.*

Here is a suggested way to do this, as well as some additional tasks for your
procedure.

(a) Have f, a, b, g, h, M, N as input to the procedure and have the approxi-
    mate value of the double integral as an output (the name for it is your
    choice). Divide $[a, b]$ into equal subintervals $[x_{i-1}, x_i]$, $i = 1, \ldots, N$,
    each of length $\Delta x = (b - a)/N$.

(b) Calculate two numbers c and d:

$$c = \min\{g(x_0), g(x_1), \ldots, g(x_N)\}$$
$$d = \max\{h(x_0), h(x_1), \ldots, h(x_N)\}.$$

This will give a rectangle $R = [a, b] \times [c, d]$ which contains D (approximately). You can use Maple's min or max commands in doing this if
you wish. See Figure 4.12. Note: c and d will vary with N.

(c) Subdivide $[c, d]$ into equal subintervals $[y_{j-1}, y_j]$, $j = 1, \ldots, M$, each
    of length $\Delta y = (d - c)/M$. This, along with the subdivision of $[a, b]$

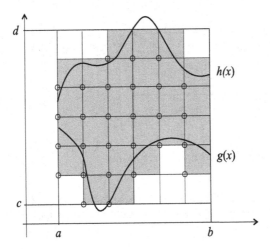

Figure 4.12: *The numbers c and d used to form a rectangle R that approximately encloses D. The subrectangles in the partition have at least one vertex inside the region D.*

above, gives a subdivision of R into subrectangles R_{ij}, $i = 1, \ldots, N$, $j = 1, \ldots, M$. Then in forming the Riemann sum, just use subrectangles R_{ij} which have at least one vertex inside the region D. See Figure 4.12. Thus, your procedure should have some logic to test each vertex of R_{ij} to see if it is inside D.

(d) Have your procedure plot all the vertices of the subrectangles that have at least one vertex inside D. Include the plot of the boundary of the region D in this picture. While the plot of the vertices is a plot of a list of points, you might find it easier to work with a set of points. Maple's union command can be used to add four more vertices to the set of vertices. After building the total set, it can easily be converted to a list for plotting.

(e) Show (by hand) how the double integral over a region D of this type can be written as an iterate of two single integrals. Show how this formula can be used to find the area of D.

(f) Test your procedure on $f(x, y) = 1$ and $f(x, y) = x^2 + y^2$, with D the triangle shown in Figure 4.13 and $a = 1, b = 3, c = 1, d = 2$.

Use $N = M = 10, 20, 100$. Compute the exact answers by hand and comment on how good your approximations are. Can you use Maple to compute the exact values ?

(g) Test your procedure on $f(x, y) = 1$ and $f(x, y) = x^2 + y^2$, with D the region bounded by: $y = 1 + \sin 5x$, $y = 3 + \cos 6x$, $x = 0$, $x = 1.5$.

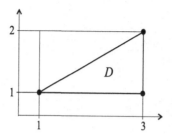

Figure 4.13: *A triangular region D.*

Use $N = M = 10, 20, 100$. Compute the exact answers by hand for $f(x, y) = 1$ and by Maple for $f(x, y) = x^2 + y^2$. Comment on how good your approximations are.

8. Generalize the procedure you wrote for exercise 6 to the case where the region D is bounded by one or more suitable curves, each with an equation of the form $H(x, y) = 0$. You may, if you wish, limit yourself to just four curves, and assume that the equations for the curves are set up so that D is the region
$$D = \{ (x, y) \quad \mathbb{R}^2 \mid H_i(x, y) \quad 0, \ i = 1, 2, 3, 4 \}.$$

Have the functions H_1, H_2, H_3, H_4 as input to the procedure. If there are fewer than four curves, just repeat one of them as necessary, e.g., H, H, H, H if there is only one curve. See Figure 4.14.

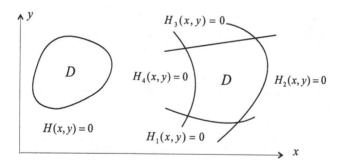

Figure 4.14: *Regions D bounded by a single curve and by several curves.*

9. **(Interiors and Exteriors of Triangles)** Suppose P, Q, R are non-collinear points in $\mathbb{R}^2$, and consider the regions determined by the lines $\overline{PQ}, \overline{QR}$, and $\overline{RP}$ as shown in Figure 4.15.

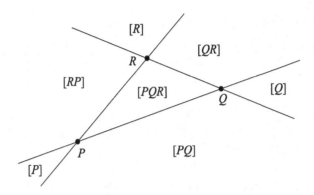

Figure 4.15: *The regions determined by three non-collinear points* P, Q, R *in* $\mathbb{R}^2$.

There are seven regions shown and, while it is not shown in the figure, it is understood that none of the points on the lines is included in any of the regions. This exercise is to write a procedure, called `checkpoint`, that takes a given point (x, y) in $\mathbb{R}^2$ and determines (1) if (x, y) lies in one of these regions and if so which one, or (2) if (x, y) lies on one of the lines, and if so if it is on a side of ΔPQR, is a vertex of ΔPQR, or is exterior to ΔPQR. The input to the procedure should be P, Q, R, x, y and the procedure should print out an appropriate designation for where the point (x, y) lies. See the Maple/Calculus Notes for some mathematical background that should help in this exercise.

4.7 Maple/Calculus Notes

4.7.1 Interiors and Exteriors of Triangles

Any triangle ΔPQR in $\mathbb{R}^2$ has an interior and exterior set of points   the interior points lie strictly inside (not on any of the three sides) and the exterior points lie strictly outside. If each side of the triangle is extended indefinitely in both directions, the resulting lines $\vec{PQ}, \vec{QR}, \vec{RP}$ further divide the exterior set of points. All together, not including the points on any of the lines, we get seven regions: $[PQR], [PQ], [QR], [RP], [P], [Q], [R]$ as shown in Figure 4.15. To define these regions precisely (and analytically), we introduce the following notation and concepts.

If $v = (v_1, v_2)$ and $w = (w_1, w_2)$ are linearly independent vectors[3] in $\mathbb{R}^2$,

---

[3]When there are just two vectors v, w, linear independence means that w is not a multiple of v.

then we say that v, w are *positively oriented* (or form *a right-handed system*) if the determinant

$$\det(v, w) = \begin{vmatrix} v_1 & v_2 \\ w_1 & w_2 \end{vmatrix} = v_1 w_2 - v_2 w_1,$$

is positive, $\det(v, w) > 0$. Otherwise, when this is negative, we say v, w are *negatively oriented* (a *left-handed system*). We say that the triangle $\triangle PQR$ is *positively oriented* if the vectors $v = \overrightarrow{PQ}, w = \overrightarrow{PR}$ are positively oriented. Otherwise, the triangle is called *negatively oriented*.

We can describe the line $\overrightarrow{PQ}$ through P, Q in terms of any given normal vector $n = (n_1, n_2)$ to the line. See Figure 4.16.

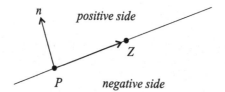

Figure 4.16: *The geometric description of the line through a point P with normal n.*

Technically all we need is one point P and a given normal n to the line. Thus, a point $Z = (x, y)$ is on the line if and only if

$$n \cdot \overrightarrow{PZ} = 0.$$

If $P = (p_1, p_2)$, then $\overrightarrow{PZ} = (x - p_1, y - p_2)$, and so the above equation is

$$n_1 x + n_2 y - c = 0, \tag{4.2}$$

where $c = n_1 p_1 + n_2 p_2$. Equation (4.2) is the implicit description of the line. Thus, letting

$$f(x, y) \quad n_1 x + n_2 y - c,$$

we get that a point (x, y) in $\mathbb{R}^2$ lies on the line if and only if $f(x, y) = 0$. Hence, a point (x, y) is not on the line, but rather is on one side or the other of it, if either $f(x, y) > 0$ or $f(x, y) < 0$. It is easy to see from the geometry in Figure 4.16 that $f(x, y) > 0$, when (x, y) is on the side of the line toward which the normal n points. This is called the *positive side* of the line and labeled $+$. The negative side of the line is the other side and is labeled $-$.

The canonical *normal vector* to v is the vector $N(v) = (-v_2, v_1)$. It is easy to see that the vectors $v, N(v)$ are positively oriented.

With this background established, we can now describe precisely the regions $[PQR], [PQ], [QR], [RP], [P], [Q], [R]$, determined by a triangle and the extensions of its sides. For this we let

$$v = \overrightarrow{PQ}, \quad w = \overrightarrow{QR}, \quad u = \overrightarrow{RP},$$

be the three vectors along the sides of the triangle (with the indicated directions), and let

$$n = N(v), \quad m = N(w), \quad k = N(u),$$

be the canonical normal vectors to these sides respectively. Further let

$$
\begin{aligned}
f(x,y) &= n_1 x + n_2 y - c \\
g(x,y) &= m_1 x + m_2 y - d \\
h(x,y) &= k_1 x + k_2 y - e,
\end{aligned}
$$

where $c = n_1 p_1 + n_2 p_2$, $d = m_1 q_1 + m_2 q_2$ and $e = k_1 r_1 + k_2 r_2$. Then the three lines determined by the sides of the triangle can be described implicitly by the three equations $f(x,y) = 0$, $g(x,y) = 0$, and $h(x,y) = 0$, respectively. Figure 4.17 shows a positively oriented triangle along with the three canonical normals to its respective sides.

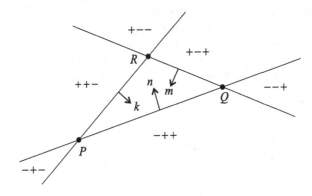

Figure 4.17: *A positively oriented triangle ΔPQR with the canonical normal vectors n, m, k to the sides $\overline{PQ}, \overline{QR}, \overline{RP}$.*

It is easy to see geometrically that a point (x, y) is in the inside of the triangle if $f(x, y) > 0$, $g(x, y) > 0$, and $h(x, y) > 0$. With this motivation we define the set $[PQR]$ (the interior of the triangle) by

$$[PQR] = \{\,(x,y) \in \mathbb{R}^2 \mid f(x,y) > 0, g(x,y) > 0, h(x,y) > 0\,\}.$$

Similarly, motivated by the geometry in Figure 4.17, we define the region $[PQ]$ by

$$[PQ] = \{\,(x,y) \in \mathbb{R}^2 \mid f(x,y) < 0, g(x,y) > 0, h(x,y) > 0\,\},$$

and the region $[P]$ by

$$[P] = \{\,(x,y) \in \mathbb{R}^2 \mid f(x,y) < 0, g(x,y) > 0, h(x,y) < 0\,\}.$$

Alternatively, you might find it easier to use a shorthand notation for each of these regions, say, by writing $[PQR] = \{+ + +\}$, $[PQ] = \{- + +\}$, and $[P] = \{- + -\}$. In any event, you can now describe the other four regions, $[QR], [RP], [Q], [R]$, in a similar fashion. *Note*: This discussion is for positively oriented triangles. For negatively oriented triangles everything is just the reverse. More specifically, each of the inequality signs must be reversed in the definition of the regions. Alternatively, you could just replace f, g, h, by $-f, -g, -h$. For programming purposes, you can cover both the positively and negatively oriented cases by using $\epsilon f,\ \epsilon g,\ \epsilon h$, where $\epsilon = \pm 1$.

4.7.2 Riemann (Double) Sums and Double Integrals

We review here the theory of integration for functions f of two variables and the definition of the double integral in terms of (double) Riemann sums.

As in the one-variable case, we simplify the discussion by restricting it to just continuous functions f. Consider first the double integral $\int\int_R f(x,y)\,dA$ of f over a rectangle $R = [a,b] \times [c,d]$, which is contained in its domain.

A *partition* of $R = [a,b] \times [c,d]$ is a set of points

$$P = \{\,(x_i, y_j)\,\}_{i=1,\dots,N}^{j=1,\dots,M},$$

where $P_x = \{x_0, x_1, x_2, \dots, x_N\}$ and $P_y = \{y_0, y_1, y_2, \dots, y_M\}$ are partitions of $[a,b]$ and $[c,d]$, respectively. (See the Maple/Calculus Notes in Chapter 3.) The points (x_i, y_j) are called the partition points of P and they divide the rectangle R into subrectangles

$$R_{ij} = [x_{i-1}, x_i] \times [y_{j-1}, y_j] = \{\,(x,y) \mid x_{i-1} \le x \le x_i,\ y_{j-1} \le y \le y_j\,\},$$

for $i = 1, \dots, N$ and $j = 1, \dots, M$. Letting $\Delta x_i = x_i - x_{i-1}$ and $\Delta y_j = y_j - y_{j-1}$, we get that the area of the i-jth subrectangle R_{ij} is:

$$\Delta A_{ij} = \Delta x_i \, \Delta y_j.$$

The fineness of the partition P is measured by how small the largest of its subrectangles is

$$\|P\| \quad \max\{\,\Delta A_{ij}\,|\,i = 1,\dots,N,\ j = 1,\dots,M\,\}.$$

For a partition P of R, suppose

$$C = \{\,(p_{ij}, q_{ij})\,|\,i = 1,\dots,N,\ j = 1,\dots,M\,\},$$

is a collection of points, one selected from each subrectangle, $(p_{ij}, q_{ij})\ \ R_{ij}$. This is called a *selection based on* P. We get a Riemann sum $S_{P,C}(f)$ corresponding to each partition P and each selection C based on P. Simply evaluate f at the point selected in each subrectangle, multiply by the area of the subrectangle, and then sum over all rectangles:

$$S_{P,C}(f) = \sum_{i=1}^{N}\sum_{j=1}^{M} f(p_{ij}, q_{ij})\,\Delta A_{ij}. \tag{4.3}$$

This is called a *Riemann (double) sum*. Just as in the one-variable case, the theory gives us here that the set of Riemann sums $\{S_{P,C}(f)|P,C\}$ approaches a unique number L as the partition fineness tends to zero. This number L has the following property. For each $\ > 0$, there is a $\ > 0$ such that

$$|S_{P,C}(f) - L| < \ , \tag{4.4}$$

for *all* partitions P with $\ \|P\| <\ $ and *all* selections C, based on the partition P. Roughly speaking this says that all the Riemann sums with fine enough partitions are within $\ $ of L, and this is so regardless of the selection C. The number L is called the (definite) double integral of f over R:

$$\int_{R} f(x,y)\,dA \quad L.$$

It is this theorem (property (4.4)) that allows us to numerically approximate the double integral in so many ways the arbitrariness in the selection C and type of partition P (as long as $\ \|P\| <\ $) gives us plenty of options. A standard option (used in this text and many calculus books) is to use only partitions with equally spaced partition points, $\Delta x_i = \Delta x = (b-a)/N$ and $\Delta y_j = \Delta y = (d-c)/M$, for every i, j. The standard choices for the selections C are:

(x_{i-1}, y_{j-1})	(lower left-hand vertices)
(x_i, y_{j-1})	(lower right-hand vertices)
(x_i, y_j)	(upper right-hand vertices)
(x_{i-1}, y_j)	(upper left-hand vertices)
$((x_{i-1} + x_i)/2, (y_{j-1} + y_j)/2)$	(centers)

The fact that we can use choices other than the standard ones is helpful in certain numerical work. For instance, if f varies more rapidly on one region of R, you will get better approximations if the partitions are finer in that region.

The extension of the above ideas to cover the case where the region of integration is not a rectangle but rather a more general region D in the domain of f is more interesting (and useful). The exact specification of what types of regions D for which $\int_D f(x, y) \, dA$ exists requires too much background to be presented here (cf. [MH, p. 451]), so the discussion will be intuitive in this aspect. Thus, we assume the basic region looks like the one shown on the left in Figure 4.14. It is the subset bounded by a suitable closed curve and is contained in a rectangle R, which itself is contained in the domain of f.

An easy way to define $\int_D f(x, y) \, dA$, one used by most calculus books, involves introducing a function F, defined by

$$F(x, y) = \begin{array}{ll} f(x, y) & \text{for } (x, y) \text{ in } D \\ 0 & \text{for } (x, y) \text{ in } R, \text{ but not in } D \end{array}$$

Since this new function F is the same as f at points in D and is zero outside of D, it seems reasonable to define $\int_D f(x, y) \, dA$, by using the double integral of F over the rectangle R:

$$\int_D f(x, y) \, dA \qquad \int_R F(x, y) \, dA.$$

However, note that F is a *discontinuous* function on R and so in order to use this approach one must have a definition of double integrals, over rectangles, of such discontinuous functions. Our definition above, as well as those in most calculus books, does not include such functions.

An alternative way to define $\int_D f(x, y) \, dA$, one that we feel is much preferable to that in the above paragraph, involves using grids that cover the region D and approximate it better and better as the grids become finer and finer. This approach is more geometric and gets at the crux of the problem. After all, no matter what definition one uses for $\int_D f(x, y) \, dA$, the special case: $f(x, y) = 1$, for all (x, y) D, will give the area of the region D (area $= \int_D 1 \, dA$).

The approach using grids is outlined and motivated in Problem 2 in the case study for this chapter. For a given partition

$$P = \left\{ (x_i, y_j) \right\}_{i=1,\ldots,N}^{j=1,\ldots,M}$$

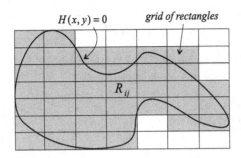

Figure 4.18: *A region D enclosed by a simple closed curve $H(x,y) = 0$ and a grid of subrectangles that overlap the region D.*

of the rectangle R, let grid(P) consist of all those subrectangles R_{ij} whose intersection with D is nonempty:

$$\mathrm{grid}(P) = \{ \, R_{ij} \mid R_{ij} \cap D \neq \emptyset \, \}.$$

See Figure 4.18.

Now, for a given selection C of points in each subrectangle, we form a Riemann sum, not as in equation (4.3), but rather by summing over just those indices i, j for which R_{ij} is in the grid:

$$S_{P,C}(f) = \sum_{i,j \text{ with } R_{ij} \in \mathrm{grid}(P)} f(p_j, q_i) \, \Delta A_{ij}.$$

With this agreement about how the Riemann sums are formed, the theory gives us that there exists a unique number L that has the property in equation (4.4). Thus, we define the double integral as

$$\iint_D f(x,y) \, dA = L.$$

4.7.3 Double Integrals in Maple

When you calculate double integrals by hand, you are really using a theorem (called Fubini's Theorem) to reduce the computation to two iterated, single integrals. For example, if f is defined and continuous on the rectangle $R = [a, b] \times [c, d]$, this theorem says that the double integral of f over R can be calculated in one of two ways by single integrals:

$$\iint_R f(x,y) \, dA = \int_c^d \left[\int_a^b f(x,y) \, dx \right] dy$$

$$= \int_a^b \int_c^d f(x,y)\, dy \ dx.$$

Because of this, double integrals can be computed in Maple in exactly the same way, namely, iterate the operator `int` for single integrals:

```
int(int(f(x,y),x=a..b),y=c..d);
int(int(f(x,y),y=c..d),x=a..b);
```

According to the theorem, doing the integration in either order should give the same result.

More generally, the theorem applies to regions other than rectangles. For example, if D is the region shown in Figure 4.11, then the double integral is reduced to the following iterated, single integrals

$$\int_D f(x,y)\, dA = \int_{x=a}^{x=b} \int_{y=g(x)}^{y=h(x)} f(x,y)\, dy \ dx.$$

To do the iterated integrals in Maple, simply use

```
int(int(f(x,y),y=g(x)..h(x)),x=a..b);
```

This assumes of course that `g,h` have previously been defined as functions in Maple.

Chapter 5

Procedures

So far the programming and algorithm building you have had to do has been relatively short and not too complex. Soon, however, either later in this book, in your other courses, or in your career, you will encounter programming tasks that are much larger and for which more skill and organization is required to construct the code.

A common approach to solving complex programming problems is to decompose the problem into a number of subproblems for which the code is easier to write. Then fitting together the various blocks of code for each subproblem gives the code for the total problem.

Breaking a task into a number of smaller pieces makes sense not only for reducing the complexity of the task but also for division of labor and reusability of code. A company writing code to track satellites in orbit about the earth can more e ciently construct the code, which can run into thousands of lines, by assigning independent pieces of the overall program to a number of different employees. This same company may also find that the code for certain subtasks in the satellite tracking program can also be used in a missile tracking program.

The means for incorporating code for a number of subproblems into the total code for a larger problem is handled differently by different programming languages. Maple uses *procedures*, which were brie"y introduced in Chapter 2. FORTRAN uses *functions* and *subroutines*, while C uses a similar facility called a *function*. Regardless of the name (procedure, subroutine, function) the purpose of this feature, or construct, in any language is the same and its syntax is similar in most languages. The exact details of the way a procedure, subroutine, or function works vary from language to language. Thus, the discussion of procedures in this chapter will give you an understanding of subroutines and functions in other languages. But beware:

you need to check the differences, because small differences can have large effects on your program.

5.1   Maple s Procedure Statement

As we saw in Chapter 2, a procedure F in Maple can have certain input parameters which are passed to the procedure when it is invoked, or called, and certain output parameters which are returned as results from the procedure. The procedure must first be *defined* before it is used, or called. The names for the input parameters, say, $I_1, I_2, \ldots, I_k$, and for the output parameters, say, $O_1, O_2, \ldots, O_m$, in the definition can be anything, and thus are called *formal parameters*, since they are only used in writing the procedure during its definition. When the procedure is used, particular names or specific values are substituted in place of the formal parameters and thus, these names or values, say, $i_1, i_2, \ldots, i_k$ and $o_1, o_2, \ldots, o_m$, are called the *actual parameters*.

The definition of any procedure in Maple is done by using code with the following format.

Procedure De nition

```
F:=proc(I_1,...,I_k,O_1,...,O_m)
    local L_1,...,L_r;
    global G_1,...,G_s;
    option O;
    description D;
    statement1; statement2; ... ; statementN
end proc:
```

Some aspects of this format were explained in Chapter 2, but here again we will discuss those parts, as well as the new ones (`local`, `global`, `option`, `description`), and in more detail.

Once the procedure has been defined, it can be used anywhere in the Maple worksheet and can also be used within the body of another procedure. To use it, just substitute the desired actual input parameters in place of the corresponding formal parameters. The procedure call has the following general format.

Procedure Call (or Invocation):

$$\mathtt{F}(i_1, \ldots, i_k, o_1, \ldots, o_m \,);$$

Thus, you use a procedure somewhat as you would use a function in mathematics: it accepts input when you assign actual values to the parameters $i_1, \ldots, i_k$ and produces output. As you have seen (and will see again below), the procedure call not only causes values to be assigned to the variables $o_1, \ldots, o_m$, but also causes the result of the last statement of the procedure to be returned as well. This last result (be it a number, an expression, or plot) appears as output in the worksheet (or a plot window) right after the procedure call. You can suppress the last result being returned by using `RETURN()` as the last statement in the procedure (this is an empty return statement).

From a mathematical point of view, a Maple procedure is a vector-valued function, with $I_1, \ldots, I_k$ being the independent variables and $O_1, \ldots, O_m$ and the result of the last statement in the procedure being the dependent (or output) variables. When values are substituted for the independent variables, the procedure/function produces values for the dependent variables.

On the other hand, all mathematical functions are procedures from Maple's point of view. Thus, when you define a mathematical function, say, $f(x) = 2 + 3x - x^2$, with the arrow notation: `f:=x->2+3*x-x^2`, Maple converts this into a procedure.

Furthermore, just about every command you have been using when working with Maple is actually a procedure. For example,

```
diff(f(x),x); plot(f(x),x=a..b); int(f(x),a..b);
```

are procedures that are part of Maple's kernel (they are read in and defined each time you start Maple). The `diff` command has two actual input parameters `f(x)` and `x` and no output parameters. Similarly `f(x)` and `a..b` are the actual input parameters to the procedures `plot` and `int`. In all three cases the output comes from the last result (command) executed in the procedure.

All built-in Maple procedures, which are not part of the kernel, are contained either in the Maple library, and can be loaded using the `readlib` command, or in Maple packages, and can be loaded using the `with(`*package*`)` command. For example, the command `with(plots)` loads and defines all the special purpose procedures that have been written for plotting (and are not part of the kernel like `plot` and `plot3d`). One goal of this course is to get you to the point where you can construct procedures as sophisticated as these.

You can look at the code for any of Maple's built-in procedures by changing the value of `verboseproc` to 2 and then using the `eval` on the name of the procedure you wish to view. For example, to see the code for Maple's `plot` procedure use the commands

```
interface(verboseproc=2); eval(plot);
```

The default setting for `verboseproc` is 1. Note that this name is a combination of the words verbose and procedure. It contains the level of verbosity in the information displayed about procedures.

5.1.1 Aspects of the Procedure Definition

The procedure definition (in the first box above) has various optional parts shown in the body of the procedure: formal parameters, local, variables, global variables, options, description. We describe some of these parts here. Additional details and explanations will occur throughout this and future chapters.

> **Formal Parameters:** A procedure need not have any parameters, but when they do occur, they form a sequence data type: $P_1, P_2, \ldots, P_n$ (in this case names separated by commas). Each name P_i can be explicitly declared as a specific data type by putting a double colon after it followed by the type you wish it to be: $P_i :: type_i$. This is useful for input parameters since it will cause Maple to generate an error message if the procedure is invoked with actual input parameters of the wrong types. It is also useful for output parameters since it allows you to make many calls to the procedure with the same output parameter names (as the examples below illustrate). There is no distinction notationally in the parameter sequence between which parameters are input and which are output. Maple can discern from the code which are which. The notation $I_1, \ldots, I_k, O_1, \ldots, O_m$ used previously was just for pedagogical reasons.
>
> **Local Variables:** These are variables $L_1, \ldots, L_k$ that you use when writing the procedure and they can be thought of as existing and having values only while the procedure is executing.[1] For this reason, local variables can have the same names as variables used outside the

---

[1]This is not strictly true, but for pedagogical purposes we will consider it as so.

procedure. Think of the local variables as dummy variables used only to write the procedure. When the procedure is called, Maple allocates storage for the local variables and keeps them distinct from variables outside the procedure.

You do not have to declare local variables, Maple will do this for you (usually correctly), but it is good programming practice to do so. The declaration has the form: `local` $L_1, \ldots, L_r$`;`. Note that the items in a sequence are separated by a comma (not a semicolon) and terminated by a semicolon or colon. As discussed below, it is important to *not* use formal parameters as if they were local variables, even though there is a natural tendency to do so.

Global Variables: These are the variables, $G_1, \ldots, G_s$, that you normally use during any interactive Maple session and are used at places outside any procedure you may define. These variables have values that are known and usable for as long as you are running Maple. Of course, as you have seen, these variables need to be reassigned values whenever you start up a new session using an old worksheet. Global variables can be used within the body of a procedure definition and this allows you to pass information to and return information from the procedure. The procedure can use the value of the global variable and can also change its value. If you use a global variable in this way, you should declare it as a global variable within the procedure. Otherwise Maple may assume that it is a local variable. Global variables are declared within a procedure by a statement of the form `global` $G_1, \ldots, G_s$`;`. On the other hand if you have a global variable A and you also want to use this name as a local variable name, in a way not connected with the global variable, you can accomplish this dual naming by making the declaration `local` A, within the procedure.

Options: There are five options that you might want to use in your procedures. They are `remember`, `operator`, `arrow`, `copyright`, and `builtin`. For example, when writing recursive procedures (see Chapter 8) you will often want to include the command: `option remember` to make the procedure more e cient.

Description: This part of the procedure definition allows you to include comments in the procedure. The comments are put in as a sequence of strings. This can be helpful if you want to explain some notation that you use or to clarify what part of the procedure does.

5.1.2   Aspects of the Procedure Call

When you make a procedure call $F(i_1, \ldots, i_k, o_1, \ldots, o_m)$, the actual parameters $i_1, \ldots, i_k, o_1, \ldots, o_m$ should match the order of the corresponding formal parameters and normally the number of actual parameters is the same as the number of formal parameters. If a procedure is invoked with too many actual parameters, Maple will use only the ones it needs, taking them from the first of the parameter list, and will ignore the extra ones at the end of the list. If too few actual parameters are used in the procedure call, you can expect to get an error message. An important exception to this is when the missing parameters are never needed in the execution of the procedure. This feature allows you to write procedures where parameters at the end of the list are optional. For example

```
plot(f(x),x=1..2); plot(g(x),x=4..9,color=blue)
```

are two calls to Maple's `plot` procedure with two and three actual parameters respectively. In essence the `plot` procedure is written to conditionally test for the presence of the extra parameter `color=blue`. If it is not there, the execution branches to a section of the procedure where this parameter is not used.

The actual input parameters $i_1, \ldots, i_k$ can be particular values (such as numbers 2, 4.31, expressions x+5*x^2, etc.) or names of global variables which have particular values. In most cases the inputs should have particular values, rather than evaluating to a name only. This is because the procedure will perhaps use these particular values in calculations and logic statements.

On the other hand, the actual output parameters $o_1, \ldots, o_m$ should evaluate to a name only. If not, you will get the error message

```
Error, (in F) illegal use of a formal parameter
```

or worse, your procedure will not work as expected. The error message usually occurs because the code in your procedure is trying to assign a value to one of the output parameters o_i (which Maple refers to as a formal parameter), and o_i itself has been evaluated to a value. This is like trying to assign the value two to two: `2:=2`. To avoid this error, the best practice is, in the *definition* of the procedure, to declare each of the output variables to be of type `evaln`, i.e.,

$$F := \text{proc}(I_1, \ldots, I_k, O_1 :: \text{evaln}, \ldots, O_m :: \text{evaln}).$$

Then if you use an actual output parameter that has a value (or values) associated with it, the procedure will not evaluate it down to that level. Note that using ::evaln to specify the type of a formal parameter O_i does not actually change, or clear out, the value (or values) that an actual parameter o_i has when the procedure is called. The code in the procedure can do this, however. For example, the procedure

```
f:=proc(a,L::evaln)
    L[1]:=a;
end proc:
```

is designed to replace the first element in a list L by the value of the variable a. Having defined the procedure, we can use it as follows.

```
L:=[1,2,3,4]:
f(0,L);
eval(L);
```

This produces two lines of output

$$0$$
$$[0, 2, 3, 4]$$

The first line is the result of the last (and only) statement executed by the procedure. The second line shows that the procedure has changed the first element of L to 0.

The distinction we have made between input and output parameters is merely for pedagogical purposes   a parameter is considered as input if its value is used in a computation by the procedure; a parameter is considered as output if it is assigned a value by the procedure. Often, a parameter can be considered as both input and output to the procedure. For example, the procedure

```
g:=proc(L::evaln)
    L[1]:=5;
end proc:
```

will replace the first element of a list L with the number 5. Thus, you can think of L as both input and output to the procedure.

5.2 Procedures - Some Details

In this section we use a number of specific examples to explain in more detail how a procedure actually works. In particular, we discuss how information

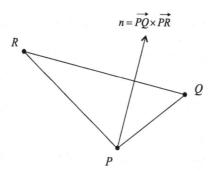

Figure 5.1: *A triangle ΔPQR with normal $n = \overrightarrow{PQ} \times \overrightarrow{PR}$.*

is passed to and returned from procedures (there are a number of different ways), and how evaluation of variables within a procedure works differently than outside a procedure.

Example 5.1 (Triangles in $\mathbb{R}^3$) Many procedures are for calculating, or determining, only one thing for output and so the easiest way to return this result is to have it be the result of the last statement in the procedure (right before the **end proc** statement). Almost all of Maple's built-in procedures work this way. In this example we write a simple procedure like this to calculate the area of a triangle in $\mathbb{R}^3$.

The procedure will have three points P, Q, R, the vertices of the triangle, as input (see Figure 5.1) and will first calculate the vectors

$$
\begin{aligned}
v &= \overrightarrow{PQ} = Q - P \\
w &= \overrightarrow{PR} = R - P.
\end{aligned}
$$

(See the Maple/Calculus Notes at the end of the chapter for a review of vectors and geometry.) Then the cross product $n = v \times w$ is computed[2] and finally, according to the theory, the area of the triangle is half the magnitude of the cross product: $area = \frac{1}{2}|n|$. Note that n is a vector that is perpendicular to the triangle, and therefore to the plane determined by the triangle.

Before writing the procedure, we decide which type of data structure: sequence, list, or array, we want to use for representing points P. These are

---

[2]Maple has a package called **linalg** with procedures to calculate cross products, dot products, and many other things. Our philosophy is for you to write your own procedures and *not* use the built-in ones.

all similar, but the mechanics of operating with each is slightly different. We
choose a list data type so that a point P has the form $P = [x_0, y_0, z_0]$. This
will be convenient for some of the graphical work below. The procedure is
defined by the following code.

```
area:=proc(P,Q,R)
  local i,v,w,n;
  for i to 3 do
    v[i]:=Q[i]-P[i];
    w[i]:=R[i]-P[i];
  end do;
  n[1]:=v[2]*w[3]-v[3]*w[2];
  n[2]:=v[3]*w[1]-v[1]*w[3];
  n[3]:=v[1]*w[2]-v[2]*w[1];
  sqrt(n[1]^2+n[2]^2+n[3]^2)/2
end proc:
```

The local variables v,w,n are used to to store the calculated components of
the two vectors as well as the components of their cross product. Note that
the procedure does the assignments and calculations *componentwise*; that
is, rather than using v:=Q-P, the do loop in the procedure calculates the
same thing as: v[1]:=Q[1]-P[1];v[2]:=Q[2]-P[2];v[3]:=Q[3]-P[3];.
We are emphasizing the componentwise approach in this chapter. The next
chapter discusses vector methods and how Maple's list data structures can
be manipulated *vectorwize*.[3]

If we now define three points: A:=[1,2,3];B:=[-1,0,1];C:=[2,1,0];
in the Maple session, then the command area(A,B,C) will return the value
of $2\,\overline{6}$ for the area of $\triangle ABC$. Note that we have written the procedure
so that the exact value, not the approximate decimal value, of the area is
returned. Also note that the actual parameter names used here are A,B,C,
while the formal, or dummy, parameter names P,Q,R are only used to write
the procedure. If you wish you can also use P,Q,R as the actual parameter
names too.

Next, let's modify the procedure to include additional output besides
the area. Some graphics are always nice, so we include code to plot both
the triangle and the vector $\frac{1}{2}n$, which is perpendicular to the triangle and
has length equal to the area of the triangle. We save the plots in variables
named **triangle**, **nvect**, which serve as output parameters. The modified
procedure is as follows.

---

[3]From the way that v,w,n are used in the procedure, Maple interprets them as tables
(even though P,Q,R are lists).

```
area2:=proc(P::list,Q::list,R::list,triangle::evaln,nvect::evaln)
  local i,v,w,n,f;
  with(plots,spacecurve);
  for i to 3 do
     v[i]:=Q[i]-P[i];
     w[i]:=R[i]-P[i];
  end do;
  n[1]:=v[2]*w[3]-v[3]*w[2];
  n[2]:=v[3]*w[1]-v[1]*w[3];
  n[3]:=v[1]*w[2]-v[2]*w[1];
  triangle:=spacecurve([P,Q,R,P],color=black,thickness=2):
  f:=t-> [P[1]+t*n[1],P[2]+t*n[2],P[3]+t*n[3]];
  nvect:=spacecurve(f(t),t=0..0.5,color=red):
  sqrt(n[1]^2+n[2]^2+n[3]^2)/2;
end proc:
```

Note that we have now explicitly declared the types of P, Q, R to be lists. Besides preventing errors in using the procedure, this also serves to remind us that the procedure was written with points being represented by lists. In other situations we might want to have points be represented by array data structures. You can see the advantage of the list data structure here. To plot the triangle we simply pass a list [P,Q,R,P] of points to the spacecurve command. Note that the first point is repeated as the last point in the list.[4] To plot the vector $\frac{1}{2}n$ with its initial point at P, the procedure first defines the vector-valued function: $f(t) = P + tn$, that parametrizes the line through P in the direction n. Then a plot of the portion of this line for t $[0, 1/2]$ gives the desired image of this vector.

We now invoke the procedure with area2(A,B,C,q1,q2); which will return the area $2\overline{}6$ as before and store the two plot strucures from the procedure in variables q1,q2. The resulting graphics can now be produced by using the commands

```
with(plots,display3d);
display3d({q1,q2},scaling=constrained,
              orientation=[-22,82],axes=normal);
```

---

[4]Generally spacecurve([P1,P2,...,Pn]) will draw a polygon in $\mathbb{R}^{\check{}}$ with sides $\overline{P_1P_2}, \overline{P_2P_3}, \ldots, \overline{P_{n\ 1}P_n}$. Thus, by including the "rst point again at the end of the list, i.e., spacecurve([P1,P2,...,Pn,P1]), you will get a closed polygon. For this to work the points must be lists: $P_i = [x_1, y_i, z_i]$. Alternatively you could use polygonplot3d([P1,P2,...,Pn]). Both spacecurve and polygonplot3d are part of the plots package.

This gives the picture shown in Figure 5.2.

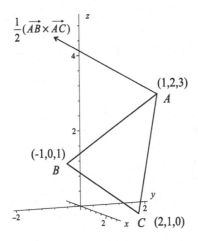

Figure 5.2: *A triangle with vertices* $A = [1, 2, 3], B = [-1, 0, 1], C = [2, 1, 0]$, *and normal vector* $\frac{1}{2}(\overrightarrow{AB} \times \overrightarrow{AC})$.

Note that `spacecurve` is part of the `plots` package, so we have included `with(plots,spacecurve)` as a command within the procedure. If this were not done, then a user of the procedure would have to make this command at some point before trying to display the plots contained in `q1,q2`.

5.2.1 Evaluation Within a Procedure

We have already discussed, in Section 2.7, Maple's evaluation rules which apply to all the work you do in an interactive session (not connected with any procedures). These rules however do *not* apply to the code that Maple sets up and executes in response to a procedure call. The differences in the rules and other aspects of the code generated by a procedure call are well worth discussing. Understanding these topics can save you hours of debugging time.

Rule 1: When a procedure is invoked, all the actual parameters are evaluated once and fully, down to their lowest levels (unless a typing statement, such as `::evaln`, specifies to do otherwise). The actual parameters are not evaluated again after this initial evaluation.

Rule 2: Maple rewrites each line of code within the procedure definition, substituting the actual (evaluated) parameters in place of the formal (dummy) parameters. Then Maple executes this rewritten code.

Rule 3: Each local variable in the rewritten code is evaluated to only one level during execution of the rewritten code. Global variables are evaluated fully inside the procedure.

Even with these rules, the processing and execution of procedures by Maple can be confusing. So we look at a number of short examples to illustrate what can happen and what can go wrong.

Suppose we write a simple program to compute the square a^2 and inverse square $1/a^2$ of a number a. The code could be as follows:

```
f:=proc(a,square,invsquare)
  square:=a^2;
  invsquare:=1/square;
end proc:
```

The procedure **f** has three input parameters and consists of only two statements, with the result of the second statement being returned as the default output. Note that **square** and **invsquare** are viewed as output parameters since the procedure assigns values to them. Having defined the procedure, we make the following assignments and call the procedure.

```
A:=h; h:=3;
f(A,B,C);
```

The output is

$$\frac{1}{B}$$

To understand this, we follow the rules. The parameter **A** evaluates fully to 3, while **B,C** evaluate fully to names (we assume they have not been assigned values elsewhere in the session. A rewrite of the two statements in the procedure definition gives

```
B:=9: C:=1/B;
```

When these are executed, the output from the procedure is $1/B$, and not the number $1/9$ (which you might have expected). This is because of Rule 1: **B** is evaluated to a name at the initial call to the procedure and is not re-evaluated again. Thus, even though **B** is assigned the value 9 by the first statement in the procedure, this is not re"ected in the default output resulting from the assignment `C:=1/B`. To the first level B evaluates to B and evaluates to 1/9 at the second level.

Note, however, that the values we wished *are* assigned correctly to B,C. This is clearly shown when, after the first procedure call, the statements B;C; are executed.

    B,C;

$$9, \frac{1}{9}$$

Since these statements are part of the interactive session, they are evaluated fully. Thus, the global variables get assigned the correct values, but this does not show up at the procedure level.

A second call `f(A,B,C)` to the procedure will result in an error message:

    Error, (in f) illegal use of a formal parameter

This (as in the example above) is because the first call assigned values to B,C and so the rewritten code for the second call is

    9:=9: 1/9:=1/9;

While the equation $9 = 9$ is mathematically correct, the Maple statement `9:=9` is not permitted. It attempts to assign to 9 the value 9. But 9 is a Maple constant and constants are not allowed to be assigned values. If this were permitted then an assignment like `2:=3;` would alter all the arithmetic done with the constant 2!! Thus, the formal parameter is being used illegally.

As an illustration of Rule 3, suppose we alter the above procedure by introducing two local variables `sq,isq` to stand for the square and inverse square.

    ff:=proc(a,square,invsquare)
      local sq,isq;
      isq:=1/sq;
      sq:=a^2;
      square:=sq;
      invsquare:=isq;
    end proc:

This seems reasonable. The second line of code merely uses the fact that the inverse square is the reciprocal of the square. However, now the statements

```
A:=h; h:=3;
ff(A,X,Y);
```

produce the output

$$\frac{1}{sq}$$

This is the default output, the result of the last statement `invsquare:=isq`. By Rule 3, the local variable `isq` only evaluates to the first level during execution of the procedute, and at the first level, `isq` has value `1/sq`. Again, the procedure works more or less as you intended, except for giving you the puzzling default output. At the interactive level, if you now make the command `X,Y;` you will see that `X` and `Y` have values 9 and 1/9.

To illustrate another problem that can occur with parameters, one that can cause the procedure not to work, we alter the above procedure as follows:

```
fff:=proc(a,square,invsquare)
  square:=a^2;
  invsquare:=1/square;
  square:=1/invsquare;
end proc:
```

If we now make the call `fff(A,y,z)`, the resulting output will be

$$\frac{1}{z}.$$

This is as expected, since the rewritten code is

```
y:=9: z:=1/y: y:=1/z;
```

So the result, $1/z$, of the last statement executed is returned by default. Since z is an actual parameter, by Rule 1, it evaluated fully (to a name in this example) and only once when the procedure was invoked. While the statements `y:=9;z:=1/y;y:=1/z`, when made at the interactive level, result in y being assigned the value of 9, this does not happen when a procedure executes these same statements.

All of this is similar to what was said in the above examples, except now something more drastic occurs   the procedure `fff` creates an infinite recursive loop, linking the global variables. Indeed, the command: `y;` results in the output

```
Error, too many levels of recursion
```

You can examine this with the **eval** command. The command

```
eval(y,1),eval(y,2),eval(y,3),eval(y,4);
```

gives the output

$$\frac{1}{z}, y, \frac{1}{z}, y$$

showing y evaluated to levels 1, 2, 3, and 4. You can see, then, that the rewritten code, y=1/z;z:=1/y, in the procedure has created an infinite number of levels and this prevents y from being evaluated fully (which is what the command y; or, equivalently, the command eval(y) is asking for). Hence, the error message.

5.2.2 Parameters and Local Variables

The procedure fff also illustrates (as the procedure f did too) that you *should not use parameters as local variables*. Often it is useful to use parameters as local variables and sometimes this does not cause trouble (as in the procedure f above). Here is another procedure that uses parameters as local variables and no harm is done.

```
h:=proc(a,b,c::evaln,ang1::evaln,ang2::evaln)
  c:=sqrt(a^2+b^2);
  ang1:=arcsin(a/c);
  ang2:=arcsin(b/c)
end proc:
```

The procedure calculates the length c of the hypotenuse in a right triangle with a,b being the lengths of the two sides about the right angle. The procedure also calculates the measures of the two acute angles ang1,ang2 in the right triangle and, for this, it is convenient to use c rather than introducing a temporary local variable. If we invoke the procedure with the call h(3,4,C,A1,A2) the output from the last result calculated in the procedure is $\arcsin(4\frac{1}{C})$, which shows that C only gets evaluated at one level during execution of the procedure (its value is a name at the first level). At the interactive level, we can inspect the actual values by using the commands C;A1;A2; and these will give us the results $5, \arcsin(\frac{3}{5}), \arcsin(\frac{4}{5})$ respectively, which are the (exact) values expected.

To further illustrate that the output parameters only get evaluated to a name at the first level during procedure execution, consider the following modification of the above example.

```
h:=proc(a,b,c::evaln,ang1::evaln,ang2::evaln)
  c:=sqrt(a^2+b^2);
```

```
  ang1:=arcsin(a/c);
  ang2:=arcsin(b/c);
  print(ang1,ang2)
end proc:
```

The procedure will work as before, giving the correct hypotenuse and angle measures, but the result of the **print** statement will not be the angle measures, but rather the names of the angles. For example, the call **h(3,4,C,A1,A2)** results in the following rewritten code.

```
C:=5:
A1:=arcsin(3/C):
A2:=arcsin(4/C):
print(A1,A2);
```

When these are executed, *as procedural code*, the output will be $A1, A2$, since the parameters **A1,A2** only evaluate to a name at their first level. Thus, the **print** statement is equivalent to **eval(A1,1), eval(A2,1);**. The two assignment statements give **A1,A2** values of **arcsin(3/C), arcsin(4/C)** at the second level, respectively. To see this replace the print statement in the procedure definition by **eval(ang1,2), eval(ang2,2);**, and re-execute everything. On the other hand if you use: **eval(ang1), eval(ang2);** in the procedure definition (instead of the print statement), this forces full evaluation and the result of the call **h(3,4,C,A1,A2)** will be $\arcsin\left(\frac{3}{5}\right), \arcsin\left(\frac{4}{5}\right)$.

While the use of **c** as a local variable causes no problems in the above examples (except for not giving evaluations to the level you wish), the use of an *output* parameter as a local variable in a conditional statement will generally result in a failure of the procedure to properly execute. The next example illustrates this.

Example 5.2 (Finding max/min Values) While Maple has two built-in procedures called **max** and **min**, consider writing a procedure, from scratch, to find the largest and smallest numbers in a list of numbers. We will write several versions of this in order to illustrate a number of points.

In each version the basic algorithm for finding the maximum and minimum values is like finding the largest and smallest apple in a basket full of apples. To find the largest one, say, pick out any one apple for starters and put it in your left hand. Then, with your right hand, begin removing each of the remaining apples, one at a time from the basket. If at any time you encounter an apple that is larger than the one in your left hand, put that apple in your left hand, tossing out the one that was there, and then proceed. By the time all the apples are removed from the basket, the largest one should be in your left hand. The process of finding the smallest apple works in a similar way.

A first attempt at writing a max/min procedure is as follows. The input will be a list A and the output will be the maximum and minimum values maxv and minv. There is a strong tendency to use these output parameters as local variables and write the procedure as follows

```
maxmin:=proc(A,maxv::evaln,minv::evaln)
  local i;
  maxv:=A[1]; minv:=A[1];
  for i from 2 to nops(A) do
    if maxv < A[i] then maxv:=A[i] end if;
    if minv > A[i] then minv:=A[i] end if;
  end do;
  RETURN()
end proc:
```

Note the use of nops(A) to pick off the number of elements (numbers) in the list A. Also, there is an empty return statement RETURN(), since we are returning the two results via the last two parameters.

This defines the procedure. However, when you use the procedure with actual parameters, say, with

```
B:=[8,160,2]
maxmin(B,M,m)
```

you will get the error message

```
Error, (in maxmin) cannot evaluate boolean
```

So the procedure does not execute. This should be expected, if you understand the points made in the above examples. To see what Maple does, suppose that prior to making the call, the list B:=[8,160,2] has been assigned. When the procedure is invoked, Maple produces the following rewritten code:

```
M:=8: m:=8:
if M < 160 then M:=160 end if:
if m > 160 then m:=160 end if:
if M < 2 then M:=2 end if:
if m > 2 then m:=2 end if:
```

When this is executed, with the procedural evaluation rules in effect, the error occurs when the first conditional statement if M < 160 is encountered. Since M only evaluates to a name at the first level, Maple cannot test the condition. Thus, even though the code is logically correct and would execute correctly outside the procedure evaluation rules, it nevertheless fails to execute when generated by a procedure call.

There are several ways to fix the maxmin procedure so that it does work. You can always introduce local variables, say, MV,mv, to take the place of the output parameters maxv,minv in the conditional statements:

```
maxmin:=proc(A,maxv::evaln,minv::evaln)
  local i,MV,mv;
  MV:=A[1]; mv:=A[1];
  for i from 2 to nops(A) do
    if MV < A[i] then MV:=A[i] end if;
    if mv > A[i] then mv:=A[i] end if;
  end do;
  maxv:=MV; minv:=mv;
  RETURN()
end proc:
```

Doing this sometimes makes the code for the procedure more confusing, especially if the code is already complicated. An alternative is to use eval to force full evaluation of maxv, minv in the conditional statements:

```
maxmin:=proc(A,maxv::evaln,minv::evaln)
  local i;
  maxv:=A[1]; minv:=A[1];
  for i from 2 to nops(A) do
    if eval(maxv) < A[i] then maxv:=A[i] end if;
    if eval(minv) > A[i] then minv:=A[i] end if;
  end do;
  RETURN()
end proc:
```

If you study carefully all the above examples, you should be able to avoid many of the pitfalls connected with evaluation rules and handling of parameters. The designers of Maple are not trying to make things di cult for you and there are specific reasons for the rules and operation of procedures. However, it is probably best not to discuss all the finer points of this behavior at this stage in your programming experience.

5.3 Groups of Related Procedures

A complicated programming task can be solved more easily by breaking it
into a number of simpler subtasks, writing procedures for all the subtasks
separately, and then putting all the procedures together. To achieve this
synthesis of the parts, you can use a main procedure to initiate calls to
the other procedures and this will produce all the required output for the
total programming task. While the main procedure initiates things, some
of the other procedures can call each other, and thus the whole group of
procedures functions as a whole to accomplish the task.

In this section we give several simple examples to illustrate how to break
a task up into simpler tasks, write the corresponding procedures, and make
them all function together. The case study in the next section offers a more
complex example of this.

Example 5.3 (Optimizing a Function) For a differentiable function f
on a closed interval $[a, b]$, you have learned how to use the first derivative f'
to find the maximum and minimum values of f. In this example we write a
procedure, to find the *approximate* maximum and minimum values of f on
$[a, b]$, the approximate places in $[a, b]$ where they occur, and also produce
some graphical output. The procedure, which does *not* use the first derivative, will also work for *nondi erentiable* functions which are continuous.

The idea behind automating the optimization process (approximately) is
to replace f by a polygonal approximation (technically by a piecewise linear,
continuous approximation).[5] This is done by dividing the interval $[a, b]$ into
N equal subintervals $[x_{i-1}, x_i]$, $i = 1, \ldots, N$, each of length $\Delta x = (b-a)/N$.
The endpoints of the subintervals, as we have seen before, are then given
by $x_i = a + i\Delta x$. The polygonal approximation to f corresponding to this
subdivision is the polygon in the plane with vertices $\{(x_i, f(x_i))\}_{i=0}^{N}$. See
Figure 5.3. As a function, the polygonal approximation is a linear function
on each subinterval $[x_{i-1}, x_i]$, $i = 1, \ldots, N$, and is a continuous function on
the whole interval $[a, b]$, hence the term *piecewise linear, continuous function*. For such functions, as Figure 5.3 suggests, the maximum and minimum
values are the largest and smallest of the values $y_i = f(x_i)$, $i = 0, 1, \ldots, N$.
For N large enough these two values will approximate the maximum and
minimum values of f.

It is a simple programming task to automate this method of finding the
approximate max/min values for f, especially since we have already written

---

[5]This idea was used also in Exercise 4, Chapter 3, where arc length is approximated
by the length of a piecewise linear function.

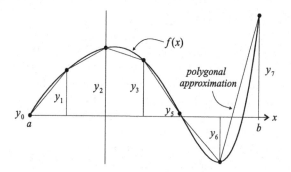

Figure 5.3: *A polygonal approximation to a function* f *on the interval* $[a, b]$. *The maximum and minimum values of this polygonal approximation are the largest and smallest of the numbers* $y_0, y_1, \ldots, y_7$.

the procedure `maxmin` to find the max and min values in a list of numbers. So, we only have to write a main procedure to calculate the subdivision points x_i, build a list $[y_0, y_1, \ldots, y_N]$, out of the corresponding y values, and then make a call to `maxmin`. We will have the procedure print out the approximate max and min values (`maxv` and `minv`) of f along with all the places (x-values) where these occur. To find these places, we scan the list of y_i's, finding the ones that are equal to `maxv` or `minv`, and build sets `xmax` and `xmin` of the corresponding x_i's. The procedure also produces graphical output consisting of a plot of the sequence of vertical line segments joining $(x_i, 0)$ to (x_i, y_i) for $i = 0, \ldots, N$. These line segments represent the y values calculated and allow us to visually check for the largest and smallest of these, as well as the places where they occur.

```
optimize:=proc(f,a,b,N,pic::evaln)
  local X,Y,L,i,A,xmax,xmin;
  X:=array(0..N);Y:=array(0..N);L:=array(0..N);
  with(plots,display);
  for i from 0 to N do
     X[i]:=evalf(a+i*(b-a)/N);
     Y[i]:=f(X[i]);
     L[i]:=plot([[X[i],0],[X[i],Y[i]]],color=black):
  end do;
  A:=[seq(Y[i-1],i=1..N+1)];
  maxmin(A,maxv,minv);
  xmax:={};xmin:={};
```

```
for i from 0 to N do
    if Y[i]=maxv then xmax:=xmax union {X[i]} end if;
    if Y[i]=minv then xmin:=xmin union {X[i]} end if;
end do;
pic:=display({seq(L[i],i=1..N)}):
print('maximum y value is',maxv,
         'and occurs at these x values',xmax);
print('minimum y value is',minv,
         'and occurs at these x values',xmin);
end proc:
```

Note how we created the list $A:=[seq(Y[i-1],i=1..N+1)]$ of numbers to pass to the maxmin procedure. A call like maxmin(Y,maxv,minv) will not work for two reasons. First, the code implictly forces Maple to make Y a table data structure, while the maxmin procedure is expecting a list data structure for its first input parameter. Secondly, maxmin assumes that A[1] is the first element in the list, and in the optimize procedure this needs to be the first y value, i.e., A[1]:=Y[0].

To test the procedure, consider the function $f(x) = 3 + 10(x^4 - x^2)e^{-x/2}$ on the interval $[-1,4]$. We make a call to the procedure using 150 points in the subdivision of $[-1,4]$:

```
f:=x->3+10*(-x^2+x^4)*exp(-x^2);
optimize(f,-1,4,150,pic);
```

This produces the output

> *maximum y value is*, 6.088066652,
> *and occurs at these x values*,$\{1.633333333\}$
> *minimum y value is*, 1.391538737,
> *and occurs at these x values*,$\{-.6333333333, .6333333333\}$

Note how the phrases, like *maximum y value is*, are printed from the procedure by using the back quote, ', to enclose them. See Chapter 6 for a discussion of this and the string data type. The command pic; produces the plot shown in Figure 5.4.

Example 5.4 (Triangles in $\mathbb{R}^3$ Again) Suppose the task is to write a procedure to determine whether a triangle $\triangle PQR$ in $\mathbb{R}^3$ is equilateral, isosceles, or a right triangle, and in the last case identify at which vertex the right angle occurs. While this would be simple enough to do with a single procedure, such a procedure would have to calculate the lengths of the three sides

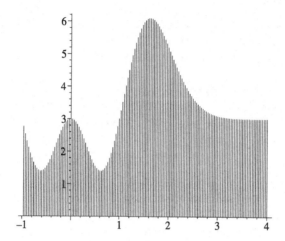

Figure 5.4: *Graphical output of the* optimize *procedure for the function* $f(x) = 3 + 10(x^4 - x^2)e^{-x/2}$ *on the interval* $[-1, 4]$ *with* $N = 150$.

and measures of the three angles of $\triangle PQR$. Having a separate procedure to just calculate these six numbers may be useful in other programming tasks, so we write the code for that first. We name the procedure triangle and we use the dot product to calculate the angle measures. Refer to Figure 5.5 for the geometric meaning of the vectors v, w, u used in the code.

```
triangle:=proc(P,Q,R,p::evaln,q::evaln,r::evaln,
               angP::evaln,angQ::evaln,angR::evaln)
   local i,v,w,u,n,p1,q1,r1;
   for i to 3 do
     v[i]:=Q[i]-P[i];
     w[i]:=R[i]-P[i];
     u[i]:=R[i]-Q[i];
   end do;
   p1:=evalf(sqrt(u[1]^2+u[2]^2+u[3]^2));
   q1:=evalf(sqrt(w[1]^2+w[2]^2+w[3]^2));
   r1:=evalf(sqrt(v[1]^2+v[2]^2+v[3]^2));
   angP:=arccos((v[1]*w[1]+v[2]*w[2]+v[3]*w[3])/(r1*q1));
   angQ:=arccos(-(v[1]*u[1]+v[2]*u[2]+v[3]*u[3])/(p1*r1));
   angR:=arccos((u[1]*w[1]+u[2]*w[2]+u[3]*w[3])/(p1*q1));
   p:=p1;q:=q1;r:=r1;
   RETURN()
```

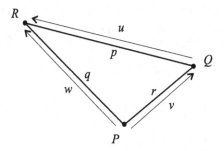

Figure 5.5: *A triangle $\triangle PQR$ with vertices P, Q, R and with p, q, r denoting the lengths of the sides opposite the respective vertices. Also shown are the vectors v, w, u used in the calculation.*

```
end proc:
```

Notice that we introduced local variables to stand for the output parameters `p,q,r`, which will be assigned the lengths of the respective sides. This is because these lengths are used in the calculation of the angles `angP,angQ,angR`.

We can now write a main procedure to simply make a call to `triangle` and the test the lengths of the sides `p,q,r` to see if they are all equal (equilateral), or if a pair of them is equal (isosceles). Similarly, a test of the angles `angP,angQ,angR` determines if the triangle is a right triangle.

It will be helpful, as a check on the classification, to have the procedure print out the lengths of the sides and the measures of the angles. Additionally, for a visual check of the classification, we have the procedure make a call to the procedure `area2` (see Example 5.1 and Figure 5.2). This will return the plots of the triangle and its normal vector, which will be displayed as the default output in the main procedure here. The main procedure is as follows:

```
checktri:=proc(P,Q,R)
   local p1,q1,r1,ap,aq,ar,c;
   triangle(P,Q,R,p1,q1,r1,ap,aq,ar);
   print('sides',p1,q1,r1);
   print('angles',evalf(ap),evalf(aq),evalf(ar));
   area2(P,Q,R,tri,normvec);
   with(plots,display3d);
   picture:=display3d({tri,normvec},
                        scaling=constrained,axes=normal);
   c:=0;
```

```
  if p1=q1 and p1=r1 then
          print('triangle is equilateral');c:=1
  elif p1=q1 or p1=r1 or q1=r1 then
          print('triangle is isosceles');c:=1
  end if;
  if cos(ap)=0 then
          print('right angle at',P);c:=1
  elif cos(aq)= 0 then
          print('right angle at',Q);c:=1
  elif cos(ar)=0 then print('right angle at',R);c:=1
  end if;
  if c=0 then
          print('not equilateral, isosceles or right');
          picture;
  else picture;
  end if;
end proc:
```

Note that the procedure uses a local variable c to keep track of whether the triangle is equilateral, isosceles, or right. If the test for any of these conditions is true then c is set to 1. Otherwise it remains 0. This helps control the logic "ow in the procedure. There is a way to structure the conditional statements so that c is not needed (exercise), however, adding additional conditional statements can often make code di cult to understand.

We now use the procedure to check the triangle in Figure 5.2, which has vertices

```
A:=[1,2,3];B:=[-1,0,1];C:=[2,1,0];
```

The procedure call checktri(A,B,C) returns the following output:

> *sides*, 3.316624790, 3.316624790, 3.464101616
> *angles*, 1.021329082, 1.021329082, 1.098934489
> *triangle is isosceles*

In addition to this, the default output will be a picture like that in Figure 5.2. You can visually check the results of the classification (that the triangle is isosceles) by rotating the figure so that the viewpoint is along the normal line, looking perpendicularly down on the plane of the triangle. You can also automate this adjustment of the viewpoint (see the exercises).

It is important to note that, while the mathematical logic in the procedure is correct, there are numerical problems that can cause some triangles to be incorrectly classified. This can occur in two ways. First, the conditional statements p1=q1, p1=r1, and q1=r1 control part of the classification, and it is possible, for example, to devise equilateral triangles whose sides have lengths that are not equal when calculated with the procedure triangle. This is because that procedure evaluates the lengths to "oating-point numbers, and roundoff error can occur. Using exact arithmetic in the procedure triangle (i.e., eliminating the evalf's) will not circumvent this problem (see the exercises).

A second problem that could cause the logic not to work correctly (as it does in the ideal world of mathematics) is in the tests for a right triangle, e.g., in the conditional statement cos(ap)=0. If you use exact arithmetic, as we have done in procedure triangle, to calculate ap (the angle at P), then when ap is /2, the conditional statement is true and the triangle is correctly classified. However, if the calculations of the angles in procedure triangle are done in "oating-point numbers, then /2 is approximated and cos(ap)=0 is false when ap:=evalf(Pi/2) (see the exercises).

These problems, and others like them, are inherent in numerical work. The sections called Numbers in Chapters 6 and 8 discuss some aspects of this in more detail. From a practical standpoint, you can restructure the procedure checktri to check equality of sides and right angles up to a specified level of tolerance (a standard engineering practice). See the exercises.

5.4 Case Study: Trig Integrals

This case study, as usual, shows you how to tackle a longer, more complex programming problem by analyzing it carefully and then designing the code to solve it. Here the solution requires breaking the problem into a number of smaller parts (three in this case), writing procedures for each, and then writing a main procedure to initiate calls to the appropriate procedures for the parts.

The case study also introduces the topic of *recursive procedures*, i.e., procedures that call themselves. This arises naturally and clearly in solving part of the problem here. However, since this topic can tend to be confusing, a more detailed discussion of recursion is provided in Chapter 8.

When studying techniques of integration in calculus, you learned how to compute integrals involving a product of a power of the sine and a power of the cosine. This was usually for small powers because doing so for large powers was tedious and often involved the repetitious use of reduction formulas. While Maple's int procedure can easily compute these indefinite integrals, the purpose of this case study is to show you how to write your own code for this (as the Maple programmers had to do).

Problem: Write a Maple procedure to calculate the trig integral

$$f(x, n, m) = \quad \sin^n x \, \cos^m x \, dx,$$

for any given, nonnegative integers n, m. You *cannot* use any of Maple's built-in procedures, especially its `int` procedure. Your procedure should be based on the techniques you learned in calculus and for a given n and m should be a function (as a Maple data structure) capable of being plotted. Here are two examples to guide your development of the general procedure.

Example 5.5 (One of n, m is odd) Suppose $n = 2$ and $m = 3$.

$$
\begin{aligned}
\sin^2 x \, \cos^3 x \, dx &= \quad \sin^2 x (1 - \sin^2 x) \cos x \, dx \\[2mm]
&= \quad u^2 (1 - u^2) \, du = \quad (u^2 - u^4) \, du \\[2mm]
&= \quad \frac{1}{3} u^3 - \frac{1}{5} u^5 \\[2mm]
&= \quad \frac{1}{3} \sin^3 x - \frac{1}{5} \sin^5 x
\end{aligned}
$$

Example 5.6 (Both n and m are even) Suppose $n = 4$ and $m = 2$.

$$
\begin{aligned}
\sin^4 x \, \cos^2 x \, dx &= \quad \frac{1}{2^2} (1 - \cos 2x)^2 \, \frac{1}{2} (1 + \cos 2x) \, dx \\[2mm]
&= \quad \frac{1}{8} \quad (1 - 2\cos 2x + \cos^2 2x)(1 + \cos 2x) \, dx \\[2mm]
&= \quad \frac{1}{8} \quad (1 - \cos 2x - \cos^2 2x + \cos^3 2x) \, dx \\[2mm]
&= \quad \frac{1}{16} \quad (1 - \cos u - \cos^2 u + \cos^3 u) \, du \\[2mm]
&= \quad \frac{1}{16} \, \frac{1}{2} u - \frac{1}{3} \sin u - \frac{1}{2} \cos u \, \sin u + \frac{1}{3} \cos^2 u \, \sin u \\[2mm]
&= \quad \frac{1}{16} x - \frac{1}{48} \sin 2x - \frac{1}{32} \cos 2x \, \sin 2x + \frac{1}{48} \cos^2 2x \, \sin 2x
\end{aligned}
$$

Note: On the fifth line, we used a reduction formula to integrate $\cos^2 u$ and $\cos^3 u$. Your procedure should do likewise. Specifically:

(1) Write a main procedure $f(x,n,m)$ that does the computation *exactly* like that in the above examples. You may call other procedures to handle the odd/even cases if you wish. In the case where one of the exponents m, n is odd, derive, by hand the exact formula for the integral. This will involve the Binomial Theorem.

(2) Write a procedure, which will be used by the main procedure, to compute, using a reduction formula, the integral

$$C(x, r) = \quad \cos^r x \, dx,$$

for any nonnegative r. The procedure should be a function (not an expression in x). Write two versions of this procedure: one with do loops and one with recursion. In the case where both exponents m, n are even, derive a formula, by hand, for the integral in terms of the various $C(x, r)$'s. *Note*: For those with the inclination to do so, derive by hand the closed form formula for $C(x, r)$ from the corresponding recurrence relation. You might even be able to find a book in the library which gives the closed form formula for $f(x, m, n)$. However, *do not* use either of these closed form formulas to code the solution of your problem. Rather, use either a do loop or a recursive procedure.

(3) Write a procedure, which will be used by your other procedures, to compute the binomial coe cients

$$b(n, k) = \frac{n!}{k!(n - k)!},$$

for integers $n \quad k \quad 0$. Do not use Maple's factorial operator ! or its built-in binomial function.

(4) Test your procedures by evaluating $f(x, 2, 3), f(x, 4, 2)$ and seeing if they give the same results as in the above examples. Use Maple's **int** command to do the eleven calculations. Are the answers different? Why?

(5) Plot the functions $C(x, 2k), C(x, 2k + 1), f(x, 2k, 2)$ for $k = 0, \ldots, 10$ and $x \quad [-2 \,, 2 \,]$.

Analysis of the Problem: First we derive the required formulas to be used in the program. This we do by cases.

Odd Case: If $m = 2k + 1$, then there is a formula for the integral:

$$\sin^n x \cos^{2k+1} x \, dx \; = \; \sin^n x \, (1 - \sin^2 x)^k \cos x \, dx$$

$$= \; u^n (1 - u^2)^k \, du$$

$$= \; \sum_{i=0}^{k} \frac{(-1)^i}{2i + n + 1} b(k, i) \sin^{2i+n+1} x. \quad (5.1)$$

Similarly if the other exponent is odd: $n = 2k + 1$, then the formula is

$$\sin^{2k+1} x \cos^m x \, dx = \sum_{i=0}^{k} \frac{(-1)^{i+1}}{2i + m + 1} b(k, i) \cos^{2i+m+1} x. \quad (5.2)$$

Even Case: If $n = 2k$ and $m = 2p$, then the formula for the integral involves

$$C(x, r) = \; \cos^r x \, dx,$$

which is computed from a reduction formula. In terms of this we can write:

$$\sin^{2k} x \cos^{2p} x \, dx \; = \; \frac{1}{2^{k+p}} \; (1 - \cos 2x)^k (1 + \cos 2x)^p \, dx$$

$$= \; \frac{1}{2^{k+p}} \sum_{i=0}^{k} \sum_{j=0}^{p} (-1)^i \, b(k, i) b(p, j) \, \cos^{i+j} 2x \, dx$$

$$= \; \frac{1}{2^{k+p+1}} \sum_{i=0}^{k} \sum_{j=0}^{p} (-1)^i \, b(k, i) b(k, j) \, C(2x, i + j)$$

$$(5.3)$$

The reduction formula for the cosine integrals is usually stated as

$$\cos^r x \, dx = \frac{1}{r} \cos^{r-1} x \sin x + \frac{r - 1}{r} \; \cos^{r-2} x \, dx. \quad (5.4)$$

Of course this assumes that $r \quad 2$. In terms of our notation this formula is

$$C(x, r) = \frac{1}{r} \cos^{r-1} x \sin x + \frac{r - 1}{r} C(x, r - 2), \quad (5.5)$$

for $r \quad 2$. Written in this latter way, we see that the reduction formula is just a recurrence relation, which recurs back to either $C(x, 0)$, when r is even, or to $C(x, 1)$, when r is odd. Note that $C(x, 0) = x$ and $C(x, 1) = \sin x$. While this recurrence relation can be solved exactly in closed form, the assignment was to code $C(x, r)$ as a procedure using the recurrence relation directly and use either a do loop or a recursive procedure to accomplish the computation.

Design of the Program: As with other code we have written earlier in the text, we will not build in any error checking on the input to the procedures. For now we want to concentrate on the main workings of the procedures, which means simple, bare bones code for accomplishing the given task.

First we write the procedure for the binomial coefficients. One way to do this involves writing the binomial coefficients in the alternative form

$$b(n, k) = \frac{n \cdot (n-1) \cdots (n-k+1)}{k \cdot (k-1) \cdots 1} = \prod_{i=0}^{k-1} \frac{n-i}{k-i}.$$

Now we can just construct the latter product with a do loop, much like we construct sums:

```
b:=proc(n,k)
   local i,prod;
   prod:=1;
   if k=0 then 1
   else for i from 0 to k-1 do
           prod:=prod*(n-i)/(k-i)
        end do;
   end if;
end proc:
```

Next we design a procedure to handle the trig integral in the odd case. For this we will have to discern which of n, m is odd and use one or the other of formulas (5.1) (5.2). However, the only differences between these two formulas is that n and m are interchanged, $\sin x$ and $\cos x$ are interchanged, and the alternating signs are $(-1)^{i+v}$ with $v = 0$ or $v = 1$. So we use v as an input parameter and write the procedure as follows.

```
oddcase:=proc(x,k,p,v)
   local s,i,u;
   if v=0 then u:=sin(x) else u:=cos(x) end if;
   s:=0;
   for i from 0 to k do
```

```
        s:=s+(-1)^(i+v)*b(k,i)*u^(2*i+p+1)/(2*i+p+1);
    end do;
  end proc:
```

The main procedure will be named f and will be invoked by f(x,n,m).
When the main procedure calls oddcase, the formal parameter p will be
replaced by either $p = n$ or $p = m$.

Writing a procedure to handle the even case involves expressing formula
(5.3) in terms of a nested pair of do loops. This is easy enough.

```
  evencase:=proc(u,k,p)
    local s,i,j;
    s:=0;
    for i from 0 to k do
        for j from 0 to p do
            s:=s+(-1)^(i)*2^(-k-p-1)*b(k,i)*b(p,j)*C(u,i+j)
        end do;
    end do;
  end proc:
```

The main procedure is now easily written and serves merely to divide
into cases and pass the appropriate values to the evencase or oddcase
procedures.

```
  f:=proc(x,n,m)
    if type(n,odd) then oddcase(x,(n-1)/2,m,1)
    elif type(m,odd) then oddcase(x,(m-1)/2,n,0)
    else evencase(2*x, n/2,m/2)
    end if;
  end proc:
```

Finally we have to design the procedure C(x,r) to compute the integral
of the rth power of the cosine function using the recurrence relation (5.5).
The do loop (i.e., nonrecursive) version of this is

```
  C:=proc(x,r)
    local u,s,i;
    if r=0 then u:=x
    elif r=1 then u:=sin(x)
    elif r=2 then u:=cos(x)*sin(x)/2+x/2
    else
        if type(r,even) then s:=2;u:=cos(x)*sin(x)/2+x/2
```

```
      else s:=1;u:=sin(x)
      end if;
      for i from s by 2 to r-2 do
          u:=cos(x)^(i+1)*sin(x)/(i+2)+(i+1)*u/(i+2)
      end do;
   end if;
end proc:
```

An alternative procedure CR(x,r) to accomplish the same thing, except with recursion, is

```
CR:=proc(x,r)
   option remember;
   local u;
   if r=0 then u:=x
   elif r=1 then u:=sin(x)
   else cos(x)^(r-1)*sin(x)/r+(r-1)*CR(x,r-2)/r
   end if;
   end proc:
```

As you can see, the code captures the essence of the recurrence relation (5.5), with the exceptional cases $r = 0, r = 1$ handled separately. Since the body of the procedure contains a statement with CR(x,r-2) in it, during execution the procedure will have to call itself. This is why it is called a recursive procedure. The command **remember** causes the procedure to keep track of the calculations it makes so that they can be used in future calls to the procedure.

Having designed the code for all the parts, we can now routinely test it on the requested examples. Thus, the commands

```
Int(sin(x)^2*cos(x)^3,x)=f(x,2,3);
Int(sin(x)^4*cos(x)^2,x)=f(x,4,2);
```

produce the output

$$\sin(x)^2 \, \cos(x)^3 \, dx = \frac{1}{3} \sin(x)^3 - \frac{1}{5} \sin(x)^5$$

$$\sin(x)^4 \, \cos(x)^2 \, dx =$$

$$-\frac{1}{48} \sin(2x) - \frac{1}{32} \cos(2x) \, \sin(2x) + \frac{1}{16}x + \frac{1}{48} \cos(2x)^2 \, \sin(2x)$$

which checks with the calculations we did by hand. (Note how the second result is ordered differently that the by-hand example.) Similarly we can use the procedure to easily produce the graphs as shown in Figures 5.6 5.7. Plotting these figures can be a real test of the difference between the recursive and nonrecursive versions of $C(x, r)$, especially if the `remember`

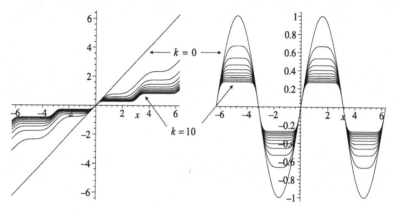

Figure 5.6: *Plots of $C(x, 2k) = \int \cos^{2k} x \, dx$ (on the left) and $C(x, 2k+1) = \int \cos^{2k+1} x \, dx$ (on the right), for $k = 0, \dots, 10$.*

option is not used in the recursive version. This is discussed, further in Chapter 8. On the other hand, the recursive version, for this example, certainly seems the clearest and most elegant way to accomplish the task.

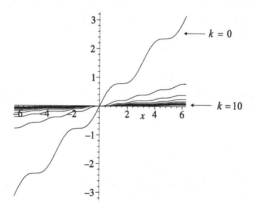

Figure 5.7: *Plots of $f(x, 2k, 2) = \int \sin^{2k} x \cos^2 x \, dx$, for $k = 0, \dots, 10$.*

Exercise Set 5

1. Rewrite the `area2` procedure so that it calculates the two angles   and   which determine the direction of the normal vector n to the plane of the triangle $\triangle PQR$. Specifically, if $n = (n_1, n_2, n_3)$, then   is the angle between n and the z-axis, while   is the angle between the vector $(n_1, n_2, 0)$ and the x-axis. Make sure you account for all the exceptional cases (such as $n_1 = 0$).

   Display the triangle and the normal vector in the same picture using a viewpoint specified by the Maple option: `orientation=[theta,phi]`, where `theta,phi` are the values calculated for , . Verify that the view is one that looks perpendicularly down on the plane of the triangle. Rewrite the `checktri` procedure so that it produces a picture of the triangle and normal vector displayed with `orientation=[theta,phi]`.

2. Rewrite the `checktri` procedure so that it uses only conditional statements to control the "ow, rather than the local control variable `c`. The output is to be one and only one of the following: (1) *the triangle is equilateral*, (2) *the triangle is isosceles*, (3) *the triangle is isosceles and a right triangle with right angle at vertex* , (4) *the triangle is a right triangle with right angle at vertex* , or (5) *the triangle is not equilateral, isosceles, or right.* You may use the fact that an equilateral triangle is never a right triangle. Keep your code as short as possible.

3. Use the code for the procedures `triangle` and `checktri` (on the CD-ROM) and apply it to the triangle $\triangle ABC$ with vertices

$$
\begin{aligned}
A &= (0,0,0) \\
B &= (\ \overline{3},\ \overline{2}, 0) \\
C &= (-\ \overline{3}(\ \overline{2}-1)/2, (3+\ \overline{2})/2, 0)
\end{aligned}
$$

   How does procedure `checktri` classify the triangle? Show by hand that this triangle is equilateral. Explain why `checktri` misses this. Change the procedure `triangle` so that it uses exact arithmetic on the calculation of the lengths of the sides (i.e., eliminate the `evalf`'s in those calculations). Now use `checktri` on $\triangle ABC$ again. Can you explain why it still fails to recognize the triangle as equilateral?

4. Alter the procedure `checktri` so that it works   correctly   up to a specified tolerance of accuracy. Specifically, include an extra input parameter `E` that the user should take to be a small positive number. Alter the logic to test for whether `abs(p1-q1)<E`, `abs(cos(ap))<E`, and so forth. Run some studies on various triangles.

5. Write a procedure called `checkquad` to classify quadrilaterals in the plane, i.e., having all four vertices P, Q, R, S in $\mathbb{R}^2$. The input to the procedure should be the four vertices. As output, your procedure should (a) print the lengths of all the sides and measures of all the angles, (b) print one and

only one of the words *quadrilateral, parallelogram, rhombus, rectangle,* or *square,* and (c) display a picture of the quadrilateral. In the classification, each quadrilateral should receive only one designation. For example, being classified a parallelogram means that it is not a rhombus, not a rectangle, and not a square. A square is classified as such, but, even though it is also a rectangle, rhombus, parallelogram, and a quadrilateral, these designations are not printed.

You may assume that the user applies the procedure only to quadrilaterals $PQRS$ where the vertices P, Q, R form a triangle, and the S lies outside this triangle in the region $[RP]$. See Exercise 7 in Chapter 4 for the definition of the region $[RP]$. Your procedure can/should use the `area2` and `triangle` procedures in the text, if you view the quadrilateral as composed of triangles $\triangle PQR$ and $\triangle PRS$.

6. Extend the procedure `checkquad` in Exercise 5 so that it requires no assumption about the positions of the four vertices P, Q, R, S of the quadrilateral. For this modify, as necessary, the procedure `checkpoint` from Exercise 7 in Chapter 4. Have the procedure `checkquad` work like this. First determine if P, Q, R are collinear. If not, determine whether the point S is in one of the regions $[PQ], [QR], [RQ]$ (see Exercise 7, Chapter 4). If so, then the quadrilateral is composed of $\triangle PQR$ and one of $\triangle PQS, \triangle QRS, \triangle RQS$, depending on the region in which S lies. Now classify the quadrilateral as in Exercise 5. In all other cases have your routine print out that the quadrilateral is either non-convex or is degenerate.

7. **(Optimizing Functions of Two Variables)** Suppose f is a function of two variables defined on a rectangle $R = [a, b] \times [c, d]$ in the plane. As in Example 5.3 for a function of one variable, write a procedure, `optimize2d`, to find the approximate maximum and minimum values of f on R and the points at which these values occur. To find the approximations, first approximate the function f by a step function as follows. Divide R into subrectangles in the standard way: $R_{ij} = [x_{i-1}, x_i) \times [y_{j-1}, y_j)$, $i = 1, \ldots, N, j = 1, \ldots, M$. Then the step function approximation is defined to have constant value $f(z_{ij})$ on R_{ij}, for $i = 1, \ldots, N, j = 1, \ldots, M$. Here $z_{ij} = (x_{i-1}, y_{j-1})$ is the lower left-hand vertex of R_{ij}. In each case have your program also produce graphical output consisting of a picture with (a) a plot of the boundary of R and (b) plots of all the vertical line segments; $L_{ij} = \overline{P_{ij}Q_{ij}}$, where $P_{ij} = (x_i, y_j, 0)$ and $Q_{ij} = (x_i, y_j, z_{ij})$, for $i = 1, \ldots, N, j = 1, \ldots, M$.

Test your `optimize2d` procedure on the functions from the following that are assigned to you.

(a) $f(x, y) = xe^{-x^2-y^2}$, on $D = [-2, 2] \times [-2, 2]$.

(b) $f(x, y) = xye^{-x^2-y^2}$, on $D = [-2, 2] \times [-2, 2]$.

(c) $f(x, y) = (x^2 + y)e^{-x^2-y^2}$, on $D = [-2, 2] \times [-2, 2]$.

(d) $f(x, y) = 5(x + y)/(1 + x^2 + y^2)$, on $D = [-5, 5] \times [-5, 5]$.

Here are some additional instructions for things to do in this exercise.

(a) As in Example 5.3, have your procedure find the approximate maximum and minimum values, places where these occur, and print these out. Return the graphical picture through an output parameter. Your procedure can call the `maxmin` procedure as the `optimize` procedure in Example 5.3 did. Now, however, you must convert the two-dimensional array of z-values $\{z_{ij}\}$ into a (one-dimensional) list.

(b) Test your procedure for values $N = M = 20, 30, 40$ and compare the results, both numerical and graphical. Annotate the figures produced with the values of N, M and all the points on the graph (x, y, z_{max}) and (x, y, z_{min}) corresponding to the maximum and minimum values. In each case, rotate the figures so that the view represents the projections of the figure on the x-z plane and the y-z plane (six figures in all). Annotate these and comment on how they help find the maximum and minimum values (and possibly how they would obscure this information).

(c) Plot the graph of the function f and its domain in the same figure.

(d) Recall from calculus that local maxima and minima of f occur at critical points, i.e., at points (x, y) that are simultaneous solutions of the equations:

$$\frac{\partial f}{\partial x}(x, y) = 0, \qquad \frac{\partial f}{\partial y}(x, y) = 0. \tag{5.6}$$

Find these critical points approximately in two ways: (i) Plot each of the curves with equations (5.6) and use the mouse to approximate where these curves intersect. (ii) Use Maple's `fsolve` command to solve equations (5.6).

(e) Use the results of part (d) to determine the maximum and minimum values of f and the points where these occur. Compare with the approximate values and points that you found in part (b).

8. **(Tetrahedra in $\mathbb{R}^3$)** A tetrahedron in $\mathbb{R}^3$ is the solid determined by four points P, Q, R, S in $\mathbb{R}^3$ and has the triangles $\triangle PQR, \triangle PQS, \triangle PSR$, and $\triangle QSR$ as its faces. See Figure 5.8. Write a program to take four points P, Q, R, S of a tetrahedron as input and to produce as output: (1) the volume of the tetrahedron, (2) the areas of its four faces, (3) the lengths of its six edges, (4) the determination of which, if any, of its faces are equilateral, isosceles, or right triangles, and (5) a plot of the tetrahedron.

9. **(Trig Integrals)** Using the case study as a guide, write a program to calculate the trig integral

$$\tan^n x \sec^k x \, dx,$$

for various nonnegative integer powers n, k. Test your program appropriately and produce plots of the functions that result from these trig integrals.

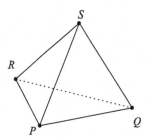

Figure 5.8: *The tetrahedron in $\mathbb{R}^3$ determined by the points P, Q, R, S.*

10. **(Lines and Planes in Space)** Many of the analytic geometry problems
    you solved in calculus III (or IV) dealt with determining certain equations
    for lines or planes in $\mathbb{R}^3$ from given information. Usually vector methods
    (using dot products, cross products, and determinants) are essential tools for
    solving such problems. This exercise is to automate the solution of some of
    these problems by writing a number of procedures. This will require that you
    understand the concepts and techniques you used in your calculus course for
    this material, and should serve to reinforce what you learned.

    Write some Maple code for the following procedures. Use the suggested
    names and parameters.

    (a) `linsolv2d:=proc(A,b,v)` Here the input is a 2×2 matrix (or two-
        dimensional array) and a vector (or one-dimensional array)

        $$A = \begin{bmatrix} a_{11} & a_{12} \\ a_{21} & a_{22} \end{bmatrix}, \qquad b = \begin{bmatrix} b_1 \\ b_2 \end{bmatrix},$$

        and the output is a vector (one-dimensional array)

        $$v = \begin{bmatrix} v_1 \\ v_2 \end{bmatrix}.$$

        The vector v is to be the solution of the equation $Av = b$, i.e., of the
        linear system

        $$a_{11}v_1 + a_{12}v_2 = b_1$$
        $$a_{21}v_1 + a_{22}v_2 = b_2.$$

        Your procedure should test to see if $\det(A) = 0$ and return the string
        `'no solution'`, or `'infinitely many solutions'` depending on which
        case applies. Otherwise your procedure should return the unique solu-
        tion v.

    (b) `planepar:=proc(a,b,c,d,P,Q,R,f)` This procedure should take the
        plane with equation $ax + by + cz = d$ (input parameters are a, b, c, d)

and return three non-collinear points P, Q, R on the plane as well as a vector-valued function f that parametrizes this plane:

$$f(s,t) = \overrightarrow{OP} + sv + tw.$$

See Figure 5.9.

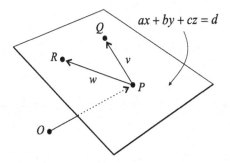

Figure 5.9: *Finding the parametric representation of the plane with equation $ax + by + cz = d$.*

(c) `linepar:=proc(a1,b1,c1,d1,a2,b2,c2,d2,g)` This procedure should return a vector-valued function $g(t) = \overrightarrow{OP} + tv$, that parametrizes the *line* of intersection of the two planes with equations

$$\begin{aligned} a_1 x + b_1 y + c_1 z &= d_1 \\ a_2 x + b_2 y + c_2 z &= d_2. \end{aligned}$$

See Figure 5.10. If the planes are parallel, your procedure should return a message saying so. Also you might want to call the `linsov2d` from within this procedure.

(d) `linesintersect:=proc(P,v,Q,u,R)` This procedure should take the data P, v, Q, w for the two lines

$$\begin{aligned} F(t) &= \overrightarrow{OP} + tv \\ G(s) &= \overrightarrow{OP} + sw, \end{aligned}$$

as input and calculate the point R of intersection of these lines. See Figure 5.11. Recall that two lines in $\mathbb{R}^3$ are parallel, skew, or intersect in a single point. Your procedure should determine which of these three possibilities occurs and return 'the lines are parallel', 'the lines are skew', or the unique point R of intersection. See the Maple/Calculus Notes at the end of the chapter.

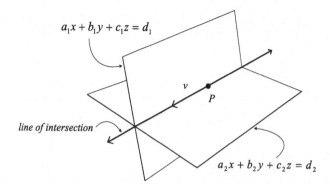

Figure 5.10: *Finding the parametric representation of the line of intersection of two planes.*

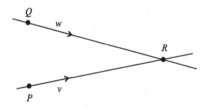

Figure 5.11: *Finding the point of intersection of two nonparallel, nonskew lines.*

(e) `linepar2:=proc(P,Q,g)` Write a procedure that takes two distinct points P, Q in $\mathbb{R}^3$ and gives a vector-valued function g which parametrizes the line through P and Q. See Figure 5.12. Use a parametrization g such that $g(0) = P$ and $g(1) = Q$. Make sure your procedure checks to see if $P \neq Q$.

General Instructions for Exercise 10:

(i) Write all of your own code. Do *not* use Maple procedures such as `solve,det, crossprod`, etc., for solving systems of equations, computing the determinant, or cross product, etc.

(ii) Test `linsolve2d` on the following systems:

$$\text{(a)} \quad \begin{array}{rcl} 2x - 5y &=& 7 \\ 3x + 6y &=& 1. \end{array} \qquad \text{(b)} \quad \begin{array}{rcl} 2x + y &=& 2 \\ -4x - 2y &=& -4. \end{array}$$

(iii) Test `planepar,linepar`, and `linepar2` by doing the following.

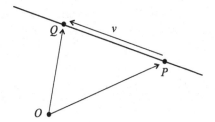

Figure 5.12: *Finding the parameric representation of the line through P and Q.*

Using the parametric equations, plot, in the same figure, the planes $x + y + z = 1, x - y = 0$, and their line of intersection. Use a greater line thickness for the line of intersection. You will have to experiment with the values for the respective parameters so that the plots are the right size. Use appropriate grid sizes for the parametric plots of the planes.

For the plane $x + y + z = 1$, use the three points P, Q, R returned by `planepar` to plot the triangle $\triangle PQR$ (use `linepar2` for this plot). Also, in the same figure, plot the plane $x - y = 0$ and the line of intersection of both planes ($x + y + z = 1$ and $x - y = 0$).

(iv) Test `linesintersect` on the following pairs of lines.

(a) $P = (4, 1, 8)$, $v = (-1, 2, 6)$ and $Q = (7, 7, 8)$, $w = (1, 4, 3)$.

(b) $P = (1, 1, 2)$, $v = (3, 2, 1)$ and $Q = (4, 2, -1)$, $w = (-1, 1, 3)$.

If the two lines intersect, plot, in the same figure, the two lines and their point of intersection.

5.5 Maple/Calculus Notes

5.5.1 Vector Methods in Geometry

It may be helpful to recall some concepts and notation from the analytic geometry part of your calculus course, especially since the notation for vectors can be slightly different from text to text.

Using vector methods in geometry not only simplifies proving geometric theorems but also makes calculations of various quantities, such as length, angle measure, area, and volume, considerably simpler. In particular, the dot product of vectors gives length and angle measure, the cross product gives area, and the scalar triple product, or determinant, gives volume. For convenience, we define these products again here.

If we denote vectors in $\mathbb{R}^3$ by v, w, u, etc.[6], then the expressions for these vectors in terms of their components are

$$v = (v_1, v_2, v_3)$$
$$w = (w_1, w_2, w_3)$$
$$u = (u_1, u_2, u_3).$$

The *dot product* is then the scalar quantity given by

$$v \cdot w = v_1 w_1 + v_2 w_2 + v_3 w_3,$$

while the *cross product* is the vector given by

$$v \times w = (v_2 w_3 - v_3 w_2, \; v_3 w_1 - v_1 w_3, \; v_1 w_2 - v_2 w_1).$$

The *scalar triple product* combines three vectors v, w, u to give the following scalar quantity

$$(v \times w) \cdot u = \det(v, w, u) = \begin{vmatrix} v_1 & v_2 & v_3 \\ w_1 & w_2 & w_3 \\ u_1 & u_2 & u_3 \end{vmatrix}.$$

The length of v is $|v| = \overline{v \cdot v} = \overline{v_1^2 + v_2^2 + v_3^2}$.

In your calculus course you should have learned the following geometric meanings for these operations

If   is the angle between the vectors v, w, then

$$v \cdot w = |v| \, |w| \, \cos .$$

Thus, v and w are perpendicular if and only if $v \cdot w = 0$.

The vector $v \times w$ is perpendicular to both v and w and $|v \times w|$ is equal to the area of the parallelogram determined by v and w. In particular $v \times w = 0$ if and only if v and w are parallel (i.e., one is a scalar multiple of the other).

The scalar triple product $(v \times w) \cdot u$ is equal to the volume of the parallelopiped determined by the vectors v, w, u.

---

[t]Some books use boldface $\mathbf{v}$, $\mathbf{w}$, $\mathbf{u}$, or arrows $\vec{v}, \vec{w}, \vec{u}$, for this in order to distinguish vectors from scalars, but we do not do that here.

While geometry deals with points, lines, and planes, as well as figures composed of these things, the use of vectors arises from the observation that between any two points in space

$$P = (x_1, y_1, z_1), \qquad Q = (x_2, y_2, z_2),$$

there is the *vector from Q to P*:

$$\overrightarrow{QP} = (\, x_1 - x_2,\, y_1 - y_2,\, z_1 - z_2\,).$$

In particular, if $Q = (0,0,0)$ is the origin, then

$$\overrightarrow{QP} = \overrightarrow{OP} = (x_1, y_1, z_1) \quad P.$$

Note: Here we have identified the vector $\overrightarrow{OP}$ with the point P. From a pedagogical point of view, it is usually bad not to make a distinction between vectors and points, but if you are going to use vector methods effectively in geometry, this identification is most convenient. So we will do it. For example, we can then express $\overrightarrow{QP}$ as a difference

$$\overrightarrow{QP} = \overrightarrow{OP} - \overrightarrow{OQ} = P - Q.$$

Some of the problems in the geometry sections of your calculus course involved writing parametric equations for lines and planes[7] based on certain information given about the lines or planes. The exercises in this chapter will deal with automating some of these calculations. So we recall a few of the details here.

A line in $\mathbb{R}^3$ is easiest to determine if you know a point $P = (x_0, y_0, z_0)$ on the line and a direction $v = (v_1, v_2, v_3)$ for the line. Then the parametrization of the line is given by the following vector-valued function.

Parametric version of the line through P in the direction of v:

$$f(t) = P + tv = (\, x_0 + t v_1,\, y_0 + t v_2,\, z_0 + t v_3 \,).$$

It is common, also, to record this same information in the following form

$$
\begin{aligned}
x &= x_0 + t v_1 \\
y &= y_0 + t v_2 \\
z &= z_0 + t v_3
\end{aligned}
$$

---

[7]Some calculus texts do not discuss parametric equations for planes.

The key to determining lines from other given information is to use that information to get a point P on the line and a direction v for the line. For example, if two points Q, P on the line are given, then $v \quad Q - P$ is a direction for the line and P or Q can be used as the given point. A parametric version is then,

$$f(t) = P + t(Q - P) = (1 - t)P + tQ.$$

Note that this is a parametrization that starts at P and runs toward Q (since $f(0) = P, f(1) = Q$).

A plane in $\mathbb{R}^3$ can be described by an equation of the form $ax + by + cz = d$, which means that the plane consists of the set of points

$$plane = \{\, (x, y, z) \,|\, ax + by + cz = d \,\}.$$

Geometrically the numbers a, b, c make up the components of a vector $n = (a, b, c)$ which is perpendicular to the plane. An alternative description for planes, one that is easier for Maple to use in plotting planes, is the parametric description. This description is entirely analogous to the parametric description of lines. Now, however, since a plane is two dimensional, two direction-vectors v, w and two parameters s, t are needed.

Parametric version of the plane through P parallel to v, w:

$$f(s, t) = P + sv + tw = (\, x_0 + sv_1 + tw_1, \ y_0 + sv_2 + tw_2, \ z_0 + sv_3 + tw_3 \,).$$

It is common, also, to record this same information in the following form

$$
\begin{aligned}
x &= x_0 + sv_1 + tw_1 \\
y &= y_0 + sv_2 + tw_2 \\
z &= z_0 + sv_3 + tw_3
\end{aligned}
$$

Note that since the plane is parallel to v, w, it follows that $n \quad v \times w$ is perpendicular to the plane. A standard problem dealing with planes is to find a parametrization of the plane passing through three given non-collinear points P, Q, R. Geometrically, this is the same as the plane through P that is parallel to $v = Q - P$ and $w = R - P$. Hence from above, a parametrization is

$$f(s, t) = P + s(Q - P) + t(R - P) = (1 - s - t)P + sQ + tR.$$

Recall that for two lines in $\mathbb{R}^3$, the lines either (a) intersect (in which case they lie in the same plane), (b) are parallel (they lie in the same plane and do not intersect), or (c) are skew (they do not lie in the same plane). We can determine which of these three possibilities occurs from the parametric equations for the lines. Thus, suppose one line passes through P in the direction of v and the other line passes through Q in the direction of w. The parametric representations for the lines are then $F(s) = P + sv$ and $G(t) = Q + tw$, respectively. Note that we have used a different parameter in each representation. This makes it easier to determine if the lines intersect or not. For the lines to intersect there must be parameter values s and t such that $F(s) = G(t)$, i.e.,

$$P + sv = Q + tw.$$

Rearranging this gives

$$sv - tw = Q - P.$$

This is a vector equation with s and t as unknowns. If we write the vector equation in terms of its components, we get the following system of three equations for the two unknowns s, t.

$$v_1 s - w_1 t = Q_1 - P_1$$
$$v_2 s - w_2 t = Q_2 - P_2$$
$$v_3 s - w_3 t = Q_3 - P_3.$$

The theory of systems of equations tells us that this system has a solution (s, t) if and only if

$$\begin{vmatrix} v_1 & -w_1 & Q_1 - P_1 \\ v_2 & -w_2 & Q_2 - P_2 \\ v_3 & -w_3 & Q_3 - P_3 \end{vmatrix} = 0.$$

Otherwise said, the two lines intersect if and only if this determinant is zero. When this is the case, we can use two of the three equations to find the parameter values s and t that give the point of intersection: $F(s) = G(t)$.

Example 5.7 Suppose the points are $P = (5, -3, 3), Q = (2, 1, 2)$ and the vectors are $v = (2, 1, 8), w = (-1, 2, 1)$. Then the parametrizations are

$$F(s) = (5, -3, 3) + s(2, 1, 8) = (5 + 2s, -3 + s, 3 + 8s)$$
$$G(t) = (2, 1, 2) + t(-1, 2, 1) = (2 - t, 1 + 2t, 2 + t).$$

Equating components gives $5 + 2s = 2 - t, -3 + s = 1 + 2t$, and $3 + 8s = 2 + t$. Writing this system in standard form gives

$$2s + t = -3$$
$$s - 2t = 4$$
$$8s - t = -1.$$

It is straightforward to check that

$$\begin{vmatrix} 2 & 1 & -3 \\ 1 & -2 & 4 \\ 8 & -1 & -1 \end{vmatrix} = 0,$$

and thus the lines intersect. If we select the top two equations in the system:

$$\begin{aligned} 2s + t &= -3 \\ s - 2t &= 4 \end{aligned}$$

and solve these, we find $s = -\frac{2}{5}$ and $t = -\frac{11}{5}$. Then we compute that the point of intersection is

$$F\left(-\tfrac{2}{5}\right) = \tfrac{21}{5}, -\tfrac{17}{5}, -\tfrac{1}{5} = G\left(-\tfrac{11}{5}\right).$$

Chapter 6

Data Structures

In the previous chapters you have seen and learned how to use basic programming constructs such as looping, branching, and procedure calls to write algorithms for a variety of programming tasks. These constructs are common to most programming languages, although the specific details vary from one to another.

Data structures, that is, the particular ways in which information is represented and accessed, are also important elements in most programming languages and a firm understanding of these structures is essential for building more complex programs.

In this chapter we will look at a *few* of Maple's data structures in more detail. You will learn that each legitimate expression in Maple has a particular data structure, and that there are specific ways to access the individual parts that make up the expression.

To a certain extent Maple's basic data structures, lists, arrays, and sets, will su ce to meet most of your programming needs. The examples below will further illustrate how these naturally occur in writing certain algorithms.

However, there is also a need to explain a few of Maple's other data types and indicate how they are useful for programming. It is important for you to know that in writing programs in the previous chapters, you used many expressions in those programs that were actually data structures. For example, in the assignment statement `eq1:=3*x+2*y=0`, the expression `3*x+2*y=0` is a Maple data structure of type `equation` and the expression `3*x+2*y` is a data structure of type `+`. The availability of additional data types such as these is necessary for Maple's symbolic manipulation capability, which is an aspect of Maple that makes it different from traditional programming languages.

Tables 6.1 and 6.2 show a basic comprehensive list of the types of data

!	And	Array	BooleanOpt
MVIndex	Matrix	NONNEGATIVE	Not
Or	Point	Range	RootOf
TEXT	Vector	$\boxed{*}$	$\boxed{+}$
$\boxed{.}$	$\boxed{=}$	$\Box$	algebraic
algext	algfun	algnum	algnumext
anyfunc	anything	arctrig	$\boxed{\text{array}}$
atomic	$\boxed{\text{boolean}}$	builtin	complex
complexcons	constant	cubic	cx_infinity
dependent	disjcyc	embedded_axis	embedded_imaginary
embedded_real	$\boxed{\text{equation}}$	even	evenfunc
expanded	$\boxed{\text{exprseq}}$	extended_numeric	extended_rational
facint	finite	$\boxed{"\text{oat}}$	$"\text{oat}[]$
$\boxed{\text{fraction}}$	freeof	function	hfarray
identical	imaginary	indexable	indexed
indexedfun	infinity	$\boxed{\text{integer}}$	intersect
last_name_eval	laurent	linear	$\boxed{\text{list}}$
listlist	literal	logical	mathfunc

Table 6.1: *Some types of data structures in Maple.*

structures that are available in Maple. The types in these tables that have boxes around them have either been discussed previously or will be discussed in this chapter. The rest you can access from Maple's help menu. (Enter ?type; to get a hyperlinked list of the types shown in Tables 6.1 and 6.2.)

All data structures in Maple are *expressions* composed of *operands* and can be represented by tree diagrams, which pictorially represent the structures for the data. We begin with a general discussion of this before looking at the details for some specific types.

6.1 Expressions and Operands

Expressions are the basic objects of Maple. They can be very simple, like the integer 7, less simple, like a 1+x^2+sin(x), or more complicated, like an equation y-1/y=1+x^2+sin(x). Lists, sets, and arrays, like [a,a,1,4], {A,B,C}, and array([[1,2],[2,4]]), are also Maple expressions.

Any expression is composed of a sequence of *operands*. The constant 7 consists of only one operand, while the expression 1+x^2+sin(x) has three operands 1,x^2,sin(x) (which mathematically are called its summands).

matrix	minus	module	moduledefinition
monomial	name	neg_infinity	negative
negint	negzero	nonnegative	nonnegint
nonposint	nonpositive	nonreal	nothing
numeric	odd	oddfunc	operator
point	polynom	pos_infinity	posint
positive	poszero	prime	procedure
protected	quadratic	quartic	radalgfun
radalgnum	radext	radfun	radfunext
radical	radnum	radnumext	range
rational	ratpoly	ratseq	real_infinity
realcons	relation	rtable	scalar
sequential	series	set	s"oat
specfunc	sqrt	string	su xed
symbol	symmfunc	table	tabular
taylor	trig	type	undefined
uneval	union	vector	verification

Table 6.2: *Some types of data structures in Maple.*

The equation `y-1/y=1+x^2+sin(x)` has two operands: the left-hand side, `y-1/y`, and right-hand side, `1+x^2+sin(x)`, of the equation.

Also any expression has a *type*, one of those shown in Table 6.1. An expression's type determines what values it can assume and what operations can be performed on it. For example, `1+x^2+sin(x)` is of type + (it is the sum of three expressions) and `y-1/y=1+x^2+sin(x)` is of type = (it is an equation).

A convenient way to visualize an expression is with an *expression tree*. At the root of the tree is the expression's type (or name, if the expression is a function). The branches from the root correspond to the operands of the expression. At the end of each branch is a name or a constant, in which case the branch terminates, or a node, in which case further branching occurs. Figure 6.1 shows the general expression tree branching from its initial root.[1]

Example 6.1 Consider the function $\sin x + 4 + x - 6x^2 + 12x^5$, which is represented in Maple by the expression

---

[1]It is customary in computer science and other disciplines to draw the tree upside down the tree grows downward by branching out from the root.

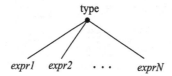

Figure 6.1: *The root and branches of an expression tree.*

```
sin(x)+4+x-6*x^2+12*x^5
```

The top of the expression tree for this expression has a node (or root node) designating its type: +, which means it is a data type consisting of a sum of other expressions (called summands in mathematics). There are five branches from this node, one for each of these other expressions (operands of the original expression): `sin(x)`, `4`, `x`, `-6*x^2`, and `12*x^5`. Some of these operands have an expression trees of their own, branching out from the operands' root node, and the root node tells the summand's type in the sum. This branching continues until a branch ends in a constant or a name. See Figure 6.2.

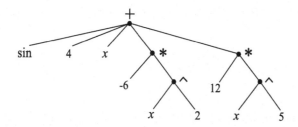

Figure 6.2: *Expression tree for $\sin x + 4 + x - 6x^2 + 12x^5$.*

When programming with expressions, you need a way to determine what the type of an expression is, how many operands the expression has, and a way of selecting these operands. There are four basic Maple commands for doing these things, which are described as follows:

---

whattype(E)

This command determines the type of the expression E. If you are in doubt as to the type of any expression, simply have Maple determine it for you with this command. For example, whattype({a,b,c}) will return the type *set*. Multiple uses of this command allow you to examine the types of all the operands in an expression. For example, whattype(3*sin(x)) returns the type *, while whattype(3) and whattype(sin(x)) return *integer* and *function* as the respective types.

$$\mathrm{type}(E, T)$$

This command returns a value of **true** if the expression E is of type T. Otherwise it returns the value **false**. This command is commonly used in conditional statements (**if-then-else**'s) to control the branching. Example 6.2 below illustrates this. Another common use is to prevent incorrect data types from being used in an algorithm you have written. For example, the code

```
if type(n,nonnegint) then A:=n! end if;
```

allows only the factorial of an nonnegative integer to be assigned to **A**.

$$\mathrm{nops}(E)$$

This command determines the number of operands of expression E. Often in programming, the number of operands for an expression is not known in advance, yet the actual number is needed within the program. For example, if a procedure has a list **L** of unknown length as input and needs to access each member of the list then a loop from **1** to **nops(L)** might be needed.

$$\mathrm{op}(i, E)$$

This command selects the ith operand of expression E. The i is optional. When it is omitted $\mathrm{op}(E)$ returns the sequence consisting of all the operands of E. The **op** command is the real workhorse for manipulating expressions in Maple. For example, suppose a square is represented by a list of its four vertices:

$$\mathtt{SQ} := [[6, 6], [8, 6], [8, 8], [6, 8]],$$

where each vertex is represented as a list consisting of the two coordinates of the vertex. Then the fourth vertex of SQ is accessed by op(4,SQ), giving [6,8], and the second coordinate of the fourth vertex is accessed by op(2,op(4,SQ)), giving 8.

Using the optional form op(E) for the op command is helpful in determining what the operands of E are and also in decomposing the expression E into its operands for use in further programming. For example, the function $f(x) = \sin x - \frac{1}{2}\sin 2x + \frac{1}{10}\sin 3x$ is the superposition of three sine waves. You can plot f and the three constituent waves with the commands

```
f:=sin(x)-sin(2*x)/2+sin(3*x)/10;
plot({f,op(f)},x=0..2*Pi);
```

Table 6.3 summarizes the whattype,type,nops, and op commands with some elementary examples of the uses of these commands.

Command	Example	Result
whattype	whattype(1+x+3*y)	+
type	type(x=y+1,equation)	true
nops	nops(1+x+3*y)	3
op	op(2,1+x+3*y)	x
op	op(1+x+3*y)	1,x,3*y

Table 6.3: *Some basic commands for manipulating expressions.*

The next example gives a more in-depth illustration of the use of these expression manipulating commands from Table 6.3.

Example 6.2 (Linear Equation Solver) Consider writing a procedure for solving a linear equation of the form

$$a_1 x + b_1 = a_2 x + b_2,$$

for x, assuming a_1, b_1, a_2, b_2 are all numeric. The input to the procedure will be an equation of this form, written in the correct Maple syntax, and we allow for the terms on each side of the equation to be in any order. For example, either $2x - 9 = 4 - x$ or $2x - 9 = -x + 4$ could be the form of the input.

This is certainly not a su ciently complicated type of equation to warrant writing an automated solver for (it's easier to solve by hand) and of course Maple's `solve` command can easily do this too. But we will write our own simple procedure and this will give us insight into how Maple does symbolic manipulation in much more complicated situations.[2]

One approach to solving equations of the stated form would be as follows. The form of the input equation `a1*x+b1=a2*x+b2` (or one of the three other possible orders) gives a simple expression tree with three levels of branching as shown in Figure 6.3.

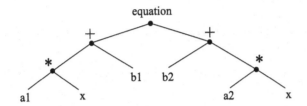

Figure 6.3: *A tree diagram for the equation* `a1*x+b1=a2*x+b2`.

We must write a routine to scan the left-hand side of the equation and identify the coe cients `a1,b1`, while allowing for the possibility that one or the other could be absent from the input (For example, an input of `3*x=5` is allowed). There are four cases for the possible type of each side: numeric, symbol, *, or +. For example, the left-hand side could be 6 (numeric type), or x (symbol type), or `7*x` (* type), or `2+9*x` (+ type). Similarly we must pick off `a2,b2` from the right-hand side of the equation.

Then by combining coe cients, `a:=a1-a2; b:=b1-b2;` the equation is reduced to one of the form `a*x+b=0`, which has either (1) a unique solution, $x = -b/a$, if $a \neq 0$, or (2) infinitely many solutions, if $a = 0, b = 0$, or (3) no solution, if $a = 0, b \neq 0$. The code is then as follows:

| Equation Solver |

```
eqnsolve:=proc(eq::equation)
  local LS,RS,a,b,a1,a2,b1,b2,i,soln;
  LS:=op(1,eq);RS:=op(2,eq);
  a1:=0;b1:=0;a2:=0;b2:=0;
  #-----------------------------------------------------------
  # If the left side has only one term, then its type must
  # be determined. If it has more than one term, then it is
```

```
# of type +, and so has more than one operand.
#-----------------------------------------------------------
if type(LS,numeric) then b1:=LS;
elif type(LS,symbol) then a1:=1;
elif type(LS,'*') then a1:=op(1,LS)
else
  for i from 1 to nops(LS) do
    if type(op(i,LS),'*') then
      a1:=op(1,op(i,LS))
    elif type(op(i,LS),symbol) then a1:=1;
    else b1:=op(i,LS)
    end if;
  end do;
end if;
#-------------------------------------------------
# The right side is checked similarly
#-------------------------------------------------
if type(RS,numeric) then b2:=RS;
elif type(RS,symbol) then a2:=1;
elif type(RS,'*') then a2:=op(1,RS)
else
  for i from 1 to nops(RS) do
    if type(op(i,RS),'*') then
             a2:=op(1,op(i,RS))
    elif type(op(i,RS),symbol) then a2:=1;
    else b2:=op(i,RS)
    end if;
 end do;
end if;
#-------------------------------------------------
a:=a1-a2;b:=b1-b2;
if a<>0 then soln:=x=-b/a
elif b=0 then soln:='x=any real number'
else soln:='no solution'
end if;
print(eq,'  solve for x . . .  ',soln);
end proc:
```

Some examples of calls to this procedure would be

```
eqnsolve(5*x+2=3-4*x);
```

$$5x + 2 = 3 - 4x, \text{ solve for } x \ldots, x = \frac{1}{9}$$

```
eqnsolve(5*x+2=5*x+2);
```

$$5x + 2 = 5x + 2, \text{ solve for } x \ldots, x{=}any \ real \ number$$

```
eqnsolve(5*x+2=5*x+1);
```

$$5x + 2 = 5x + 1, \text{ solve for } x \ldots, no \ solution$$

6.2   Quotes and Strings

In the last example above, part of the output for the procedure is the expression: *solve for x ....* This was accomplished by enclosing this phrase in *back quotes*. A back quote in this book appears as , while on your keyboard it will appear differently. Enclosing a phrase in back quotes makes that phrase into a name with data structure type: `symbol`. You can assign values to these, e.g.,

```
'luckynumber':=13;
```

   luckynumber := 13

The back-quoted name `'luckynumber'` is the same as the name `luckynumber`, without the the back quotes. So in most cases the back quotes are not necessary. The exception to this is when the phrase you want to make into a name includes spaces (and other nonvalid naming symbols). Then back quotes are the only way to go:

```
'my friend s lucky number is':=77;
```

   my friend s lucky number is := 77

The whole phrase here, *my friend s lucky number is*, is a name for a variable which gets assigned the value 77. This is no different, in principle, than the assignment `x:=77`.

   We have found it convenient, in various place in the book, to use back quotes on phrases that we want to print as output (especially error messages). As an alternative to this, we could turn the phrase into a *string*.

   An ordered collection of Maple characters, which stands only for itself and cannot be assigned any value, is called a *string* data type. A string can be formed by enclosing characters in *double quotes*. For example,

```
print("Error, n must be an integer");
```

produces output that looks like

```
"Error, n must be an integer"
```

Note that the double quotes " appear in the output. For printing out phrases like this, the double quotes are essentially equivalent to back quotes.[3]

A string has only one operand (itself), and can be of any length up to the maximum allowed by your machine. Unlike the symbol data structure, the string data structure permits certain operations which are convenient for some programming tasks (but will not be used or discussed here).

The use of *single quotes* around an expression, *expression* , delays the evaluation of that expression. Often this can be used to produce output that has a special form. For example, the following procedure definition and invocation:

```
improperint:=proc(f)
   int(f,x=0..infinity) =int(f,x=0..infinity)
end proc:
improperint(exp(-x^2));
```

produce the output

$$\int_0 e^{-x^2}\, dx = \frac{1}{2}\ -\ .$$

Example 4.4 also gives another example of where using single quotes to suppress evaluation is valuable (necessary)   the plotting of functions whose definitions involve boolean expressions.

6.3   Numbers

The earliest computers were created to do arithmetic calculations. Since a machine can store only a finite amount of information and the set of real numbers is infinite, there are natural limitations on the arithmetic capabilities of any computer. Each programming language typically has distinct data structures for storing and doing arithmetic on integers and rational numbers. Computer algebra systems like Maple can augment this arithmetic by performimg symbolic or exact arithmetic. We will discuss some aspects of Maple's arithmetic in this section. Chapter 8 has further discussion of numbers, both mathematically and computationally.

---

[3]The Maple documentation accessed by ?**name**; says In the past, names formed by using backquotes were often used in the sense of character strings instead of names.

Maple has three basic data types for representing real numbers, depending on whether they are integers or rational numbers. These types are: `integer`, `fraction`, and `float`. Of course, an integer can be represented in any of these types (for example, $2 = 2/1 = 2.0$), while a rational number can only be represented in the later two types (for example, $3/2 = 1.5$).[4]

The `integer` data type has several subtypes, like `nonneqint`, `posint` (see Tabls 6.1-6.2). A nonnegative integer is represented by a sequence of decimal digits, and a negative integer is represented by a sequence of decimal digits preceded by a minus sign. The maximum length of an integer depends on the system. A 32-bit system limits the length to 524,289 digits.

A rational number is represented, in the `fraction` data type, as a pair of integers (numerator, denominator) with all common factors removed and with a positive denominator. The data type `rational` includes the `integer` and `rational` data types.

A rational number $x = c/d$ can be represented, mathematically, in the decimal (base 10) system, by a $\pm$ sign, followed by a finite sequence of decimal digits, a decimal point, and a second (possibly infinite) sequence of decimal digits (see Chapter 8).

$$x = \pm a_s \cdots a_1 a_0 . b_0 b_1 \cdots$$

This mathematical representation leads to the reperesentation of x in the "oating-point, or "oat, data type. For this, write x in scientific notation as

$$x = \pm (.a_s \cdots a_1 a_0 b_0 b_1 \cdots) \times 10^{s+1}.$$

Then internally, x is stored in the `float` data type as a pair of integers, (m, e), where $m = \pm a_s \cdots a_1 a_0 b_0 b_1 \cdots$, which is called the *mantissa*, and $e = s + 1$, which is called the exponent. Of course, this storage may require rounding of $.a_s \cdots a_1 a_0 b_0 b_1 \cdots$ so that the integer m is within the maximum length. If e exceeds the maximum length, then storage (internal representation) of x is not possible.

Integers and rational numbers can be input to Maple in their decimal or fractional representations, e.g., `243`, `45/673`, and are then stored as `integer` and `fraction` data types respectively.[5] A number, integer or

---

[4]In Maple, di erent variables can have the same numerical value, even though their types are di erent. For example, `a:=2` and `b:=2.0` result in a and b having the same value. So the conditional a=b is true. However, `type(b,integer)` returns a value of false.

[5]The `rational` data type includes both the `integer` and `fraction` data types. Thus if `a:=2`, then the type statements: `type(a,integer)`; `type(a,fraction)`; `type(a,rational)`; return values of *true, false, true*, respectively.

rational, can also be input using a decimal point in the notation or using scientific notation. For example, a:=-56843.6734, b:=-568436734E-4, and c:=-5.68436734e4, give the same numerical value to a,b, and c. (Note that the exponential part can be specified using either E or e.) Numbers input in such a way then become "oating-point numbers in Maple   they are stored as float data types.[6]

An integer has one operand, itself. A fraction has two operands, its numerator and denominator. A fraction is automatically put in reduced form by Maple and the sign goes with the numerator. A "oat x has two operands, a mantissa m and exponent e, where $x = m \times 10^e$. For example the number -56843.6734 has operands -568436734 and -4.

We emphasize again that Maple is capable of representing only rational numbers and only finitely many (though a tremendously large number) of these. Let us say that these are the numbers that Maple knows. Thus infinitely many rational numbers and all irrational numbers must necessarily be approximated in Maple by the numbers that Maple knows. For instance, $\overline{2}$ is approximated with 10 digits by 1.414213562.

Being a computer algebra system, Maple can perform *symbolic* computations as well. In particular, it can do symbolic or exact arithmetic when all the numbers involved are either integers or fractions. Exact arithmetic also includes symbolic calculations that involve the usual symbolic representations for real numbers such as , square roots of positive integers such as $\overline{2}$, and symbolic function evaluations such as ln(2) or cos(1). The result of an exact arithmetic calculation is a symbolic representation of a real or complex number which is in fact the exact answer as it would appear in a manual calculation. We will limit our discussion to situations where the exact answer represents a real number. For example,

```
1+1/2+1/3+1/4+1/5+1/6+1/7+1/8;
```

indicates an exact arithmetic calculation and has exact answer

$$761/280.$$

Another example is the following;

```
5^(1/4)+14/3*sin(Pi/3);
```

---

[t]The exact value of some rational numbers will be lost when they are input and stored as "oats. For example, a:=1/3 and b:=1/3.0, gives a the exact value of one third, while b gets the lesser value of 0.3333333333.

and the exact answer is

$$5^{\frac{1}{4}} + \frac{7}{3}\ \overline{3}.$$

Maple also does "oating-point arithmetic. This is just the usual arithmetic performed on the decimal representations of numbers (subject to the storage limitations of the computer). Generally, if one of the numbers in an arithmetic operation is a "oat then all the other numbers in the calculation will be converted to "oats and the arithmetic will be done in "oating point.[7] Converting a non-"oating-point number to a "oat means representing it, or an approximation of it, in decimal (base 10) form. This conversion is what the `evalf` command does. You can always force Maple to give a decimal answer by using the `evalf` command.

The calculation `2/5.0` results in the exact answer .400000000. The calculation `2/3.0` gives an approximation .6666666667. In Maple's "oating-point arithmetic, the result of any single basic arithmetic operation (addition, subtraction, multiplication, division), if it is not too big or too small, is correctly rounded according to the value set by the `Digits` command. The number of digits used to represent a "oat is set using the `Digits` command. The default value is `Digits:=10`. Thus the approximation .6666666667 of 2/3 is correctly rounded to the tenth decimal place. The difference between the exact value of a number and the approximation is called the roundoff error.

Since many numbers in Maple are approximations, there will usually be roundoff error. These roundoff errors accumulate if approximate values are used in further calculations and can give final answers that are not acceptable. The accumulation of roundoff error in computer calculations is a topic of numerical analysis and is beyond the scope of this book (see [FK], [YG]). We can hope to minimize the roundoff error in our calculations by increasing the precision of "oating-point numbers using the `Digits` command. If you want more digits used in "oating-point arithmetic, you can reset the value of `Digits`:

```
a:=(.135467817)^2;
```

$$a:=.01835152944$$

```
Digits:=20;
b:=(.135467817)^2;
```

---

[7]However, calculations involving Maple constants, such as `Pi`, give answers in terms of the constant. So, `Pi/2.0` yields .5000000000π even though there is the decimal 2.0 in the command.

$$Digits := 20$$
$$b := .018351529442745489$$

Maple calculations will be done using 20 decimal digits for the remainder of the session unless you reset **Digits** to some other value. If you only want one calculation done using 20 digits the command

```
b:=evalf[20]((.135467817)^2);
```

could be used. The value of **Digits** will not be changed from what it was before this command and all further calculations will be done using that number of digits.

In this example of a "oating-point calculation, it is easy to verify that **b** is assigned the exact value (do the calculation by hand), while **a** is only approximate. There was a single arithmetic operation performed   exponentiation and this led to a roundoff error of $b - a = .00000000000745489$. Since this error is quite small it might be considered of no importance, i.e., for some purposes the error can be assumed to be zero. Just keep in mind that ignoring roundoff error in a result can cause problems when this result is used elsewhere. For instance, when comparing the the fraction 1/3 and the "oat .3333333333 (which mathematically compare as $1/3 > .3333333333$), Maple gives

```
is(1/3>.3333333333);
```

false

What happens is this. When comparing a fraction and a "oat, Maple first converts the fraction 1/3 to a "oat, getting .3333333333, then makes the comparison .3333333333 > .3333333333, and gets *false*, as indicated.

We see that Maple can do two types of arithmetic: exact and "oating-point arithmetic. Exact arithmetic has the advantage that there is no round-off error. The exact answer, however, may be long and complicated or not particularly instructive. For example, consider the exact expansion of $(\sqrt{3} + \frac{13}{47} +)^5$:

```
expand((sqrt(2)+13/47+Pi)^5);
```

$$\frac{120135685}{4879681} + \frac{27127949}{4879681}\sqrt{2} + \frac{1366151813}{229345007} + \frac{1192620}{103823}\sqrt{2}$$
$$+ \frac{1744990}{103823}\sqrt{2} + \frac{49250}{2209}\sqrt{2}^2 + \frac{260}{47}\sqrt{2}^3 + 5\sqrt{2}^4 +$$

$$\frac{45870}{2209} \quad _3 + \frac{65}{47} \quad _4 + \quad _5$$

There is no roundoff error here but the answer is unwieldy and may not be useful in further calculations.

6.4   Lists: Vector Methods in Geometry

In this section we will continue our study of the list data structure with some geometrical applications. The problems we look at are quite naturally described using vectors and the operations defined on them. A review of the material in section 5.5.1 might be helpful at this point.

The general form for a list is

$$[\ expr1, expr2, \ldots, exprN \ ]$$

where *expr1, expr2, ... , exprN* stand for any Maple expressions. The elements of a list are ordered and need not be distinct. A list is a Maple expression and its operands are its elements.

The simplest type of list is a list of numbers, e.g., L:=[1,2,16,8,16]. We can select the third element using either L[3] or op(3,L). On the other hand, the four operands of the list T:=[1,x^2,x^3+x,2*x^4] are themselves expressions and, hence, we can use multiple nested op commands to select operands from them. For example, op(1,op(3,T)) yields x^3. The command op([3,1],T) accomplishes the same thing.

The list data structure is the most elementary and useful of all the data structures, and in a certain sense there is not much more to explain about lists. However, because points and vectors in $\mathbb{R}^2$ and $\mathbb{R}^3$ can be conveniently represented by lists and easily manipulated in Maple, it will be a good reinforcement of your prior study of analytic geometry to discuss vector methods and some elementary programs connected with these methods.

We have seen that in order to draw the segment connecting two points (a,b), (c,d) in $\mathbb{R}^2$, we can use the plot command with the list [[a,b],[c,d]]:

    plot([[a,b],[c,d]],scaling=constrained,axes=normal);

In this way we associate the point (x,y) with the list [x,y]. Similarly, the point (x,y,z) can be associated with the list [x,y,z]. Recall that a point (x,y) can also be identified with the vector having initial point $(0,0)$ and terminal point (x,y). Sometimes to distinguish points from vectors, we use the mathematical notation (x, y) for the point and x, y for the vector. See Figure 6.4.

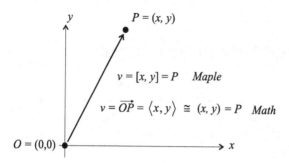

Figure 6.4: *Mathematical and Maple notation for points and vectors.*

In Maple, a list with numerical elements can be treated as a point sometimes (for the purposes of drawing line segments) and can be treated at other times as a vector, with the operations of addition, subtraction, and scalar multiplication being available. For instance, if $u = [3, 1]$ and $v = [1, 2]$, we can picture u and v as vectors in the plane which start at the origin $(0, 0)$ and end at the points $(3, 1)$ and $(1, 2)$, respectively. The following commands plot the vectors $2u, v$, and $2u + v$ (see Figure 6.5).

```
u:=[3,1];v:=[1,2];
plot({[[0,0],2*u],[[0,0],v],[[0,0],2*u+v]});
```

Note: Multiplication of a list by a scalar is done automatically only if the scalar is numeric. Thus, `3/4*[1,2]` yields `[3/4,3/2]` but `sqrt(2)*[1,2]` or `t*[1,2]` is left as is. To force Maple to do these scalar multiplications you can use the **expand** command: `expand(t*[1,2])` results in the list `[t,2*t]`

This dual use of the list data structure to represent both vectors and points can be confusing at first and we must be careful to pay attention to what is intended in each particular case.

Some of the exercises in the previous chapter illustrate the use of vector methods in geometry, as they are traditionally encountered in the calculus sequence. As another typical example, we consider how the formula for the midpoint of the line segment from point A to point B is easy to derive using vectors. The midpoint is the point halfway from A to B. Viewing A and B

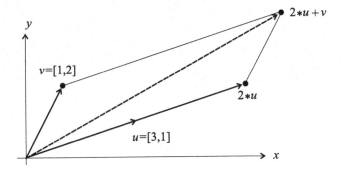

Figure 6.5: *The vectors $u, 2u, v$ and $2u + v$.*

as vectors[8] the parametric form of the equation for the line segment $\overline{AB}$ is

$$f(t) = (1 - t)A + tB,$$

for $0 \le t \le 1$. The midpoint corresponds to $t = 1/2$, giving the formula for the midpoint: $\frac{1}{2}(A + B)$. The point one third of the way from A to B corresponds to $t = 1/3$ and is given by $\frac{2}{3}A + \frac{1}{3}B$.

The midpoint formula allows us to also derive a well-known geometric property of triangles. Let A, B, and C be the vertices of a triangle and P and Q the midpoints of the sides $\overline{AC}$ and $\overline{BC}$, respectively (see Figure 6.6). Then in terms of points $P = \frac{1}{2}A + \frac{1}{2}C$ and $Q = \frac{1}{2}B + \frac{1}{2}C$. This gives us that

$$\overrightarrow{PQ} = Q - P = \frac{1}{2}(B + C) - \frac{1}{2}(A + C)$$
$$= \frac{1}{2}(B - A) = \frac{1}{2}\overrightarrow{AB}.$$

This proves, using vector methods, the well-known geometric fact: *the line segment joining the midpoints of two sides of a triangle is parallel to the third side and half as long.*

Example 6.3 (Altitudes of a Triangle) Let points A, B, C be the vertices of a triangle in the x-y plane. For each vertex there is an altitude. For example, the altitude at C is $\overline{CP}$ where P is a point on the line $\overrightarrow{AB}$ such that $\overline{CP}$ and $\overrightarrow{AB}$ are perpendicular. See Figure 6.7. Using algebra, the coordinates of the endpoint of the altitude can be calculated. Vector

---

[8]That is, make the identi"cations $A \sim \overrightarrow{OA}$ and $B \sim \overrightarrow{OB}$, where O is the origin.

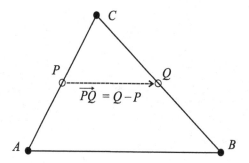

Figure 6.6: *Triangle △ABC and midpoints P and Q.*

methods can also be used to determine these coordinates. Consider Figure 6.7 for the purposes of our discussion.

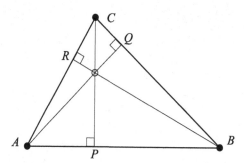

Figure 6.7: *The altitudes of a triangle.*

There are two conditions which define the foot, point P, of the altitude from vertex C: (1) $\angle CPA$ is a right angle and (2) P lies on the line through A and B. Using vectors methods, we get an equivalent set of conditions:

(1) $(P - C) \cdot (B - A) = 0$

(2) $P = (1 - t)A + tB$ for some real number t.

Hence, to determine P we need to find the value of t. This can easily be done by substituting the expression for P in (2) into the equation in (1).

We will write a procedure to find P and also plot $\triangle ABC$ and the altitude CP. For convenience, we first write a short procedure, `dotprod`, to calculate the dot product of two two-dimensional vectors. We will call the dot product in the main procedure.

```
dotprod:=proc(v,w)
  v[1]*w[1]+v[2]*w[2]
end proc:
```

In Figure 6.7, the point P lies between A and B. For general triangles this may not be the case and drawing the altitude may require extending the side AB. For a given triangle ABC and the altitude from C, we will determine which is the case. This decision can be made by examining the value of t in the definition of P. P will lie between A and B only if t is between 0 and 1. If $t < 0$ then A is between P and B. If $t > 1$ then B is between A and P. The following code checks these cases and produces a picture of the triangle and the altitude $\overline{CP}$.

```
altitude:=proc(A,B,C)
  local P,eq,t;
  P:=expand((1-t)*A+t*B);
  eq:=dotprod(P-C,B-A)=0;
  t:=solve(eq,t);
  if t<0 then plot([[A,B],[B,C],[C,A],[C,P],[A,P]],
                    color=black,scaling=constrained);
  elif t>1 then plot([[A,B],[B,C],[C,A],[C,P],[B,P]],
                    color=black,scaling=constrained)
  else  plot([[A,B],[B,C],[C,A],[C,P]],
                    color=black,scaling=constrained) end if;
end proc:
```

Figure 6.8 shows the output from the call `altitude(A,B,C)`, where `A:=[0,1]`, `B:=[2,0]`, and `C:=[4,2]`.

Notice, in the code for the procedure, that to express P in coordinate form, which is required for use of **dotprod** in the next line, we use the **expand** command. At this stage in the procedure, t has no value and so Maple does not automatically do the vector operations.

Example 6.4 (Scalar Triple Products and Volume) Recall that the volume of a parallelepiped can be computed using the scalar triple product of vectors. Suppose that the parallelepiped is oriented in space so that one of its vertices is at the origin. The edges emanating from that vertex determine three vectors, say, u, v, and w. The volume is then given by the absolute value of the scalar triple product of these vectors $(u \times v) \cdot w$. The procedure in this example has as input the three vectors u, v, and w. Each edge of the parallelepiped is a line segment connecting two of the vertices and all of the

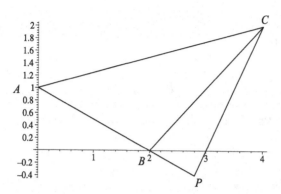

Figure 6.8: *Triangle* $\triangle ABC$ *with altitude from* C.

edges can be determined from u, v, and w. In addition to calculating the
volume we also want to produce a picture of the parallelepiped, so we will
have two output parameters: one, `vol`, for the volume and a second `graph`
which will contain the set of plot structures for graphing the edges.

After calling the procedure we display this set of plot structures to obtain
a picture of the parallelogram.

```
Volume:=proc(u,v,w,vol::evaln,graph::evaln)
  local edges,vcrossw,i;
  edges:=array(1..12,1..2);
  edges[1,1]:=[[0,0,0],w];edges[1,2]:=red;
  edges[2,1]:=[w,u+w];edges[2,2]:=black;
  edges[3,1]:=[u+w,u];edges[3,2]:=black;
  edges[4,1]:=[[0,0,0],u];edges[4,2]:=blue;
  edges[5,1]:=[w,v+w];edges[5,2]:=black;
  edges[6,1]:=[u+w,v+u+w];edges[6,2]:=black;
  edges[7,1]:=[u,v+u];edges[7,2]:=black;
  edges[8,1]:=[[0,0,0],v];edges[8,2]:=green;
  edges[9,1]:=[v,v+w];edges[9,2]:=black;
  edges[10,1]:=[v+w,v+u+w];edges[10,2]:=black;
  edges[11,1]:=[v+u,v+u+w];edges[11,2]:=black;
  edges[12,1]:=[v+u,v];edges[12,2]:=black;
  ucrossv:=[u[2]*v[3]-u[3]*v[2],u[3]*v[1]-u[1]*v[3],
           u[1]*v[2]-u[2]*v[1]];
  vol:=abs(ucrossv[1]*w[1]+ucrossv[2]*w[2]+ucrossv[3]*w[3]);
```

```
   with(plots,spacecurve):
   graph:={seq(spacecurve(edges[i,1],color=edges[i,2]),'i'=1..12)}:
   RETURN();
end proc:
```

A call to this procedure:

```
Volume([5,1,1],[0,6,1],[1,0,2],volume,parallelepiped);
```

will produce no printed output because of the RETURN statememt but the
volume is stored in the name volume and the set of plot structures for the
individual edges is stored in the name parallelepiped. Thus, the following
commands will give the volume and the picture of the parallelepiped.

```
with(plots,display):
volume;
display(parallelepiped,axes=normal,scaling=constrained,
        orientation=[-50,80]);
```

Figure 6.9 shows the graphic displayed (with additional elements added).

When you invoke this procedure you decide what names to use for the
output parameters vol and graph in the procedure definition. Remember
that these are evaluated to names, i.e., they can be assigned values each
time the procedure is called.

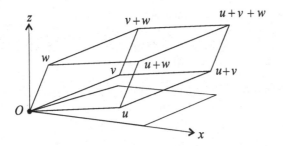

Figure 6.9: *Parallelepiped with volume* $|(u \times v) \cdot w|$.

6.5 Arrays and Tables

Arrays and tables are similar types of data structures   a table can be
thought of as a generalization of an array. Both types have their uses,

but it is the array type that you will find more commonly used. Despite their similarities, arrays and tables have different operand structures and, for programming purposes, it is important to be aware of these differences.

A table is a collection of indexed names. The indices do not have to be integers but can be any expression. A table can be created implicitly by assignment to indexed names. For instance, if we wanted to store some important information about Alice, we could create a table named `Alice` in the following way.

```
Alice[birthday]:='Oct 5,1966'; Alice[dresssize]:=10;
Alice[shirtsize]:='med'; Alice[shoesize]:=6.5;
Alice[favoritecolor]:='teal';
```

This information then can be conveniently accessed.

```
Alice[birthday];
```

> *Oct 5,1966*

```
Alice[shoesize];
```

> 6.5

The table command can also be used to define the table `Alice`.

```
Alice:=table([birthday = 'Oct 5,1966', favoritecolor = 'teal',
shoesize = 6.5, shirtsize = 'med', dresssize = 10]);
```

Maple will automatically create a table data structure for a variable when the variable is indexed in the program code and is not explicitly declared as an array. A good example to illustrate this is the one from Chapter 3 involving the partial sums

$$s_k = \sum_{n=1}^{k} \frac{1}{n(n+1)},$$

of the series $\sum_{n=1}^{} \frac{1}{n(n+1)}$. We used the following code to compute the partial sums `s[1],s[2],`... as long as the kth term $a_k = 1/(k(k+1))$ is larger than a specified :

```
a:=n->1/(n*(n+1));
epsilon:=0.001;
k:=0; s[0]:=0;
while a(k+1)>epsilon do
    k:=k+1:
    s[k]:=s[k-1]+(-1)^(k+1)*a(k);
end do:
```

Since we do not know in advance how many partial sums will be computed in the do loop, we cannot dimension s as an array. As it is, the code implicitly defines s as a table data structure.

The general form of the `table` command is

$$\texttt{table}(F, L)$$

where L is a list of equations or a list of entries. If L is a list of equations, the left-hand side of each equation is the index and the right-hand side the table entry. If L is a list of entries, the indices are taken to be $1, 2, \ldots$. In the `table` command F is optional and is an indexing function. An indexing function is a procedure for the indexing method to be used to set up the table.

A table has only one operand, the table itself. To get a list of the entries in a table T we could use the command `op(op(T))` or the command `op(eval(T))`. For example, if A is the table created implicitly by the pair of do loops:

```
for i to 2 do for j to 2 do A[i,j]:=i^2+j^3 end do:
```

then `op(A)` gives the following output:

$$\texttt{table}([\ (1,\ 1) = 2,\ (1,\ 2) = 9,\ (2,\ 1) = 5,\ (2,\ 2) = 12\ ])$$

An *array* is a special case of a *table* in Maple. An array is a table with specified dimensions and each dimension has an integer range. An array must be explicitly declared using the **array** command, which has the general form:

$$\texttt{A:=array}(F, R, L)$$

where (1) F is an indexing function (as for tables), (2) R is a sequence of ranges $a_1..b_1, a_2..b_2, \ldots, a_k..b_k$, where $a_i < b_i$ are all integers, and (3) L is a list of corresponding entries for the array. Here k is the dimension of the array. For example, the declaration of A as a three-dimensional array generally has the form:

```
A:=array(a1..b1,a2..b2,a3..b3)
```

Note that this declaration only dimensions the array, i.e., specifies that it is three-dimensional and establishes the ranges for its three indices (typically denoted by i, j, k). The declaration does not place any entries in the array. This would have to be done in other ways, for example, either by explicity including a list L in the declaration (which is impractical for arrays with large ranges), or by using do loops, say:

```
for i from a1 to b1 do
  for j from a2 to b2 do
    for k from a3 to b3 do
      A[i,j,k]:=i-j+k
    end do;
  end do;
end do;
```

To select an entry of A, we can just specify the indices: A[i,j,k]. To see all the entries, we can use either the **print** or the **eval** commands.

In general, an array is a Maple expression and has three operands. The first operand is the indexing function F, if there is one. The sequence of ranges R is the second operand, and the list L of its data entries is the third. So we could use the commands op(op(A)) or op(eval(A)) to obtain a sequence of these operands.

Here is an example where each entry of an array is a plot data structure.[9] Let p be the $3 \times 3 \times 3$ array which stores the plots of the vectors $[i, j, k]$ for $i = 1, \ldots, 3$, $j = 1, \ldots, 3$, $k = 1, \ldots, 3$. To plot the vector $[i, j, k]$ we use the **spacecurve** command with the list [[0,0,0],[i,j,k]] consisting of the two endpoints of the vector. See Figure 6.10.

```
p:=array(1..3,1..3,1..3);
with(plots):
for i from 1 to 3 do
  for j from 1 to 3 do
    for k from 1 to 3 do
```

---

[9]Plot data structures are discussed in Chapter 7.

```
      p[i,j,k]:=spacecurve([[0,0,0],[i,j,k]]);
    end do;
  end do;
end do;
display({seq(seq(seq(p[i,j,k],i=1..3),j=1..3),k=1..3)},
         color=black,axes=box,scaling=constrained,
         orientation=[-45,70],tickmarks=[2,2,2],
         projection=0.8);
```

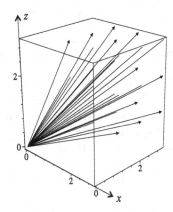

Figure 6.10: *Some 3-dimensional vectors.*

6.6 Sets: The Cantor Set and Limiting Covers

In this section we will study a few more examples of programming algorithms that naturally involve the set data structure. These lead to constructions of certain sets by taking limits. Working with infinitely many sets, which might be new to you, is a fundamental aspect of modern analysis that you may encounter in later courses.

The general form for a set data structure is

$$\{ \ expr1, expr2, \ldots, exprN \ \}$$

where *expr1, expr2,* ... , *exprN* stand for any Maple expressions. The elements of a set data structure are distinct and the order of these elements is not determined by the user. A set is a Maple expression and its operands are its elements.

Maple allows the usual operations on sets: union, intersection, minus, member. See Table 6.4 for the set commands.

Operation	Maple	Math
union	A union B	A B
intersection	A intersect B	A B
minus	A minus B	$A - B$
member	member(x,A)	x A
in	x in A	x A

Table 6.4: *Set commands.*

The member command `member(x,C)` determines if `x` is a member of the set `C`. This command can also be used with lists. Note that Maple treats the integer 2 and the "oat 2.0 as different elements, so if $C = \{1, 2, 6, 3, 9\}$, then the statement `member(2.0,C)` evaluates to false.

To generate the set of prime numbers between 500 and 600 the `union` command is useful.

```
primes:={}:
for k from 500 to 600 do
 if is(k,prime) then primes:=primes union {k} end if;
end do:
primes;
```

$$\{503, 509, 521, 523, 541, 547, 557, 563, 569, 571, 577, 587, 593, 599\}$$

```
nops(primes);
```

$$14$$

Example 6.5 (The Cantor Set) There is a very well-known set in mathematics called the Cantor set which has many interesting properties (Cf. [Sch], [Hof], [MH]). This set is obtained by a limiting process as follows. Start with the interval $[0, 1]$ and remove the middle third $\left(\frac{1}{3}, \frac{2}{3}\right)$ leaving two intervals $[0, \frac{1}{3}]$ and $[\frac{2}{3}, 1]$. We will let

$$S_1 = \left\{[0, \tfrac{1}{3}], [\tfrac{2}{3}, 1]\right\}$$

$$T_1 = \{(\tfrac{1}{3}, \tfrac{2}{3})\}.$$

The next step is to take each interval in S_1 and remove the middle third to obtain two subintervals. Let S_2 be the set consisting of the four resulting intervals and T_2 be the set of the two intervals which were removed:

$$S_2 = \{[0, \tfrac{1}{9}], [\tfrac{2}{9}, \tfrac{1}{3}], [\tfrac{2}{3}, \tfrac{7}{9}], [\tfrac{8}{9}, 1]\}$$

$$T_2 = \{(\tfrac{1}{9}, \tfrac{2}{9}), (\tfrac{7}{9}, \tfrac{8}{9})\}.$$

Similarly, S_3 will be the set of intervals obtained by taking each interval in S_2 and removing the middle third. Continuing in this fashion, S_k is the set of intervals obtained by taking each interval in S_{k-1} and removing the middle third. The set T_k is the set of all the middle thirds removed at that stage. Since the number of remaining intervals doubles in going from one stage to the next, we see that S_k has 2^k elements and has the form

$$S_k = \{\, I_1^k, I_2^k, \ldots, I_{2^k}^k \,\},$$

where I_j^k, $j = 1, \ldots, 2^k$, are the intervals that arise from this construction. Also T_k has 2^{k-1} elements and is of the form

$$T_k = \{\, J_1^k, I_2^k, \ldots, J_{2^k}^k \,\}.$$

Now let C_k be the union of the intervals in S_k:

$$
\begin{aligned}
C_0 &= [0, 1] \\
C_1 &= [0, \tfrac{1}{3}] \quad [\tfrac{2}{3}, 1] \\
C_2 &= [0, \tfrac{1}{9}] \quad [\tfrac{2}{9}, \tfrac{1}{3}] \quad [\tfrac{2}{3}, \tfrac{7}{9}] \quad [\tfrac{8}{9}, 1] \\
&\;\;\vdots \\
C_k &= I_1^k \quad I_2^k \quad \cdots \quad I_{2^k}^k.
\end{aligned}
$$

See Figure 6.11.

Using limits we can define a set $C = \lim_{k \to} C_k$, as a subset of $[0, 1]$. This limiting set is the *Cantor set*, i.e., what is left of $[0, 1]$ after continually removing middle thirds in the indicated way. For large k, the set C_k is an approximation of the Cantor set.

The following procedure calculates the sets S_k of intervals used in the Cantor set construction. Each set S_k is stored as an element of an array named S. Thus, S[k] refers to the kth set of intervals. The sets T_k of middle thirds removed at the kth step in the construction are also calculated and stored in an array T.

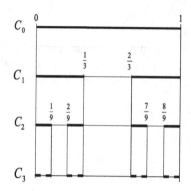

Figure 6.11: *Cantor set construction.*

```
cantor:=proc(m,S::evaln,T::evaln)
  local k,i,I,newS,mid;
  S:=array(0..m);T:=array(0..m);
  S[0]:={[0,1]};
  T[0]:={};
  for k from 1 to m do
    S[k]:={};T[k]:={};
    for I in S[k-1] do
    newS:={[I[1],I[1]+1/(3^k)],[I[1]+2/(3^k),I[2]]};
    mid:={[I[1]+1/(3^k),I[1]+2/(3^k)]};
    S[k]:=S[k] union newS;
    T[k]:=T[k]   union mid;
    end do;
  end do;
  RETURN()
end proc:
```

Calling this procedure will generate sets of intervals in the approximations to the Cantor set.

```
cantor(3,S,T);
'S[1]':=S[1];'T[1]':=T[1];'S[2]':=S[2];'T[2]':=T[2];
'S[3]':=S[3];'T[3]':=T[3];
```

$$S[1] := [0, \tfrac{1}{3}], [\tfrac{2}{3}, 1]$$

$$T[1] := (\tfrac{1}{3}, \tfrac{2}{3})$$

$$S[2] := \quad [\tfrac{8}{9}, 1], [\tfrac{2}{3}, \tfrac{7}{9}], [\tfrac{2}{9}, \tfrac{1}{3}], [0, \tfrac{1}{9}]$$

$$T[2] := \quad (\tfrac{1}{9}, \tfrac{2}{9}), (\tfrac{7}{9}, \tfrac{8}{9})$$

$$S[3] := \quad [\tfrac{8}{9}, \tfrac{25}{27}], [\tfrac{26}{27}, 1], [\tfrac{20}{27}, \tfrac{7}{9}], [\tfrac{8}{27}, \tfrac{1}{3}], [\tfrac{2}{27}, \tfrac{1}{9}], [\tfrac{2}{9}, \tfrac{19}{27}], [\tfrac{2}{27}, \tfrac{7}{27}], [0, \tfrac{1}{27}]$$

$$T[3] := \quad (\tfrac{1}{27}, \tfrac{2}{27}), (\tfrac{7}{27}, \tfrac{8}{27}), (\tfrac{19}{27}, \tfrac{20}{27}), (\tfrac{25}{27}, \tfrac{26}{27}) \quad .$$

Note: The ordering of the elements in these sets is determined by Maple.[10]

There is also a Cantor function: $f : [0,1] \quad [0,1]$ (Cf. [Sch], [Hof]). The definition of this function is quite complicated, but we can easily define some approximations. This is why we had our **cantor** procedure also calculate the sets T_k of the middle thirds removed at the kth step.

We will denote the mth approximation to f by f_m. It will be a step function which will be zero on each interval in S_m, and have specified positive values on the middle thirds removed up through the mth step in the Cantor set construction. For example, $f_1, f_2,$ and f_3 are defined as follows. Note the pattern.

$$f_1(x) = \quad \begin{array}{ll} 0 & \text{if } x \quad C_1 \\ \tfrac{1}{2} & \text{if } x \quad (\tfrac{1}{3}, \tfrac{2}{3}) \end{array}$$

$$f_2(x) = \quad \begin{array}{ll} 0 & \text{if } x \quad C_2 \\ \tfrac{1}{2} & \text{if } x \quad (\tfrac{1}{3}, \tfrac{2}{3}) \\ \tfrac{1}{4} & \text{if } x \quad (\tfrac{1}{9}, \tfrac{2}{9}) \\ \tfrac{3}{4} & \text{if } x \quad (\tfrac{7}{9}, \tfrac{8}{9}) \end{array}$$

$$f_3(x) = \quad \begin{array}{ll} 0 & \text{if } x \quad C_3 \\ \tfrac{1}{2} & \text{if } x \quad (\tfrac{1}{3}, \tfrac{2}{3}) \\ \tfrac{1}{4} & \text{if } x \quad (\tfrac{1}{9}, \tfrac{2}{9}) \\ \tfrac{3}{4} & \text{if } x \quad (\tfrac{7}{9}, \tfrac{8}{9}) \\ \tfrac{1}{8} & \text{if } x \quad (\tfrac{1}{27}, \tfrac{2}{27}) \\ \tfrac{3}{8} & \text{if } x \quad (\tfrac{7}{27}, \tfrac{8}{27}) \\ \tfrac{5}{8} & \text{if } x \quad (\tfrac{19}{27}, \tfrac{20}{27}) \\ \tfrac{7}{8} & \text{if } x \quad (\tfrac{25}{27}, \tfrac{26}{27}) \end{array}$$

Figure 6.12 shows the graphs of f_2 and f_3.

---

[10]Also note that mathematically the intervals in the sets $T[1], T[2],$ and $T[3]$ are *open* intervals (a, b), and are shown as such in the above display. In Maple, these intervals are stored as lists $[a, b]$, and would appear like that in output.

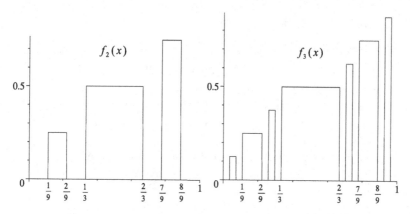

Figure 6.12: *The approximations f_2 and f_3 to the Cantor function f.*

In general, for any $m > 1$, the mth approximation f_m to the Cantor function is defined by the requirements:

(1) it is zero on C_m;

(2) it agrees with f_{m-1} on the intervals removed at the $(m-1)$st stage; and

(3) It has values on the intervals $J_1^m, J_2^m, \ldots, J_{2''}^m$ in T_m given by

$$f_m(x) = \frac{2i-1}{2^m},$$

for x J_i^m and $i = 1, \ldots, 2^{m-1}$. *Note:* This assumes that these intervals are listed in increasing order.

The following code defines the mth approximating function. To define this approximation, the intervals in the T_k must be in increasing order. The procedure below first calls `cantor(m,S,T)` which calculates the S_k's and T_k's. The intervals in the T_k's are then sorted in increasing order, using Maple's `sort` command, and are stored in an array `OT`. Thus, `OT[k]` is the list consisting of intervals in `T[k]` but in increasing order. The function's value is then easily defined using the pattern above.

```
cantorfctn:=proc(x,m)
  local n,val,i,OJ,k,Sfirst,first,OT;
  OT:=array(1..m);
  cantor(m,S,T);
```

```
val:=0;
for k from 1 to m do
  n:=nops(T[k]);
  Sfirst:=[];
  first:= [seq(T[k][i][1],i=1..n)];
  Sfirst:=sort(first);
  for i from 1 to n do
    OJ[i]:=[Sfirst[i],Sfirst[i]+1/(3^k)];
  end do;
OT[k]:=[seq(OJ[i],i=1..n)];
  for i from 1 to n do
    if OT[k][i][1]<x and x<OT[k][i][2] then
            val:=(2*i-1)/(2^k)
    end if;
  end do;
end do;
val;
end proc:
```

Figure 6.13 shows the graphs of the approximations f_4 and f_5 to the Cantor function f.

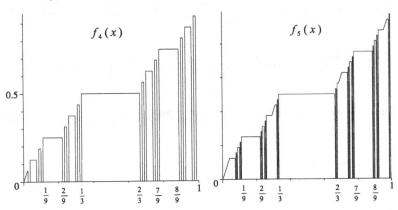

Figure 6.13: *The approximations f_4 and f_5 to the Cantor function f.*

You can perhaps begin to see the shape of the Cantor function f emerging from this sequence of plots.

We can now be more specific about the definition f. Let D_k be the union of all the intervals in $T_1, T_2, \ldots, T_k$. This is the set of all the middle thirds removed up through step k in the Cantor set construction. Using limits we can define the set $D \quad \lim_{k \to} D_k$. This is just the set $[0,1] \setminus C$, i.e., the numbers in the interval $[0,1]$ which are not in the Cantor set. The theory shows that there is a unique continuous function $f : [0,1] \quad [0,1]$ which agrees with f_m on D_m for every $m = 1, 2, 3, \ldots$. This function f is the Cantor function.

Note: Our approximating functions for the Cantor function are not the ones usually seen in the mathematics literature, and were chosen for simplicity. Normally, instead of letting f_m be zero on C_m, one first defines f_m on the intervals $T[1], \ldots, T[m]$ which have been removed and then uses linear interpolation to fill in the values (Cf. [Sch]). So the second approximating function using this scheme, say, f_2, would look like Figure 6.14. One advantage of these approximating functions is that they are continuous.

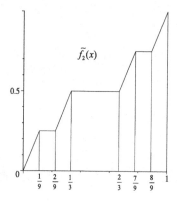

Figure 6.14: *The approximation f_2 to the Cantor function f.*

In the case study and exercises from Chapter 4, we discussed Riemann sum approximations of double integrals over several types of domains. We will consider this problem again in the next example.

Example 6.6 (More Riemann Sums for Double Integrals) In this example we will discuss a program for approximating the double integral

$$\int_D f(x,y)\, dA$$

of a function over a region D of the form $D = \{(x,y) | H(x,y) \geq 0\}$, where H is a suitable function. A program like this was developed in solving Problem 2 in the case study from Chapter 4. Here we want to improve upon the approximation method so that it is more efficient. The improved method also gives a good illustration of the convenience of the set data structure in Maple.

The previous approximation scheme was based upon enclosing the region D within a rectangle R, subdividing this rectangle into subrectangles, and then producing a grid G_0 consisting of all those subrectangles that overlap (i.e., intersect) D. See Figure 6.15.

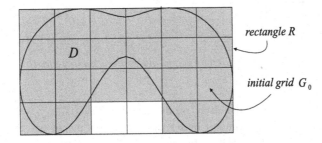

Figure 6.15: *An initial grid G_0 that covers the region D.*

While G_0 is *not* a rectangle it will cover[11] D and be a better approximation to D than the rectangle R, at least if the initial subdivision of R is fine enough. Then the double integral can be approximated as:

$$\iint_D f(x,y)\, dA \approx \sum_{i,j \text{ with } R_{ij} \subseteq G_0} f(x_{i-1}, y_{j-1})\, \Delta A_{ij}.$$

This was the previous scheme. What we did not discuss previously was how to go about making finer and finer approximations. Do we just make a finer subdivision of R, recalculate G_0, and then recompute the above Riemann sum? This could be computationally expensive, since the recalculation of the grid G_0 would involve testing a great many vertices.

A more efficient approach would be as follows. Having calculated the initial grid G_0, we produce a finer grid G_1 by dividing each rectangle in G_0 into four subrectangles, testing each of these four to see if it intersects D,

---

[1.] By a *cover* of D we mean any set C that contains D as a subset. While $G_\iota = \{R_{\iota}\}_{(i,j) \in \kappa}$ is taken to be a set of subrectangles R_{ι} as described, saying that G_ι covers D means that union of all the rectangles in G_ι covers D. Specifically, let $D_\iota = \cup\{R_\iota | R_{ij} \in G_0\}$. Then $D \subseteq D_0$. So D_0 covers D.

retaining those that pass the test and tossing out those that do not. The new grid G_1 will still cover D and will be a better approximation to D. The corresponding Riemann sum over the rectangles in G_1 should give a better approximation to $\int_D f(x,y)\,dA$ as well. See Figure 6.16.

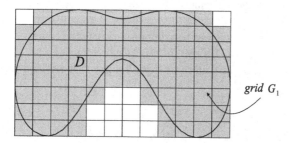

Figure 6.16: *A better approximating Riemann sum arises from a refinement G_1 of the initial grid G_0.*

This scheme can be continued indefinitely, producing a sequence $\{G_n\}_{n=0}$ of grids which cover D and approximate it better and better as n increases. If we let D_n be the union of all the rectangles in G_n, then in theory $\lim_{n\to\infty} D_n = D$. Thus, D is realized as the limit of a decreasing sequence $D_0 \supseteq D_1 \supseteq D_2 \supseteq \ldots \supseteq D$ of sets that cover it. The corresponding sequence of areas of the covers D_n will converge to the area of D, as we would expect since

$$\text{area of } D = \int_D 1\,dA.$$

To automate the calculation of the grids and the approximating Riemann sums, we establish an array data structure G to store the sequence of grids: G[0],G[1],G[2],... (or at least finitely many of them). Each grid G[n] will be a set whose elements are the rectangles in the grid. In selecting a data structure for the rectangles there are several possible choices. Since the subdivision process will be uniform, producing rectangles of the same length and width at each stage, we can represent each rectangle in G[n] by its *lower left-hand vertex* [x,y], provided we keep track of the length dx and width dy at each stage. See Figure 6.17. We will use an array Rsum to store the approximating Riemann sums, with Rsum[n] being the approximation formed by summing over all the rectangles in G[n].

We design the program to consist of a main procedure int2d, and two procedures test,initgrid that are called from the main procedure.

The procedure test merely tests if a rectangle, represented by its lower

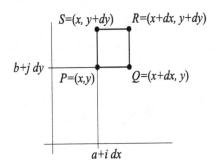

Figure 6.17: *A rectangle PQRS is determined by its lower left-hand vertex P and the lengths dx, dy of its sides.*

left-hand vertex P and side lengths dx, dy has at least one of its four vertices inside the region $H(x, y)$ 0. It returns a Boolean value of true or false.

It is *important to note* that the finite nature of the computer prevents us from designing the procedure **test** so that it determines if a rectangle intersects D. This would require testing infinitely many points (not just the four vertices). Convince yourself of this (exercise). Thus, the grids produced by the algorithm in **test** may end up not covering D.

The code for the procedure **test** is:

```
test:=proc(H,P,dx,dy)
  if H(P[1],P[2])<=0 or H(P[1]+dx,P[2])<=0 or
      H(P[1]+dx,P[2]+dy)<=0 or H(P[1],P[2]+dy)<=0
    then RETURN(true) else RETURN(false)
  end if;
end proc:
```

The procedure **initgrid** will generate the initial grid G[0] The logic to generate this grid is different from that for the other grids and so it seems best to separate this calculation from the main procedure. The input to **initgrid** will be the function H of two variables that defines the region D, the function f of two variables, the numbers a, b, c, d giving a rectangle $R = [a, b] \times [c, d]$ that encloses D (which must be chosen by the user), and the positive integers N, M in the initial subdivision of R into NM equal subrectangles. The output variables will be the arrays G, Rsum, even though only G[0], Rsum[0] are produced by this procedure. Its code is

```
initgrid:=proc(H,f,a,b,c,d,N,M,G,Rsum)
  local dx,dy,P,i,j;
```

```
dx:=(b-a)/N;dy:=(d-c)/M;G[0]:={};Rsum[0]:=0;
for i from 0 to N-1 do
  for j from 0 to M-1 do
    P:=[a+i*dx,c+j*dy];
    if test(H,P,dx,dy) then
      G[0]:=G[0] union {P};
      Rsum[0]:=Rsum[0]+evalf(f(P[1],P[2])*dx*dy);
    end if;
  end do;
end do;
end proc:
```

The main procedure int2d will have the same input as the procedure initgrid, with an additional input: a positive integer K indicating the number of grids $G_0, G_1, \ldots, G_K$ to produce. CAUTION: at each step the new subdivided grid has 4 times as many rectangles as the previous one. Some of these are cast out, but after K steps, we can expect roughly 4^K as many rectangles in G[K] as there were in G[0]. Hence, your computer will run out of memory if K is too large. Even for small values of K, the computations can take a relatively long time to execute. The code is:

```
int2d:=proc(H,f,a,b,c,d,N,M,K,G,Rsum)
  local dx,dy,i,j,k,P,NP;
  dx:=(b-a)/N;dy:=(d-c)/M;
  initgrid(H,f,a,b,c,d,N,M,G,Rsum);
  for k from 1 to K do
    dx:=dx/2;dy:=dy/2;G[k]:={};Rsum[k]:=0;
    for P in G[k-1] do
      for i from 0 to 1 do
        for j from 0 to 1 do
          NP:=[P[1]+i*dx,P[2]+j*dy];
          if test(H,NP,dx,dy) then
            G[k]:=G[k] union {NP};
            Rsum[k]:=Rsum[k]+evalf(f(NP[1],NP[2])*dx*dy);
          end if;
        end do;
      end do;
    end do;
  end do;
  RETURN()
end proc:
```

We test the procedure with $f(x, y) = 1$ and the region D shown in the previous figures. This region is determined by

$$H(x, y) = 1 - (2x^2 + y) \exp(1 - x^2 - y^2).$$

We take $R = [-1.66, 1.66] \times [-0.64, 1.13]$ and divide this into 250 subrectangles by taking $N = 25$ and $M = 10$. The call to the main procedure is:

```
int2d(H,f,-1.66,1.66,-0.64,1.13,25,10,2,grid,Rsum);
```

To produce the picture in Figure 6.18, we use the commands:

```
curve:=implicitplot(H(x,y)=0,x=-0.8..1.2,y=-1.8..1.8,
                numpoints=1000,color=black):
p1:=plot(grid[1],style=point,color=black):
display({curve,p1},tickmarks=[2,2], scaling=constrained);
```

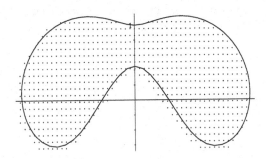

Figure 6.18: *Lower left-hand vertices of rectangles in the approximating grid G_1 produced by the Maple code. The approximate area of the region D is 4.128171000.*

6.7 Polynomials

Mathematically, a polynomial in one variable, x, has the following general form:

$$a_n x^n + a_{n-1} x^{n-1} + \cdots + a_1 x + a_0, \tag{6.1}$$

where the coefficients $a_n, a_{n-1}, \ldots, a_0$ are constants. Polynomials are a data type in Maple and there are several useful commands that are available for working with polynomials.

In the general form above the summands are arranged in decreasing powers of the variable, so the polynomial is called *sorted*. For example $3x^4 - 2x^3 + 6x^2 - 8$ is sorted and $5x^2 - 7x^3 - 2x + 7$ is not. Polynomials in Maple are not necessarily stored in sorted form but the **sort** command can be applied to a polynomial to rewrite it in decreasing powers of x. Applying the **ops** command to a polynomial will give a sequence of its summands.

```
p:= 3*x^4-2*x^3+6*x^2-8;
op(p);
```

produces the sequence

$$3x^4, -2x^3, 6x^2, -8.$$

Of course the monomial $3x^4$ is also a polynomial but it is of type *, not +. Thus, if the **op** command is applied to `3*x^4` the sequence $3, x^4$ is given.

Maple allows the usual arithmetic operations of addition, subtraction, multiplication, and exponentiation for polynomials. These operations can be used in conjunction with the **expand** command to operate on polynomials and produce polynomials in general form. For example,

```
p:=6*x^3+x^2-2;
q:=p^2;
```

produces the output

$$p := 6x^3 + x^2 - 2$$

$$q := (6x^3 + x^2 - 2)^2.$$

Apply the **expand** command `expand(q)` to get

$$q := 36x^6 + 12x^5 - 24x^3 + x^4 - 4x^2 + 4.$$

Table 6.5 gives some of the useful operations on polynomials.

We see that Maple has many built-in commands for working with polynomials. In Maple a polynomial in x is said to be sorted if it is in the form of (6.1). i.e., if it is written in decreasing powers of x. For a polynomial p the command `coeffs(p)` produces a list of coe cients of p listed in increasing powers of x.

The next example illustrates how to write your own procedure to do essentially the same thing, but with the added advantage of including zeros for missing powers of x.

command	operation
+,-	addition and subtraction
*,^	multiplication and exponentiation
quo	quotient of two polynomials
rem	remainder upon division of two polynomials
coeff	extract a coe cient of a polynomial
coeffs	construct a sequence of all the coe cients
degree	the degree of a polynomial
subs	evaluate a polynomial
factor	factor a polynomial over a field
collect	group coe cients of like terms together
expand	distribute products over sums
sort	sort a polynomial

Table 6.5: *Operations with polynomials.*

Example 6.7 (A Coe cients Procedure) We construct a procedure that will take a polynomial as input, with its terms arranged in any order, and produce as output a list of its coe cients in increasing order. The list will include zeros for the coe cients of the missing terms. The logic in the procedure has to allow for all the possible types of polynomial expressions. In addition, we write our own code to determine the degree of the polynomial, rather than using Maple's **degree** command.

Coe cients of Polynomials

```
cfs:=proc(p)
  local n,i,a,degr,T,m,q;
  n:=nops(p);degr:=0;
  #--------------------------------------------------------
  if type(p,'+') or type(p,symbol) then
    for i from 1 to n do
      T:=op(i,p);
      if type(T,'*') then
        if type(op(2,T),symbol) then m:=1;a[m]:=op(1,T)
          else m:=op(2,op(2,T));a[m]:=op(1,T) end if;
        if m>degr then degr:=m end if;
      elif type(T,'^') then
        m:=op(2,T);a[m]:=1;
        if m>degr then degr:=m end if;
```

```
  elif type(T,symbol) then m:=1;a[m]:=1;
    if m>degr then degr:=m end if;
  else a[0]:=T;
  end if;
end do;
for i from 0 to degr do
  if not type(a[i],numeric) then a[i]:=0 end if;
end do;
#------------------------------------------------------------
elif type(p,'^') then m:=op(2,p);a[m]:=1;degr:=m;
  for i from 0 to m-1 do
      a[i]:=0
  end do;
#------------------------------------------------------------
elif type(p,'*') then a[0]:=0;
      if type(op(2,p),symbol) then
        degr:=1;a[1]:=op(1,p);
      else degr:=op(2,op(2,p));
            for i from 1 to degr-1 do
              a[i]:=0;
            end do;
            a[degr]:=op(1,p);
      end if;
  else a[0]:=p;
  end if;
#------------------------------------------------------------
  [seq(a[i],i=0..degr)];
end proc:
```

The call cfs(-x+2+5*x^5) would result in the list $[2, -1, 0, 0, 0, 5]$.

It is also interesting and instructive to write our own procedure for multiplying two polynomials. Multiplication of polynomials is just the repeated application of the distributive law of multiplication over addition. Letting $p = a_0 + a_1 x + a_2 x^2 + \cdots + a_n x^n$ and $q = b_0 + b_1 x + b_2 x^2 + \cdots + b_m x^m$, we need a method for calculating the product pq as a polynomial arranged in increasing powers of x. A change of notation will help us see things more clearly. Write $p = \sum_{i=0}^{n} a_i x^i$ and $q = \sum_{j=0}^{m} b_i x^j$ so that

$$pq = \sum_{i=0}^{n} a_i x^i \sum_{j=0}^{m} b_j x^j = \sum_{i=0}^{n} \sum_{j=0}^{m} (a_i b_j) x^{i+j}.$$

Now we need to collect the terms involving like powers, i.e., all the terms involving x^{i+j} with $i + j = k$, for $k = 0, 1, \ldots, n + m$. For each such k, we sum the coefficients $a_i b_j$ with $i + j = k$ and this will give the coefficient of x^k in the product pq. The formula for the product is thus seen to be:

$$pq = \sum_{k=0}^{n+m} \left(\sum_{\substack{1 \le n, j \le n \\ i+j=k}} a_i b_j \right) x^k.$$

We would like to simplify this formula to

$$pq = \sum_{k=0}^{n+m} \left(\sum_{i+j=k} a_i b_j \right) x^k.$$

This formula will in some cases refer to coefficients a_i, b_j that are not present in the original polynomials. However, if we redefine p and q by inserting extra terms with zeros as coefficients:

$$p = a_0 + a_1 x + \cdots + a_n x^n + 0 \cdot x^{n+1} + 0 \cdot x^{n+2} + \cdots + 0 \cdot x^{n+m}$$

$$q = b_0 + b_1 x + \cdots + b_m x^m + 0 \cdot x^{m+1} + 0 \cdot x^{m+2} + \cdots + 0 \cdot x^{n+m},$$

then we can use the simpler formula. Next, for convenience in coding the double sum, we reindex the inner sum. All we need to do is change the summation index as follows:

$$pq = \sum_{k=0}^{n+m} \left(\sum_{i=0}^{k} a_i b_{k-i} \right) x^k.$$

Now the coding can be done with a pair of nested do loops. In this procedure we will use Maple's **degree** command but instead of the built-in command **coeff** we will use **cfs** from the previous example.

Product of Polynomials

```
polyprod:=proc(p,q)
  n:=degree(p);m:=degree(q);
  for i from 0 to n+m do
    if i<=n then a[i]:=cfs(p)[i+1]
    else a[i]:=0
    end if;
  end do;
```

```
for i from 0 to n+m do
  if i <=m then b[i]:=cfs(q)[i+1]
  else b[i]:=0 end if;
end do;
c:=array(0..n+m);
for k from 0 to n+m do
  sum:=0;
  for i from 0 to k do
    sum:=sum+a[i]*b[k-i];
  end do;
  c[k]:=sum;
end do;
pq:=0;
for k from 0 to n+m do
  pq:=pq+c[k]*x^k;
end do;
pq;
end proc:
```

The call

```
polyprod(6*x^3-5*x^2+3,-8*x^5+2);
```

gives

$$6 - 10x^2 + 12x^3 - 24x^5 + 40x^7 - 48x^8$$

6.8   Case Study: Partial Fractions

An interesting application of working with polynomials occurs when we write
a procedure to give the partial fraction decomposition of a rational function.
This comes up in calculus when we want to integrate quotients of polynomial
functions:

$$\frac{p(x)}{q(x)}\,dx.$$

Example 6.8 Recall that the quotient $p(x)/q(x)$ of two polynomial functions $p(x)$ and $q(x)$ is called a rational function. If the rational function is
proper, i.e., $\deg(p(x)) < \deg(q(x))$, and $q(x)$ factors over the reals into a
product of factors each having degree at least one, then according to the
theory $p(x)/q(x)$ can be decomposed into (or written as) a sum of simpler

rational functions. These simpler rational functions can be integrated using logarithms, power functions, and inverse trig functions.

Here are two rational functions and their respective decomposition into partial fractions.

$$\frac{x+5}{(x+1)(2x-3)} = \frac{-4/5}{x+1} + \frac{13/5}{2x-3}$$

$$\frac{2x^2+3x-5}{(x+4)(5x^2-3x+2)} = \frac{15/94}{x+1} + \frac{\frac{113}{94}x - \frac{125}{94}}{5x^2-3x+2}$$

In general, for a rational function (left-hand sides of the above equations) the problem is to calculate the partial fraction decompositions (right-hand sides). This involves setting up the correct form of the decomposition with the coe cients in the numerators as the unknowns. Setting this decomposition equal to the original rational function produces an equation which when simplified yields a system of linear equations involving the unknown coe cients. One then must solve this system of equations to obtain the partial fraction decomposition.

Problem: Write a Maple procedure to calculate the partial fraction decomposition of the rational function

$$\frac{p(x)}{q_1(x)q_2(x)\cdots q_n(x)},$$

which is assumed to be proper: $\deg(p(x)) < \deg(q_1(x)) + \deg(q_2(x)) + \cdots + \deg(q_n(x))$. Assume that each $q_i(x)$ is a linear or quadratic factor with integer coe cients which is completely factored over the reals and that all the $q_i(x)$'s are distinct.

Analysis of the Problem: To get started let's look at a particular example of how we do partial fraction decomposition by hand.

Example 6.9

$$\frac{3x^2+2x-6}{(3x+1)(2x^2-3x+4)}.$$

The correct form to assume for the partial fraction decomposition in this case is

$$\frac{3x^2+2x-6}{(3x+1)(2x^2-3x+4)} = \frac{A}{3x+1} + \frac{Bx+C}{2x^2-3x+4},$$

where A, B, and C are constants. The problem is to find A, B, and C. Simplifying the above equation we have

$$3x^2 + 2x - 6 = (3B + 2A)x^2 + (B - 3A + 3C)x + 4A + C.$$

Equating the corresponding coefficients gives a system of equations for us to solve:

$$3B + 2A = 3$$
$$B - 3A + 3C = 2$$
$$4A + C = -6.$$

Upon doing this we find that $A = -57/47, B = 85/47, C = -54/47$ so that the partial fraction decomposition is

$$\frac{3x^2 + 2x - 6}{(3x + 1)(2x^2 - 3x + 4)} = \frac{-57/47}{3x + 1} + \frac{(85/47)x - 54/47}{2x^2 - 3x + 4}.$$

From the example we see that there are basically three steps involved in this analysis:

☐ The first thing we need to do is set up the partial fraction decomposition of

$$\frac{p(x)}{q_1(x)q_2(x)\cdots q_n(x)}.$$

We must look at each $q_i(x)$ and determine its degree. If $\deg(q_i(x)) = 1$, we do one thing and if $\deg(q_i(x)) = 2$, we do something else. In Figure 6.19, we illustrate this branching scheme and have included some checks along the way. Diagrams such as this help to organize our thinking before we get involved in too many details.

☐ Next we need to sum up the partial fractions from step 1, and set the sum equal to $\frac{p(x)}{q_1(x)\cdots q_n(x)}$ to get an equation for the partial fraction decomposition. Simplifying this equation and equating corresponding coefficients gives a system of equations.

☐ Our final task is to solve the system of equations from step 2 and then substitute the solutions into the partial fraction decomposition equation from step 2.

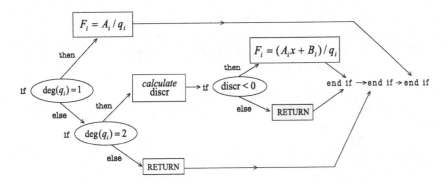

Figure 6.19: *Flow of control in the partial fraction code.*

The Maple code for this procedure can now be written by following the
above outline. We will use many of Maple's commands for working with
polynomials.

Design of the Code: Our procedure which we call `pfrac` will have two
input parameters: the numerator p and the denominator q of our rational
function. We assume that $q(x)$ is entered in factored form as described in
the statement of the problem. We first need to check that p/q is proper
and if not to return an error message. If the rational function is proper
we will need to store the factors of the denominator so that we can easily
access them. We can form a list of the factors which we call `Q` using the `seq`
command. We also need the number of factors of q and we can determine
this using a call to `nops`. The first few lines of the procedure would be

```
pfrac:=proc(p::polynom,q::polynom)
  local Q,m,A,B,F,i,e,e2,e3,e4,Cfs,k,ans,eqns,r,a,b,s;
  if degree(p)>=degree(q)
    then RETURN("Error - p/q is not proper")
  end if;
  Q:=[seq(i,i=q)];
  m:=nops(q);
```

Now we have the factors of the denominator q stored in the list `Q` and we
know there are m of them. With the notation $q = q_1(x)q_2(x)\cdots q_m(x)$ then
`Q[i]`$= q_i(x)$ and $i = 1,2,3,\ldots,m$. Next we need to set up the partial
fraction for each q_i using a do loop. We will store these in a array called `F`.
For the coe cients of the numerators of the partial fractions we need two

arrays which we name A and B. It is a good idea to put in a check to make sure any quadratic factor q_i is irreducible. We can do this by checking to see if the discriminant is negative. Since all the factors of q should be linear or quadratic we will return an error message otherwise.

```
A:=array(1..m);B:=array(1..m);F:=array(1..m);
for i from 1 to m do
  if degree(Q[i])=1 then
        F[i]:=A[i]/Q[i]
  else if degree(Q[i])=2 then
        a:=coeff(Q[i],x,2);
        b:=coeff(Q[i],x,1);
        c:=coeff(Q[i],x,0);
        discr:=b^2-4*a*c;
        if discr<0 then F[i]:=(A[i]*x+B[i])/Q[i]
        else RETURN("Error-factor",Q[i],"is not irreducible")
        end if;
      else RETURN("Error - degree of factor", Q[i],
                  "should be 1 or 2")
      end if;
    end if;
end do;
```

We now need to sum the partial fractions. If we give this sum the name e then we need to consider the equation $p/q = e$ or, equivalently, the equation $p - e \cdot q = 0$. Once the left-hand side is simplified and like terms collected, we set each of the coefficients equal to zero. Storing the coefficient of x^k in the kth location of an array which we name Cfs, we can generate the set of equations we need to solve.

```
e:=0;
for i from 1 to m do
  e:=e+F[i];
end do;
e2:=simplify(p-e*q);
e3:=collect(e2,x);
Cfs:=array(0..degree(e3,x));
for k from 0 to degree(e3,x) do
  Cfs[k]:=coeff(e3,x,k)
end do;
eqns:={seq(Cfs[r]=0,r=0..degree(e3,x))};
```

All that remains is to solve the system, substitute the solution values into e, and output the partial fraction decomposition in a nice form.

```
ans:=solve(eqns);
e4:=p/q=subs(ans,e);
end proc:
```

Here are several calls to pfrac and the results:

```
pfrac(3*x^2+2*x-6,(3*x+1)*(2*x^2-3*x+4));
```

$$\frac{3x^2 + 2x - 6}{(3x + 1)(2x^2 - 3x + 4)} = -\frac{57}{47}\frac{1}{3x + 1} + \frac{\frac{85}{47}x - \frac{54}{47}}{2x^2 - 3x + 4}$$

```
pfrac(x+5,(x+1)*(2*x-3));
```

$$\frac{x + 5}{(x + 1)(2x - 3)} = -\frac{4}{5}\frac{1}{x + 1} + \frac{13}{5}\frac{1}{2x - 3}$$

```
pfrac(x+5,(x+1)*(2*x^2-3));
```

$$\text{"Error-factor"}, \ 2x^2 - 3, \ \text{"is not irreducible"}$$

Exercise Set 6

1. **(Medians)** A median of a triangle is a segment from a vertex to the midpoint of the opposite side. A basic fact is that the three medians of a triangle intersect at a point P, called the *centroid* of the triangle. For triangle $\triangle ABC$, use vector methods to find a formula for the coordinates of the point P of intersection of the medians. Write a Maple procedure which has vertices A, B, C as input and which has as output the point P and a plot of the triangle with the three medians.

2. **(Medians)** Suppose $\triangle ABC$ is any triangle and let $\triangle A_1 B_1 C_1$ be the triangle formed using the midpoints A_1, B_1, C_1 of $\triangle ABC$. See Figure 6.20.

   Use vector methods to show that $\triangle A_1 B_1 C_1$ is similar to $\triangle ABC$. Specifically, use the dot product and vector algebra to show that the corresponding angles are equal: $\_A = \_A_1$, $\_B = \_B_1$, and $\_C = \_C_1$. Are the other three triangles shown in Figure 6.20 also similar to $\triangle ABC$? Justify your answer. Write a procedure to produce a picture of $\triangle ABC$ and $\triangle A_1 B_1 C_1$, with each shown in a different color.

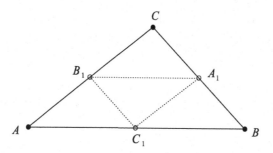

Figure 6.20: *The triangle $\triangle A_1 B_1 C_1$ formed from the midpoints of the sides of triangle $\triangle ABC$.*

3. **(Medians)** As in the previous exercise, suppose $\triangle ABC$ is any triangle and let $\triangle A_1 B_1 C_1$ be the triangle formed using the midpoints A_1, B_1, C_1 of $\triangle ABC$. See Figure 6.20. The process of going from $\triangle ABC$ to $\triangle A_1 B_1 C_1$ can be repeated. Namely, let $\triangle A_2 B_2 C_2$ be the triangle formed using the midpoints A_2, B_2, C_2 of $\triangle A_1 B_1 C_1$. This process can be continued indefinitely, and at the nth step, we have a triangle $T_n = \triangle A_n B_n C_n$, and get the next triangle $T_{n+1} = \triangle A_{n+1} B_{n+1} C_{n+1}$ by using the midpoints A_{n+1}, B_{n+1}, and C_{n+1} of T_n. As you can readily see by sketching a few of these triangles, the successive triangles get smaller and smaller. This exercise is to study the sequence of triangles $\{T_n\}_{n=0}$. *Note:* You need to know some linear algebra to work this exercise. It also involves the idea of iterated maps.

(a) Write some code to draw, for a given N, all the triangles $T_0, T_1, \ldots, T_N$ (in the same figure).

(b) Show that each sequence of vertices $\{A_n\}_{n=0}$, $\{B_n\}_{n=1}$, and $\{C_n\}_{n=0}$, converges to the point P which is the centroid of the initial triangle T_0 (see Exercise 1). For example,

$$\lim_{n \to} A_n = P.$$

There are several ways to show this. Here is one suggested way.

(i) Show that the vertices of T_{n+1} are computed from those of T_n by

$$
\begin{aligned}
A_{n+1} &= \tfrac{1}{2} B_n + \tfrac{1}{2} C_n \\
B_{n+1} &= \tfrac{1}{2} A_n + \tfrac{1}{2} C_n \\
C_{n+1} &= \tfrac{1}{2} A_n + \tfrac{1}{2} B_n
\end{aligned}
$$

Show that these three equations can be written as a single matrix equation:

$$
\begin{matrix}
A_{n+1} \\
B_{n+1} \\
C_{n+1}
\end{matrix}
=
\begin{bmatrix}
0 & \tfrac{1}{2} & \tfrac{1}{2} \\
\tfrac{1}{2} & 0 & \tfrac{1}{2} \\
\tfrac{1}{2} & \tfrac{1}{2} & 0
\end{bmatrix}
\begin{matrix}
A_n \\
B_n \\
C_n
\end{matrix}
. \qquad (6.2)
$$

(ii) Let $x_n = (A_n, B_n, C_n)$. Then equation (6.2) can be written as $x_{n+1} = x_n$, where is the 3×3 matrix in equation (6.2). Show from this that $x_n = ^n x_0$, for all n.

(iii) Show that

$$\lim_{n \to} ^n = \begin{pmatrix} \frac{1}{3} & \frac{1}{3} & \frac{1}{3} \\ \frac{1}{3} & \frac{1}{3} & \frac{1}{3} \\ \frac{1}{3} & \frac{1}{3} & \frac{1}{3} \end{pmatrix}. \tag{6.3}$$

Use Maple to compute a few of the powers: ^n, $n = 2, 3, 4, \ldots$, of . This will give empirical evidence to the above result. To *prove* the result, find the eigenvalues $_1, _2, _3$, and corresponding linearly independent eigenvectors, v_1, v_2, v_3, of . Then let M be the 3×3 matrix formed with v_1, v_2, v_3 as columns. Show that $ M = MD$, where D is a diagonal matrix with $_1, _2, _3$ as its diagonal entries. The powers of D are easy to compute and the limit: $\lim_{n \to} D^n$ is readily discerned. Compute the inverse of M and use it to find the limit of the powers of :

$$\lim_{n \to} ^n = \lim_{n \to} MD^n M^{-1}.$$

4. (**Altitudes**) This exercise pertains to the discussion in Example 6.3 on altitudes of trangles $\triangle ABC$.

   (a) Use the conditions (1) and (2) in Example 6.3 to find a formula for the point P which is the foot of the altitude from vertex C onto side $\overline{AB}$.

   (b) Alter the procedure `altitude` to get a new procedure, `altitude2`, that draws the triangle $\triangle ABC$ in black with thickness 2, the altitude $\overline{CP}$ in red with thickness 1, and the extension of the side $\overline{AB}$ (where necessary) in black with thickness 1. Also have the foot (point P) of the altitude as an output parameter to the new procedure.

   (c) Write a procedure `orthocen` that calls the procedure `altitude2` and uses its output to (1) draw a figure like that in Figure 6.7 showing the triangle and all three of its altitudes in the same picture, and (2) calculate the points U, V, W where the pairs of altitudes: $\overline{CP}$ and $\overline{BR}$, $\overline{BR}$ and $\overline{AQ}$, $\overline{AQ}$ and $\overline{CP}$, intersect. *Note:* It is a theorem that all three altitudes intersect in a common point, i.e., $U = V = W$, so your output will show this when you apply the procedure to any triangle. Extra Credit: Prove this property. The point where the three altitudes intersect is called the *orthocenter* of the triangle.

5. Let A, B, C, D be the four vertices of a tetrahedron. The altitude of this tetrahedron from vertex A is the line segment from A to the point P where P lies in the plane containing vertices B, C, and D and the line segment from A to P is perpendicular to this plane. Use vector methods to find a formula for the point P. Write a procedure which calculates P and draws the tetrahedron $ABCD$ and the altitude AP. See Figure 6.21.

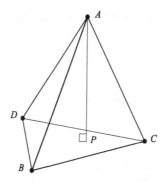

Figure 6.21: *Tetrahedron ABCD with altitude AP.*

6. Write a procedure to generate the nth Taylor polynomial for a function f.
   Call the procedure `Tpoly` and have input parameters the function f, the
   order n of the Taylor polynomial to be generated, and the expansion point
   a.

7. Write a procedure that takes a polynomial written in increasing powers of x
   and reverses the ordering, i.e., outputs the polynomial written in decreasing
   powers of x.

8. Explain the logic for the procedure `cfs` in Example 6.7 by drawing a "ow
   chart.

9. Write a procedure named `degr` that inputs a polynomial and computes its
   degree.

10. Find the sums of lengths of all the intervals in the complement of the Cantor
    set.

11. Write a procedure `cantorfctn2(x,m)` to produce the continuous approxima-
    tions f_m to the Cantor function. Figure 6.14 is a graph of f_2. The precise
    definition of f_m is as follows.

    Let $OD[m]$ be the ordered list of all the intervals removed up through stage
    m, say $OD[m] = [OT[1], OT[2], \ldots, OT[2^{m-1}]]$. For x in some interval in
    $OD[m]$ let $f_m(x) = f_m(x)$ where f_m is the step function approximation to
    the Cantor function given in Example 6.5. So the graph of f has the same
    steps as the graph of f.

    To complete the definition of f, linearly interpolate between the values on
    consecutive steps. For this, let $S[m]$ be the set of intervals remaining at stage
    m. If x is in an interval in $S[m]$, then x is in the complement of $OD[m]$, and
    so either x is to the left of the first interval in $OD[m]$, or x is in the interval
    between two consecutive intervals in $OD[m]$, or x is to the right of the last
    interval in $OD[m]$. Specifically: either

(a) $0 \leq x \leq OT[1][1]$, or

(b) x is in the interval between two consecutive elements of $OD[m]$, say $OT[i]$ and $OT[i+1]$, so $OD[i][2] \leq x \leq OD[i+1][1]$, or

(c) $OT[2^{m-1}][2] \leq x \leq 1$.

If (a) is the case, define $f_m(x)$ be $L(x)$, where L is the line though the point $(0,0)$ and the left endpoint of the first step $(OT[1][1], f_m(OT[1][1]))$.

If (b) is true, let $f_m(x)$ be $L(x)$, where L is the line though the right endpoint of the ith step $(OT[i][2], f_m(OT[i][2]))$ and the left endpoint of the $(i+1)$th step $(OT[i+1][1], f_m(OT[i+1][1]))$.

If (c) is the case, then $f_m(x)$ is defined to be $L(x)$, where L is the line though the right endpoint of the last step $(OT[2^{m-1}][2], f(OT[2^{m-1}][2]))$ and the point $(1,1)$.

Use your procedure to plot the graphs of f_2, f_3, f_4 and f_5.

12. In this problem you will look at a two dimensional analogue of the Cantor set and produce some pictures of this new set. Start with the square $[0,1] \times [0,1]$ and remove the middle square $[\frac{1}{3}, \frac{2}{3}] \times [\frac{1}{3}, \frac{2}{3}]$. Divide the part of the original square that remains into 8 congruent squares each having side of length $\frac{1}{3}$:

$$[0, \tfrac{1}{3}] \times [0, \tfrac{1}{3}], \quad [0, \tfrac{1}{3}] \times [\tfrac{1}{3}, \tfrac{2}{3}], \quad [0, \tfrac{1}{3}] \times [\tfrac{2}{3}, 1], \quad [\tfrac{1}{3}, \tfrac{2}{3}] \times [\tfrac{2}{3}, 1]$$

$$[\tfrac{2}{3}, 1] \times [\tfrac{2}{3}, 1], \quad [\tfrac{2}{3}, 1] \times [\tfrac{1}{3}, \tfrac{2}{3}], \quad [\tfrac{2}{3}, 1] \times [0, \tfrac{1}{3}], \quad [\tfrac{1}{3}, \tfrac{2}{3}] \times [0, \tfrac{1}{3}]$$

From each of these smaller squares remove the middle square. See Figure 6.22. Then the remaining part of each of the smaller squares can be divided into 8 congruent squares each having side of length $1/9$. Remove the middle square from each of these smaller squares.

Figure 6.22 shows this process at the second and fourth removal stages. Continue this pattern ..., indefinitely. What remains of the original square is called the Sierpinski carpet.

To do this problem, first write two preliminary procedures: `middlesquare` and `newsquares`.

The `middlesquare` procedure has as input two points, the lower left corner and upper right corner of a square which has horizontal and vertical sides, and produces as output a list of the four vertices of the middle square.

The `newquares` procedure accepts as input four points: `A,B,C,D` where `A` and `B` are the lower left corner and upper right corner of a square and `C` and `D` are the lower left corner and upper right corner of the middle square. The output of `newsquares` is a list of lists of vertices of the eight squares remaining when the middle square is removed from the first input square.

Next write the main procedure named `Sierpinski:=proc(N)`, where the parameter is the number of stages of removing middle squares that are to be completed. The output will be a display of the initial square and all the middle squares removed. The `Sierpinski` procedure should make a call to `middlesquare` to produce the new set of middle squares to be removed and then call `newsquares` to create the next set of squares whose middles will be removed in the next stage.

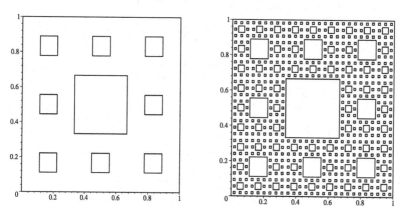

Figure 6.22: *Constructing the Sierpinski carpet*

13. In Example 6.6, the Riemann sum approximations to the double integral

$$\int_L f(x,y)\,dA,$$

were constructed by a scheme using grids of rectangles with each rectangle having at least one vertex inside D. Modify the procedures there so that the scheme uses grids of rectangles with each rectangle having *all four* vertices inside D. Such grids will not cover D, but will (for suitable regions) lie inside D and will approximate the region D fairly well when the subdivisions are fine enough. Have your code produce the results and graphics like that in Example 6.6. Test it on the region D used in that example and with functions $f(x,y) = 1$ and $f(x,y) = x^2 + y^2$. Also use these same two functions and the region D defined by $H(x,y) = x^2 + y^2 - 1$ to study the approximations to $V = \int_L f(x,y)\,dA$. Compute V exactly, in each case, by hand. Compare these studies with the corresponding ones you get by using the scheme in Example 6.6.

14. Without using Maple, determine the result of the following commands and explain your reasoning.

    (a) `pfrac(2*x-1,(x+3)*(4*x-1))`

(b) `pfrac(x^3+5,(x+1)*(2*x^2-3))`

(c) `pfrac(x+5,(x+1)*(2*x^2))`

15. Write a procedure to take a rational function $f(x) = p(x)/q(x)$ and determine whether it is proper (degree of p is less than degree of q) and if not to write f as the sum of a polynomial and a proper rational function. This involves dividing $p(x)$ by $q(x)$.

   (a) Do this exercise using Maple's built-in commands for polynomials.

   (b) Write your own procedure for dividing $p(x)$ by $q(x)$.

16. Suppose for the rational function $f(x) = p(x)/q(x)$ that q is factorable over the integers into a product of distinct linear and quadratic factors. Write a procedure to factor q and then find the partial fraction decomposition. This will be a short procedure using Maple's `factor` command and then a call to the procedure `pfrac` (which you have placed in your worksheet) to determine the partial fraction decomposition of f.

17. Write a program to find the partial fraction decomposition of

$$\frac{p(x)}{q_1(x)^{k_1} q_2(x)^{k_2} \cdots q_n(x)^{k_i}},$$

where $\deg(p(x)) < \deg(q_1(x)^{k_1}) + \cdots \deg(q_n)^{k_i}$ and the $q_i(x)$'s are distinct, completely factored linear or quadratic polynomials.

18. Evaluate the following integrals. The only help from Maple that is allowed is the use of the procedures in this chapter.

   (a)

$$\frac{x^3}{x^2 + 1}\, dx.$$

   (b)

$$\frac{3 - 11x}{(x - 2)^3(x^2 + 1)(2x^2 - 5x + 7)^2}\, dx.$$

Chapter 7

Graphics Programming

In this chapter we discuss programming techniques involved in producing graphic images for mathematical objects, such as curves and surfaces. We will also examine the plot structures, PLOT and PLOT3D, that Maple uses to render these images on plot devices such as the monitor screens or printers.

Our goals are threefold. We want you to learn how to write algorithms to generate the data that comprise various types of graphics (2-D and 3-D plot structures). Often, the analysis and design of such algorithms can be quite challenging. Also, we want you to obtain a deeper understanding of the geometry involved in representing functions of one or two variables. Understanding the geometry and spatial relationships will be necessary in order for you to solve the programming tasks. On the other hand, the solution of the programming tasks will make you appreciate more fully many aspects of geometry and multivariable calculus that you have previously studied. In addition, we want to show you further examples of how integral formulas for certain quantities arise from discrete approximations (Riemann sums) and a passage to a limit.

Most programming languages have high-level and low-level features for producing graphical output. At the high level, a single command will do all the work for you, while at the low level you will have to build the graphic out of basic elements, such as points and lines. You are already familiar with many of the high-level commands in Maple. For example,

```
plot(exp(-x)*sin(2*Pi*x),x=0..2,style=point,numpoints=400);
plot3d(x*exp(-x^2-y^2),x=-2..2,y=-2..2,style=patchcontour);
```

are high-level commands that produce plots of $f(x) = e^{-x}\sin 2\ x$ and $f(x,y) = xe^{-(x^2+y^2)}$, respectively. Below, you will write your own code to produce such plots. For functions of a single variable, these plots, which represent curves, will be constructed out of line segments (similar to some of the activities you encountered in previous chapters), and for functions of two variables, the plots which represent the surfaces will be constructed out of triangles. The actual graphic image will be only an approximation to the true mathematical curve or surface (which are ideal objects or abstractions). This is inherent in the finiteness of the computer. Thus, it is important for you to realize that in Maple displays:

All curves are polygons   polygonal approximations to the true curves.
All surfaces are triangular complexes   triangulations of the true surfaces.

Below we will examine the various elements in the plot structures that Maple creates for each of these and this examination will allow us to build plot structures in these and a variety of other circumstances. Programming like this, with the basic elements at the low level, allows precise control over how the graphic is produced. However, we should mention that any graphic produced at the low level can also be produced by using high-level commands and some appropriate programming.

7.1   Preliminary Examples

In Maple all high-level commands (Maple procedures) that produce graphic output do so by first creating either a `PLOT` structure or a `PLOT3D` structure. By default, the information in the plot structure is passed to a device driver that interprets it and displays it on your monitor. In the next section, we will describe the makeup of the `PLOT` and `PLOT3D` structures in general, but here we look at two specific examples of these structures.

Example 7.1 (Polygonal Approximations) As we have mentioned previously, a Maple plot of any real-valued function $y = f(x)$ of a single real variable is actually just a polygonal approximation to the mathematical (ideal) graph of f. This is readily apparent if we specify a small number of points to use in the plot. For instance, consider the following:

```
f:=x->exp(-x)*sin(2*Pi*x);
plot(f(x),x=0..2,color=black,numpoints=11,adaptive=false);
```

The resulting plot is shown in Figure 7.1. As you can see, the plot is a

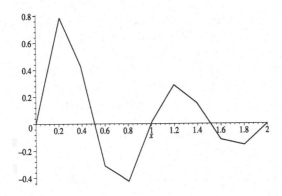

Figure 7.1: *Plot of the function* $f(x) = e^{-x} \sin 2\, x$ *on the interval* $[0, 2]$ *using* 11 *points.*

polygon with 10 sides and 11 vertices.[1] One way to view the PLOT structure that Maple uses to create the display is to assign the above plot command to a named variable and have the results displayed on the screen:

```
ps:=plot(f(x),x=0..2,numpoints=11,adaptive=false);
```

The output of this command looks like the following

$$
\begin{aligned}
ps := \mathrm{PLOT(CURVES}([&[0,0], \\
&[.2092526800000000, .7847492911058794], \\
&[.3913223180000001, .4266585765323162], \\
&[.5960786680000000, -.3127712397707445], \\
&[.8021922760000001, -.4244526317143005], \\
&[1.007326262000000, .01680477453751293], \\
&[1.197511726000000, .2856725693114512], \\
&[1.394437758000000, .1526752737646165], \\
&[1.598098942000000, -.1169344445098951], \\
&[1.801106982000000, -.1566759698860453], \\
&[2., -.6629516857096724\, 10^{-16}]]), \\
&\mathrm{COLOUR}(RGB, 1.0, 0, 0)), \mathrm{AXESLABELS}(\ x\ ,\ ), \\
&\mathrm{VIEW}(0..2., DEFAULT))
\end{aligned}
$$

This PLOT structure has three arguments. First there is the object: CURVES, which itself is comprised of a list of the 11 calculated points (the object in-

---

[1]Note that the option adaptive=false must be used in order to get Maple to calculate and plot only 11 points. By default, Maple has an adaptive routine that will calculate extra points where needed in order to have the plot appear to be smooth.

formation) and the local object information: COLOUR(RGB,1.0,0,0), which specifies that the color of the plot is red.[2] The second and third arguments contain global information about the plot. AXESLABELS("x",'') specifies that the horizontal axes be labeled with an x and that there be no label on the vertical axis. VIEW(0..2.,DEFAULT)) specifies a view with horizontal range $0\ldots2$ and the default for the vertical range (which is based on the calculated values of the function).

The production of the plot shown in Figure 7.1 actually involves two steps. When you use the high-level command

```
plot(f(x),x=0..2,color=black,numpoints=11,adaptive=false);
```

Maple first produces the PLOT structure shown on the right-hand side of the *ps:=* output above and then passes this structure to a device driver that uses the information in the data structure to render the graphic on the screen (or produce a printed version). Every high-level command in Maple that produces a graphic will create an intermediate PLOT or PLOT3D structure like this.

Now that you see how a basic PLOT structure is created by the high-level plot command, you can build your own PLOT structures piece by piece using programming at a lower level. There are always a number of different approaches to such tasks, some of which involve using high-level commands as well, and generally we want you to write algorithms to generate the data (usually points in $\mathbb{R}^2$ or $\mathbb{R}^3$). Then you can use either a high-level or low-level plot command to display the graphic.

For example, to produce the plot in Figure 7.1 by generating our own data points, we simply divide the interval $[0,2]$ into $N=10$ equal subintervals, calculate the division points $x_i = 2i/N$, $i = 0, 1, \ldots, N$, and get the data points (x_i, y_i), where $y_i = f(x_i)$, $i = 0, 1, \ldots, N$. Some Maple code for this is as follows.

```
N:=10:
X:=array(0..N):Y:=array(0..N):pts:=array(0..N):
for i from 0 to N do
        X[i]:=2.0*i/N:
        Y[i]:=evalf(f(X[i])):
        pts[i]:=[X[i],Y[i]]:
end do:
polyapprox:=[seq(pts[i],i=0..N)]:
```

---

[2]Note that Maple uses the British spelling *colour* for the word color. It will also accept COLOR(RGB,1.0,0,0), if you want to use this in programming at the low level.

Here `polyapprox` is a list consisting of the same points shown above as the object information for the `CURVES` object.

To create the plot of this polygonal approximation with a blue color, we could use the high-level command

```
plot(polyapprox,color=blue);
```

Alternatively, we can do the same thing using the following low-level command:

```
PLOT(CURVES(polyapprox,COLOUR(RGB,0,0,1)));
```

In summary then, you see that you can use high-level commands or low-level commands to produce a specific graphic. If you prefer, you can use a combination of high-level and low-level commands for accomplishing the same thing.

As a second introductory example of the type of graphics programming discussed in this chapter, we consider a solid of revolution and its approximation by cylindrical disks.

Example 7.2 (Volumes of Solids of Revolution) Suppose f is a given function on an interval $[a, b]$ and consider the solid S obtained by revolving the graph of f about the x-axis. See Figure 7.2.

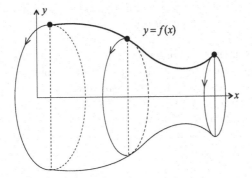

Figure 7.2: *The solid of revolution S obtained by revolving the graph of $y = f(x)$ about the x-axis.*

A standard formula for the volume V of this solid of revolution is

$$V = \int_a^b f(x)^2 \, dx. \tag{7.1}$$

This formula is associated with the *disk method* for measuring the volume of such a solid and is motivated in your calculus book by the following argument. As in Chapter 3 (Example 3.1 and Exercise 2), we approximate f by a step function using a subdivision of $[a, b]$ into N equal subintervals, each of length $\Delta x = (b - a)/N$ and having the steps determined by the values of f at the right-hand endpoints of these intervals. The subdivision points are $x_i = a + i(b - a)/N$, $i = 0, \ldots, N$, and the step heights are $y_i = f(x_i)$, $i = 1, \ldots, N$. Then the steps (or treads) in the step function will be the line segments with endpoints (x_{i-1}, y_i) and (x_i, y_i), for $i = 1, \ldots, N$. See Figure 7.3.

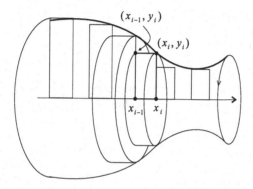

Figure 7.3: *A step-function approximation to f and the corresponding cylindrical disk approximation to the solid of revolution.*

The figure shows several cylinders that are generated when several treads (line segments) in the approximating step function are revolved about the x-axis. When the entire approximating step function is revolved, we obtain a sequence of cylinders that approximate the solid of revolution S. This is the *cylindrical disk approximation*, and the volume of this solid is easy to compute. It is just the sum of the volumes of the N cylindrical disks that comprise it. The volume of a cylinder is the area of its circular base times its height, $r^2 h$, and as shown in the figure the ith disk has radius $r = y_i = f(x_i)$ and height (or thickness of the disk) $h = \Delta x = x_i - x_{i-1}$. Thus, the ith volume is $f(x_i)^2 \Delta x$, and so the volume of the total cylindrical disk approximation is

$$V_N = \sum_{i=1}^{N} f(x_i)^2 \Delta x. \tag{7.2}$$

For very thin disks, i.e. for $\Delta x \to 0$ (or equivalently for large N), this approximate volume will be close to the volume V for the solid of revolution S generated by f. This comes from the theory which says that the limit

$$V = \lim_{N \to} \sum_{i=1}^{N} f(x_i)^2 \, \Delta x = \int_a^b f(x)^2 \, dx, \qquad (7.3)$$

gives us the integral formula for this actual volume.

This is the motivation for the integral formula used in the disk method. Using these ideas, we want to write a procedure that not only computes the approximate volumes in equation (7.2), but also produces a picture of the corresponding cylindrical disk approximation.

To create the graphic, we will simply plot the cylindrical surface of each disk (the exercises will consider embellishments to this), and for this we represent each surface parametrically. If you have not studied parametric representations of surfaces, you should read the Maple/Calculus Notes at the end of this chapter. For the cylindrical surfaces here, we can easily determine a parametrization using Figure 7.4.

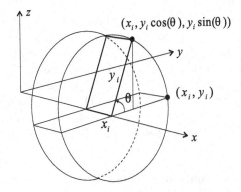

Figure 7.4: *Parametrizing a cylindrical surface.*

As shown in the figure, a point on the surface of the ith cylindrical disk is determined by two numbers (or parameters): a number x in the interval $[x_{i-1}, x_i]$ and an angle which indicates its inclination with the x-y plane. From the geometry shown, it is clear that any point (x, y, z) on the surface is given by

$$\begin{aligned} x &= x \\ y &= r \cos \\ z &= r \sin , \end{aligned}$$

where $r = y_i$. These equations constitute the *parametric representation* of this surface. Alternatively, one considers this as giving a vector-valued function $G : \mathbb{R}^2 \quad \mathbb{R}^3$ of two variables,

$$G(x, \ ) = (x, r \cos \ , r \sin \ ),$$

with $(x, \ )$ in $[x_{i-1}, x_i] \times [0, 2\ ]$. Maple's `plot3d` applies to functions of two variables, and when they are vector-valued functions, it gives a plot of the parametric surface. Note that in the present example, the radius of the cylinder is $r = y_i = f(x_i)$. Assuming that these values have been previously computed and stored in an array `Y`, then for an assigned value of i we can plot the ith cylindrical surface with the command:

```
plot3d([x,Y[i]*cos(theta),Y[i]*sin(theta)],x=X[i-1]..X[i],
       theta=0..2*Pi,grid=[2,51])
```

The option `grid=[2,51]` is used to control the appearance of the surface. (We only want two of the x-coordinate curves on the surface. See the Maple/Calculus Notes on coordinate curves.)

    With this background, we can now easily write a procedure to compute the approximate volume in equation (7.2) and create a picture of the corresponding cylindrical disk approximation. The procedure uses a do loop to subdivide the interval $[a, b]$ in the standard way, and sum the approximating terms. Within this do loop, controlled by $i = 1, \ldots, N$, the plot of the ith cylinder is created and assigned to an array element `ps[i]`. This is then used to display all the cylinders in a single figure. This graphic is returned through an output parameter with formal name `figure`. The following is the code for the procedure:

Procedure for Approximating Solids of Revolution

```
diskmethod:=proc(f,a,b,N,figure::evaln)
  local X,Y,s,x,dX,i,disk;
  with(plots,display3d);
  X:=array(0..N);Y:=array(1..N);disk:=array(1..N);
  dX:=evalf((b-a)/N); X[0]:=a;s:=0;
  for i to N do
     X[i]:=a+i*dX; Y[i]:=evalf(f(X[i]));
     s:=s+evalf(Pi*Y[i]^2*dX);
     disk[i]:=plot3d([x,Y[i]*cos(theta),Y[i]*sin(theta)],
              x=X[i-1]..X[i],theta=0..2*Pi,grid=[2,51]):
  end do;
```

```
    figure:= display3d({seq(disk[i],i=1..N)},
            axes=box,projection=0.8,orientation=[-60,70]);
    print('The approximate volume using',N,'disks is V =', s);
end proc:
```

We use this procedure on the function $f(x) = e^{-x}\sin(2\ x)$ from Example
7.1, with $N = 10$. The call: **diskmethod(f,0,2,10,fig);** produces the
output

> *The approximate volume using*, 10, *disks is*, .7477548010

To see the figure, we enter **fig;**, and this gives the cylindrical disk approximation shown in Figure 7.5.

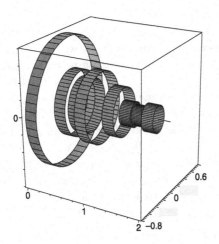

Figure 7.5: *A cylindrical disk approximation to the solid of revolution obtained by revolving* $f(x) = e^{-x}\sin 2\ x$, *on* $[0,2]$, *about the x-axis.*

The example uses only high-level plotting commands to produce the cylindrical surfaces. The exact form of the **PLOT3D** structure that Maple creates
for each of these surfaces will be discussed below.

7.2   Maple s Plot Structures

Maple has numerous special-purpose procedures (high-level commands) for
producing plots, graphs, and other visual displays. For example, the **plots**
package consists of the following 51 procedures for exhibiting graphical data
in two and three dimensions:

```
animate, animate3d, animatecurve, changecoords, complexplot,
complexplot3d, conformal, contourplot, contourplot3d,
coordplot, coordplot3d, cylinderplot, densityplot, display,
display3d, fieldplot, fieldplot3d, gradplot, gradplot3d,
implicitplot, implicitplot3d, inequal, listcontplot,
listcontplot3d, listdensityplot, listplot, listplot3d,
loglogplot, logplot, matrixplot, odeplot, pareto, pointplot,
pointplot3d, polarplot, polygonplot, polygonplot3d,
polyhedra_supported, polyhedraplot, replot, rootlocus,
semilogplot, setoptions, setoptions3d, spacecurve,
sparsematrixplot, sphereplot, surfdata, textplot,
textplot3d, tubeplot
```

Each procedure is designed to conveniently accomplish a certain plotting task and typically includes required input parameters as well as optional ones. We do not want to discuss the specifics of these procedures here (see the Appendix and Maple's help menu), but rather wish to emphasize that each of the procedures that produces graphic output does so by first producing either a PLOT structure or a PLOT3D structure. By default, the information is passed to a device for display, resulting in a figure that is displayed on your monitor.

The PLOT or PLOT3D structure that is produced at the intermediate step is a data structure, called a *plot data structure* or *plot structure*. These are unevaluated function calls having one of the following generic forms:

```
PLOT(objects, global-information)
```

which is for graphics in two dimensions, and

```
PLOT3D(objects, global-information)
```

which is for graphics in three dimensions. Both PLOT and PLOT3D are functions with arguments (or parameters) which are shown above in two groups: objects and global-information.

Here objects stands for a sequence of parameters that specify the objects to be plotted and global-information stands for a sequence of parameters that specify how the total picture, comprised of all these objects, is to be rendered. The specific types of objects that Maple allows are:

☐ Objects for the PLOT command:

CURVES, POINTS, POLYGONS, TEXT, ANIMATE,

☐ Objects for the PLOT3D command:

CURVES, POINTS, POLYGONS, TEXT, ANIMATE, GRID, MESH

Each of these objects is itself a function and has arguments (or parameters)
that are either object information, i.e., the data comprising the object,
or local information, i.e., descriptions of how the individual object is to
be rendered (color, scaling, etc.) in the absence of global information. For
example,

```
PLOT(CURVES([[0,0],[8,2],[2,6],[0,0]],COLOUR(RGB,0,0,1)),
     CURVES([[4,1],[5,4],[1,3],[4,1]],COLOUR(RGB,1,0,0)),
     SCALING(CONSTRAINED))
```

has two parameters that are objects, each of which is of type CURVES, and one
parameter specifying global information: SCALING(CONSTRAINED), which
renders both objects with a constrained scaling. Each object also has local
information specified about its color. When this unevaluated function call is
evaluated by a plotting device, the result is two polygonal curves, one blue
and one red, each with three sides.

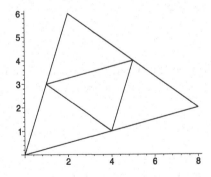

Figure 7.6: *A pair of triangles.*

As shown in Figure 7.6, the resulting figure is actually a pair of triangles
(color not shown) with the smaller one formed from the midpoints of the
sides of the larger one.

The specific parameters for global information are as follows:

☐ Global Information for PLOT and PLOT3D:

SCALING, VIEW, STYLE, AXESSTYLE, AXESLABELS, AXESTICKS,
THICKNESS, LINESTYLE, SYMBOL, COLOUR, FONT, TITLE

Some of these, such as THICKNESS, LINESTYLE, SYMBOL, COLOUR, can also
occur as local object information since they control attributes of the object
rather than attributes of the picture with all its objects.

Each of these options for global information is a function with parameters (usually only one) indicating the global information. For example,
SCALING(CONSTRAINED) and SCALING(UNCONSTRAINED) specify that the objects in the picture are to be rendered either with the same scale for each
axis or a variable scale (which does not maintain proportions).

Example 7.3 As a simple example of creating a graphic image from the
basic objects discussed above, consider the triangle and median shown in
Figure 7.7.

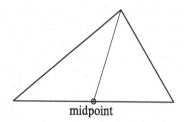

Figure 7.7: *A triangle and one of its medians.*

The triangle has vertices $(0,0), (6,0), (4,3)$ and the median shown has
endpoints $(3,0), (4,3)$. We can produce the graphic shown in Figure 7.7
using the following command.

```
PLOT(CURVES([[0,0],[6,0],[4,3],[0,0]],COLOUR(RGB,0,1,0),
        THICKNESS(2)),
    CURVES([[3,0],[4,3]],COLOUR(RGB,1,0,0)),
    POINTS([3,0],SYMBOL(CIRCLE),COLOUR(RGB,1,0,0)),
    TEXT([3,-0.1],'midpoint',ALIGNBELOW,
        FONT(TIMES,ROMAN,16),COLOUR(RGB,1,0,0)),
    SCALING(CONSTRAINED),AXESSTYLE(NONE));
```

The first CURVES object shown is the triangle, rendered in green and with
a thickness of 2 units. The second CURVES object is the median, rendered

in red and with a default thickness of 1 unit. The POINTS object creates a single point at $(3, 0)$, rendered in red and with the circle style. The TEXT object creates the label for the midpoint, rendered in red and aligned below the point $(3, -0.01)$, which is slightly below the midpoint. The font for this label is 16 pt, Times Roman.

Note how the local object information is included as parameters in the objects CURVES, POINTS, and TEXT. This is necessary since each of these has different attributes. The global object information SCALING(CONSTRAINED) and AXESSTYLE(NONE) applies to the whole picture.

You can get more information on Maple's PLOT and PLOT3D structures from the help menu by using the command ?plot,structure;. This explains many of the options for the local and global object information. The rest of this chapter also will help you learn the syntax and details of these plot structures. More important, it will help you understand the mathematics and geometry that underlie these structures.

7.3 Approximating Curves and Surfaces

This section discusses techniques for approximating curves and surfaces and the use of these approximations to represent the curves and surfaces graphically. From a mathematical standpoint, approximation techniques are central to computing approximate numerical values for quantities that are not known exactly. For example, the step-function approximation to the curve $y = f(x)$ in Example 7.2 led to the procedure approxsolid that can be used to calculate approximate values for the volume

$$V = \int_a^b f(x)^2 \, dx,$$

of a solid of revolution. While this integral formula for the volume can often be evaluated exactly, the cases where it cannot be done exactly clearly require some numerical scheme for finding an approximate value.

From a graphical standpoint, approximation techniques are essential for constructing representations of curves and surfaces in $\mathbb{R}^2$ and $\mathbb{R}^3$. Some aspects of this have already been discussed in previous chapters, examples, and exercises.

7.3.1 Approximating Curves

We have used step-function approximations to curves $y = f(x)$ that are graphs of real-valued functions of a single variable. While these approximations are useful in connection with Riemann sums and the definition of the definite integral, they are rather poor for representing the curve graphically. Furthermore, the step-function approximations rely on the special nature of the curve, viz., that it is the graph of a function, and thus they are not applicable for more general curves in the plane and do not apply at all to curves in $\mathbb{R}^3$.

On the other hand, any curve in $\mathbb{R}^2$ or $\mathbb{R}^3$ is more suitably approximated by a polygon (polygonal curve) formed by selecting a sequence of points on the curve and connecting these points with line segments, which form the sides of the polygonal approximation. Selecting the sequence of points on the curve is easy if the curve is given parametrically. We discuss this first. When the curve is not given parametrically, the determination of points on the curve is more challenging, and we will not discuss this case (see [FK, p. 136] and [YG, p. 160]).

A parametric representation of a curve in $\mathbb{R}^3$ can be considered as a vector-valued function of a single variable, $F : I \quad \mathbb{R}^3$, where I is an interval, say, $I = [a, b]$ (cf. the Maple/Calculus Notes). Then for each $t \quad I$, the value of F at t is a point

$$F(t) = (\, f(t),\, g(t),\, h(t)\,),$$

in $\mathbb{R}^3$ that lies on the curve. Indeed, the curve itself is conceived of being traced out by $F(t)$ as the parameter t moves continuously through the interval $[a, b]$. Thus, it is easy to select a sequence of points on the curve to use for the polygonal approximation. We merely select a sequence $t_0 < t_1 < \cdots < t_N$ of parameter values in $[a, b]$, with $t_0 = a, t_N = b$, to get a sequence of points $\{F(t_i)\}_{i=0}^{N}$ on the curve. A common selection of the parameter values is

$$t_i = a + i(b - a)/N, \qquad (i = 0, 1, \ldots, N),$$

which arises from dividing $[a, b]$ into N subintervals of equal length. This equally spaced selection of parameter values works well if the curve varies uniformly with the parameter value. However, if the curve varies rapidly for, say, t near a and less so for t near b, then you can achieve a better approximation by selecting more of the t_i's near a and fewer near b.

Example 7.4 Consider the vector-valued function defined by

$$F(t) = (\, t \cos t,\, t \sin t,\, t\,),$$

for $t \quad [-10, 10]$. The curve with F as the parameter map can be plotted in Maple as follows:

```
F:=t->[t*cos(t),t*sin(t),t];
ps:=spacecurve(F(t),t=-10..10,color=black,numpoints=11,
             scaling=constrained,orientation=[-120,60],
             axes=box,projection=0.8,thickness=2,
             labels=['x','y','z'],tickmarks=[3,3,3]);
ps;
```

We have included a number of options in the **spacecurve** command to control characteristics of the resulting picture. In particular we have chosen to divide the domain $[-10, 10]$ of F into $N = 10$ equal subintervals with division points

$$t_i = -10 + i(10 - (-10))/10 = -10 + 2i, \quad (i = 0, \ldots, 10).$$

The output from the second command above (the one defining **ps**) looks like

$ps := \text{PLOT3D(CURVES}([[8.390715291, -5.440211109, -10.],$
$[1.164000270, 7.914865973, -8.],$
$[-5.761021720, -1.676492989, -6.],$
$[2.614574483, -3.027209981, -4.],$
$[.8322936731, 1.818594854, -2.],$
$[0, 0, 0],$
$[-.8322936731, 1.818594854, 2.],$
$[-2.614574483, -3.027209981, 4.],$
$[5.761021720, -1.676492989, 6.],$
$[-1.164000270, 7.914865973, 8.],$
$[-8.390715291, -5.440211109, 10.]],$
$\text{COLOUR}(RGB, 0, 0, 0)), \text{AXESTICKS}(3,3,3),$
$\text{AXESSTYLE(BOX)}, \text{SCALING}(CONSTRAINED),$
$\text{THICKNESS}(2), \text{AXESLABELS}(x, y, z),$
$\text{PROJECTION}(-120., 60., .8))$

The **spacecurve** command thus creates a single **CURVES** object which has object data consisting of the 11 calculated points on the curve and a specification of a black color for the plot. The other arguments of **PLOT3D** consist of global object information indicating how the picture is to be rendered. These correspond to the options used in the **spacecurve** command.

The plot produced by the command **ps;** will be a ten-sided polygonal approximation to the curve. This is shown in Figure 7.8, along with a better approximation using $N = 100$ sides. You can perhaps discern from the better approximation that this curve lies on the surface of a double cone.

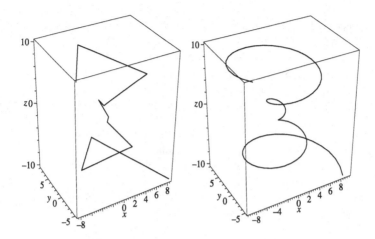

Figure 7.8: *Plots of approximations to the curve $F(t) = (t \cos t, t \sin t, t)$ using polygons with $N = 10$ and $N = 100$ sides.*

Example 7.5 (Distance to a Plane Curve) This example illustrates why you might need to know the details of the PLOT structure created by one of Maple's plotting commands.

Suppose C is a curve in $\mathbb{R}^2$ and $P = (x_0, y_0)$ is a point not on C. The *distance* from P to the curve C is by definition the least of all the distances

$$D(x, y) = \overline{(x - x_0)^2 + (y - y_0)^2},$$

between P and points $Q = (x, y)$ on the curve. See Figure 7.9.

The determination of this minimum distance is a standard optimization problem from calculus and this problem can be solved exactly when the curve C is fairly simple. (See the exercises.) Here we write an elementary procedure to approximate the distance between P and C.

The procedure is easy to write if the curve C is given parametrically (see the exercises), and so we look at the more complicated case where C is given implicitly by an equation $H(x, y) = 0$. We assume that either the entire curve, or the portion that we are interested in, lies in the rectangle $R = [a, b] \times [c, d]$. To approximate the distance between P and C we simply generate a su ciently good polygonal approximation to C and then find the least of all the distances from P to the vertices of the polygon.

However, it is a challenging programming problem to write a routine to plot polygonal approximations to implicitly defined curves, and so, instead

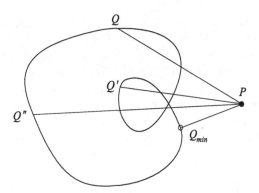

Figure 7.9: *The distance between point P and curve C is the minimum of all the distances $\overline{PQ}$ between P and points Q on the curve.*

of writing such a program from scratch, we use Maple's `implicitplot` command instead. The `PLOT` data structure returned by `implicitplot` has the coordinates of the vertices contained in it and we need to determine how to access this information.

To see the details of the plot structure, we look at a simple example. Consider the circle given implicitly by $x^2 + y^2 - 1 = 0$. We define the corresponding function H, create a plot structure, and assign it to the variable named `curve` by using

```
H:=(x,y)->x^2+y^2-1:
curve:=implicitplot(H(x,y)=0,x=-1.5..1.5,y=-1.5..1.5,
                    style=point,grid=[3,3]);
```

This results in the following output:

$$curve:=\text{PLOT(CURVES }([[0,-.67],[-.67,0]],[[-.67,0],[-.33,.33]],$$
$$[[0,.67],[-.33,.33]][[0,-.67],[.33,-.33]],$$
$$[[.67,0],[.33,-.33]],[[.67,0],[0,.67]],$$
$$\text{COLOUR}(RGB,1,0,0)),\text{STYLE}(POINT),$$
$$\text{AXESLABELS}(x,y))$$

As you can see, `PLOT` has three arguments, `CURVES`, `STYLE`, and `AXESLABELS`, and these form the three operands of `curve`.[3] The first operand, `CURVES`, has object information consisting of a sequence of six lists, L_i, $i = 1, \ldots, 6$,

---

[3]To make the output more readable, we have rounded the numbers in it to two decimal places.

with each list consisting of a pair of points $[x_i, y_i]$, $[\overline{x}_i, \overline{y}_i]$, $i = 1, \ldots, 6$. Each of the two points are the endpoints of a line segment. The plot of all the line segments gives the polygonal approximation to the circle. Thus, the implicitplot command creates a sequence of pairs of points which is used to draw the polygon. The number of pairs (six in this example) that Maple returns is not predictable from the grid size, but will be one less than the number of operands of the **CURVES** object. The last operand is the color information. Figure 7.10 shows the polygonal approximation to the circle which is produced an implicitplot command like that above.

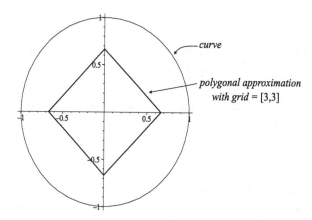

Figure 7.10: *A polygonal approximation to the circle $x^2 + y^2 - 1 = 0$.*

The approximation appears as a four-sided, diamond-shaped polygon even though the data returned by Maple consist of six line segments. This is so since two pairs of these segments lie on the same line. As you can see, this polygonal approximation is not very good.

We can select the information we need from the plot structure **curve** by using the **op** command and the selection brackets **[ ]** as follows. First we select the **CURVES** object using

```
pairs:=op(1,curve);
```

which gives

$pairs := \text{CURVES}([[0, -.67], [-.67, 0]], [[-.67, 0], [-.33, .33]],$
$[[0, .67], [-.33, .33]], [0, -.67], [.33, -.33]],$
$[[.67, 0], [.33, -.33]], [[.67, 0], [0, .67]],$
$\text{COLOUR}(RGB,1,0,0))$

If we want to select the third pair of points from the object information in CURVES, we can use

```
op(3,pairs);
```

which gives

$$[[0, .67], [-.33, .33]]$$

Then the x and y coordinates of the second point in this pair of points can be selected by

```
op(3,pairs)[2][1]; op(3,pairs)[2][2];
```

which gives

$$-.33, .33$$

With this understood, we can now write the code for a procedure to approximate the distance between a given point and a given, implicitly defined, curve. Inputs to the procedure are (1) the coordinates x0,y0 of the given point, (2) the function H defining the curve, and (3) the numbers a,b,c,d giving the ranges for the implicitplot command. Note that for a bounded curve, like a circle, you will want to take a,b,c,d so that the rectangle $[a, b] \times [c, d]$ is only slightly larger than needed to enclose the curve.

As output from the procedure, we return, through parameters, (1) the approximate distance, dist, between the curve and the given point $P = (x_0, y_0)$ and (2) the coordinates x1,y1 of the approximate point $Q_{min} = (x_1, y_1)$ that is nearest to P. The local variables X,Y are used to store the coordinates of Q_{min} before they are assigned to x1,y1. Two do loops are used to scan the sequence of pairs of points returned by the implicitplot. Within the loops, the coordinates of each point Q are assigned to XX,YY and the distance dd between P and Q is computed and compared to the previously computed least distance d1.

We also have the default output be a figure showing the curve rendered in the point style and the line segment between P and Q_{min}.

Procedure for an Approximate Distance from a Point to a Curve

```
distance:=proc(x0,y0,H,a,b,c,d,dist::evaln,x1::evaln,y1::evaln)
   local curve,pairs,N,X,Y,d1,dd,i,n,XX,YY,line;
   with(plots,implicitplot,display):
   curve:=implicitplot(H(x,y)=0,x=a..b,y=c..d,style=point):
```

```
pairs:=op(1,curve):N:=nops(pairs)-1;
X:=op(1,pairs)[1][1];Y:=op(1,pairs)[1][2];
d1:=((X-x0)^2+(Y-y0)^2)^.5;
  for n to N do
    for i to 2 do
    XX:=op(n,pairs)[i][1];YY:=op(n,pairs)[i][2];
    dd:=((XX-x0)^2+(YY-y0)^2)^.5;
    if dd < d1 then d1:=dd;X:=XX;Y:=YY end if;
    end do;
  end do;
  x1:=X;y1:=Y;dist:=d1;
  line:=plot([[x0,y0],[X,Y]],color=black):
  display({curve,line},scaling=constrained);
end proc:
```

As an example of the use of the procedure suppose the given point is $P = (0,0)$ and the curve is

$$\left(x^3 + 4x^2y^2\right)e^{1-x^2-y^2} - 1 = 0.$$

Figure 7.11 shows the resulting graphical output.

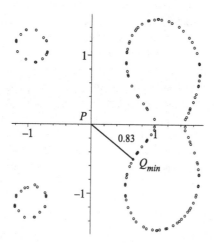

Figure 7.11: *The approximate distance between the point $P = (0,0)$ and the curve $\left(x^3 + 4x^2y^2\right)e^{1-x^2-y^2} - 1 = 0$.*

The procedure also finds that

$$Q_{min} = (.6612206704748800, -.5012206704748806),$$

and that the approximate distance from P to Q_{min} is 0.8297197934.

The procedure uses the default 25×25 grid size in the `implicitplot` command. For some curves this might not be fine enough to give a good polygonal approximation to the curve, and hence a good approximation to the distance. However, it is easy enough to modify the procedure to include input parameters for the grid size.

7.3.2 Approximating Surfaces

In Chapter 4, we used step-function approximations to surfaces $z = f(x, y)$ that are graphs of real-valued functions of two variables and saw how these approximations serve to motivate and define the double integral in terms of limits of Riemann sums. However, from a graphical standpoint surfaces are best approximated by triangular complexes, which are the analogs of the polygonal approximations to curves discussed in the last subsection.

Triangular complexes and their use in approximating surfaces are concepts that are not typically discussed in calculus books and so you may be learning these topics here for the first time. The ideas are not complicated and are best illustrated for the special case of surfaces $z = f(x, y)$ which are graphs.

Example 7.6 (Surfaces that are Graphs) Suppose $f : R \quad R$ is a real-valued function defined on a rectangle $R = [a, b] \times [c, d]$. Recall that the graph of f is the set of points

$$S = \{ (x, y, f(x, y)) \,|\, (x, y) \quad R \},$$

in $\mathbb{R}^3$. We approximate S by a sequence of contiguous triangles in space (a triangular complex) by first dividing the rectangle R into a mesh of contiguous triangles, called a *triangulation* of R. There are many ways to do this and Figure 7.12 shows one possible way that consists of first dividing the rectangle R into subrectangles and then dividing each subrectangle into two triangles using one of its diagonals. Corresponding to any triangle in this triangulation of R, we get a triangle in space that has its three vertices on the surface S, and hence we get a triangulation of S as indicated in Figure 7.12.

We proceed to automate this triangulation process and produce a graphic showing the triangular complex that approximates the surface. For this we partition R into subrectangles by using the standard partitions of $[a, b]$ and $[c, d]$ into N subintervals and M subintervals, respectively. Figure 7.13 shows a triangulation for the case $N = 5$ and $M = 4$. Note the choice of diagonals

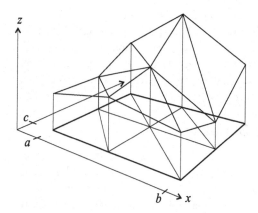

Figure 7.12: *A triangulation of the domain* $R = [a, b] \times [c, d]$ *of* f *and the corresponding triangular complex that approximates the graph of* f.

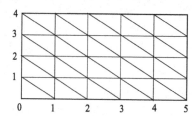

Figure 7.13: *A triangulation of* R *with* $N = 5$ *and* $M = 4$.

for the subrectangles and that groups of these diagonals form parallel line segments. We will use prior indexing schemes to denote the vertices of the subrectangles and from this we get the indexing of the vertices of the corresponding triangles.

The vertices of the subrectangles in the partition of R are $P_{i,j} = (x_i, y_j)$, where $x_i = a + i\Delta x$ and $y_j = c + j\Delta y$, for $i = 0, \ldots, N$, $j = 0, \ldots, M$. Here $\Delta x = (b - a)/N$ and $\Delta y = (d - c)/M$. To these there are corresponding points on the surfaces S:

$$Q_{i,j} = (x_i, y_j, f(x_i, y_j)).$$

The i-jth subrectangle R_{ij} is (by definition) the rectangle with vertices $P_{i-1,j-1}, P_{i,j-1}, P_{i,j}, P_{i-1,j}$. Corresponding to these vertices are four points $Q_{i-1,j-1}, Q_{i,j-1}, Q_{i,j}, Q_{i-1,j}$ on the surface S. See Figure 7.14. From these points we get two triangles T_{ij}, T_{ij} which approximate the part of the surface

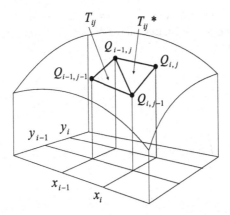

Figure 7.14: *Four points on the surface S and two triangles T_{ij}, T_{ij} determined by them.*

S that lies over the rectangle R_{ij}. The points $Q_{i-1,j-1}, Q_{i,j-1}, Q_{i-1,j}$ are the vertices of the triangle T_{ij} and $Q_{i-1,j}, Q_{i,j}, Q_{i,j-1}$ are the vertices of triangle T_{ij}, as shown in Figure 7.14.

There are several ways to draw the triangular complex

$$\{ T_{ij}, T_{ij} \mid i = 0, \ldots, N, j = 0, \ldots, M \}.$$

Drawing each triangle separately would be ine cient since there are numerous overlapping sides. Instead we view the triangular complex as made up of polygonal paths of three types: vertical, horizontal, and diagonal. This terminology comes from the corresponding triangulation of the rectangle R (cf. Figure 7.13), and is perhaps not appropriate for polygons in $\mathbb{R}^3$, but nevertheless is easy to remember. We write out explicitly the vertices for the polygons of each type as follows. (For convenience the diagonal type is divided into two further types.)

(1) **(Vertical)** This type has vertices where all the x-coordinates are the same. Specifically, for a given $i = 0, \ldots, N$,

$$\texttt{vert[i]} = [\, Q_{i,0}, Q_{i,1}, Q_{i,2}, \ldots, Q_{i,M} \,].$$

(2) **(Horizontal)** This type has vertices where all the y-coordinates are the same. Specifically, for a given $j = 0, \ldots, M$,

$$\texttt{horiz[j]} = [\, Q_{0,j}, Q_{1,j}, Q_{2,j}, \ldots, Q_{N,j} \,].$$

(3) **(Diagonal Type 1)** This type has vertices where the x-coordinates decrement by Δx while the y-coordinates increment by Δy, starting with $y_0 = c$ and a given x_i. Specifically, for a given $i = 1, \ldots, N$,

$$\texttt{diag1[i]} = [\,Q_{i,0}, Q_{i-1,1}, Q_{i-2,2}, \ldots, Q_{i-r_i, r_i}\,],$$

where $r_i = \min\{i, M\}$.

(4) **(Diagonal Type 2)** This type has vertices where the x-coordinates decrement by Δx while the y-coordinates increment by Δy, starting with $x_N = b$ and a given y_j. Specifically, for a given $j = 1, \ldots, M - 1$,

$$\texttt{diag2[j]} = [\,Q_{N,j}, Q_{N-1,j+1}, Q_{N-2,j+2}, \ldots, Q_{N-s_j, j+s_j}\,],$$

where $s_j = \min\{N, M - j\}$.

Figure 7.15 shows these polygonal paths in space arise from corresponding vertical, horizontal, and diagonal sequences of vertices in the domain R of f.

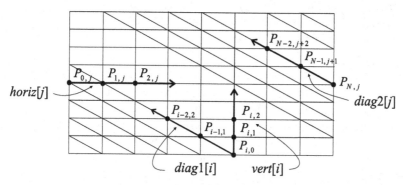

Figure 7.15: *Sequences of vertical, horizontal, and diagonal vertices form straight-line polygonal paths in the domain R of f. The corresponding sequences of points on the surface S form polygonal paths that approximate curves on the surface.*

Thus, plotting all of these types of polygonal paths using the `spacecurve` command will give us a representation of the triangular complex that approximates S for these values of N and M. This representation is called the *wire frame* style since only the sides of each triangle are drawn. The *patch* style arises from rendering the interiors of the triangles but not their sides (see the exercises).

With the above analysis complete, it is now easy to write a procedure that draws the wire frame representation. The input to the procedure is the function f, the numbers a, b, c, d that determine the rectangle R on which f is defined, and the numbers N, M of subintervals in the partitions of $[a, b], [c, d]$ respectively. The procedure is as follows:

A Surface Plot Procedure

```
surfplot:=proc(f,a,b,c,d,N,M)
  local i,j,X,Y,Q,r,s,vert,horiz,diag1,diag2;
  X:=array(0..N):Y:=array(0..M):Q:=array(0..N,0..M):
  vert=array(0..N):horiz:=array(0..M):
  diag1:=array(0..N): diag2:=array(0..M):
  X[0]:=evalf(a):Y[0]:=evalf(c):
  for i to N do X[i]:=evalf(a+(b-a)*i/N) end do:
  for j to M do Y[j]:=evalf(c+(d-c)*j/M) end do:
  for i from 0 to N do for j from 0 to M do
     Q[i,j]:=[X[i],Y[j],f(X[i],Y[j])]  end do; end do;
  for i from 0 to N do vert[i]:=[seq(Q[i,j],j=0..M)] end do:
  for j from 0 to M do horiz[j]:=[seq(Q[i,j],i=0..N)] end do:
  for i from 1 to N do
     diag1[i]:=[seq(Q[i-r,r],r=0..min(i,M))] end do:
  for j from 1 to M-1 do
     diag2[j]:=[seq(Q[N-s,j+s],s=0..min(N,M-j))] end do:
  with(plots,spacecurve);
  spacecurve({seq(vert[i],i=0..N),seq(horiz[j],j=0..M),
              seq(diag1[i],i=1..N),seq(diag2[j],j=1..M-1)},
             color=black,orientation=[-45,65],axes=frame,
             scaling=constrained,projection=.8);
  end proc:
```

As a test of the procedure we use the function $f(x, y) = x^2 - y^2$, whose graph is the well-known hyperbolic paraboloid. We restrict the function to the square $[-1, 1] \times [-1, 1]$ and use subdivisions with $N = 8, M = 8$. Then the command `surfplot(f,-1,1,-1,1,8,8)` produces the plot shown in Figure 7.16. The figure appears to indicate the presence of straight lines that lie on the surface   the lines corresponding to the diagonals in the triangulation of the parameter domain. This is indeed the case since the surface is a hyperbolic paraboloid, but this will not occur for surfaces in general.

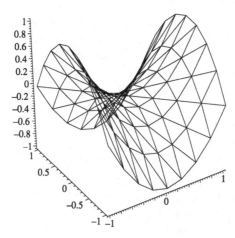

Figure 7.16: *A triangular complex that approximates the graph of* $f(x, y) =$ $x^2 - y^2$ *on the square* $[-1, 1] \times [-1, 1]$.

7.4 The GRID and MESH Objects

We have just seen how to represent (and approximate) a surface by a triangular mesh built from horizontal, vertical, and diagonal polygonal curves in $\mathbb{R}^3$. Each of these polygonal curves has its vertices on the surface.

When you use the `plot3d` command to plot a surface corresponding to a function of two variables, Maple will generate the data for these polygonal curves automatically (the diagonal ones are not shown by default) and this becomes the object data for the `GRID` object or the `MESH` object. In this section we discuss the details of these two `PLOT3D` objects.

7.4.1 The GRID Object

We begin with a simple example that illustrates the `PLOT3D` structure that Maple creates when you plot the graph of a function, say, $f(x, y) = x^2 + y^2$ on $R = [0, 2] \times [0, 2]$. We use a grid specification of `grid=[3,3]` to keep the output small and readable.[4]

```
f:=(x,y)->x^2+y^2;
ps:=plot3d(f(x,y),x=0..2,y=0..2,grid=[3,3],
            axes=normal,orientation=[-66,76]);
```

---

[4]The default grid size in Maple is `grid=[25,25]`.

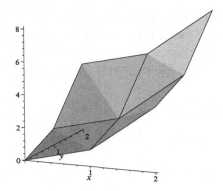

Figure 7.17: *A plot of* $f(x,y) = x^2 + y^2$, *on* $R = [0,2] \times [0,2]$, *using a grid of* $3 \times 3 = 9$ *points.*

The output from the `plot3d` command is

Maple V Ouput:

$ps :=$ PLOT3D(GRID(0..2.,0..2.,[[0,1,4],[1,2,5],[4,5,8]]),
AXESSTYLE(*NORMAL*),AXESLABELS($x,y,$),
PROJECTION(-66.,76.,1))

Maple 6 Output:

$ps :=$ PLOT3D(GRID(0. .. 2.,0. .. 2.,
Array(1..3, 1..3, [[...],[...],[...]], datatype = "oat[8],
storage = rectangular, order = C_order)),
AXESSTYLE(*NORMAL*),AXESLABELS($x,y,$),
PROJECTION(-66.,76.,1))

Note that the Maple V and Maple 6 **PLOT** data structures appear differently in the output. The Maple 6 output suppresses the display of the numbers in the lists of data and also contains some new options in its parameter sequence. Internally, the lists of data are the same, so the discussion below will be for the Maple V version, since it is more explicit.

Figure 7.17 shows the picture that results from using the command `ps;`. As you can see from the output, Maple creates a **GRID** object

```
GRID(0..2.,0..2.,[[0,1,4],[1,2,5],[4,5,8]])
```

which contains the information for the plot. The first two arguments, `0..2` and `0..2`, of **GRID** are the x and y ranges, giving the size of the rectangle

$R = [0,2] \times [0,2]$. Three equally spaced points $x_0 = 0, x_1 = 1, x_2 = 2$ are chosen in the x interval $[0,2]$, dividing it into two equal subintervals. Similarly, equally spaced points $y_0 = 0, y_1 = 1, y_2 = 2$ are chosen in the y interval.[5] Thus, the grid consists of $3 \times 3 = 9$ points (x_i, y_j), $i, j = 0, 1, 2$. The third argument of GRID is

```
[[0,1,4],[1,2,5],[4,5,8]]
```

and this gives the z values corresponding to each point in the grid. Thus, if we let

$$z_{ij} = f(x_i, y_j),$$

then the z values are ordered as

$$[[z_{00}, z_{01}, z_{02}], [z_{10}, z_{11}, z_{12}], [z_{20}, z_{21}, z_{22}]].$$

Besides plotting the horizontal and vertical polygons shown in Figure 7.17, Maple also fills in the interiors of each rectangle with colored shadings. You should realize that each rectangle in Figure 7.17 is actually composed of two triangles as is apparent from grayscales shown in the figure. In order to render the figure, Maple creates a triangular complex exactly like we did in Example 7.6 and has a default scheme for assigning a range of colors to the rectangles. *Note:* The figure that Maple displays on your monitor will not look like Figure 7.17. Besides being in color, it will *not* exhibit clearly the triangles which comprise each rectangle.[6] You can see the triangles clearly on the monitor if you specify GRIDSTYLE(TRIANGULAR) in the plot. For example, the command

```
PLOT3D(GRID(0 .. 2.,0 .. 2.,[[0,1,4],[1,2,5],[4,5,8]]),
        AXESSTYLE(NORMAL),AXESLABELS(x,y,' '),
        PROJECTION(-66.,76.,1),GRIDSTYLE(TRIANGULAR));
```

produces the plot shown in Figure 7.18.

You should realize that by default Maple does not use the triangular grid style, i.e., does not draw in the diagonal polygons like we did in Example 7.6. However, that does not mean the triangles are not there (as is apparent from the PostScript figure shown in Figure 7.17). You should also know that each of the rectangles that comprise the picture shown in Figures

---

[5]It is just coincidental in the example that $x_i = i$ and $y_i = i$, for $i = 0, 1, 2$.

[6]The triangles appear clearly in Figure 7.17 only because of the way Maple assigns grayscales to the colors when the "gure is exported to PostScript for printing in this book. If you look at the same "gure displayed on a computer monitor, the triangles will not be apparent.

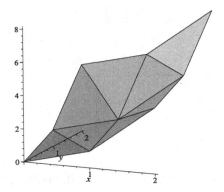

Figure 7.18: *A plot of $f(x,y) = x^2 + y^2$, on $R = [0,2] \times [0,2]$, rendered with a triangular grid style.*

7.17 7.18, need not be an actual rectangle in $\mathbb{R}^3$, i.e., its four sides may not lie in the same plane. This is easier to see in the following example.

If we create a plot consisting of a grid of just four points by using the following command:

```
PLOT3D(GRID(0..1,0..1,[[1,2],[0,4]]),
       PROJECTION(-10,80,1),AXESSTYLE(NORMAL));
```

we get the four points $(0,0,1), (0,1,2), (1,0,0), (1,1,4)$ in $\mathbb{R}^3$ as the vertices of two triangles, and Maple draws the picture shown in Figure 7.19.

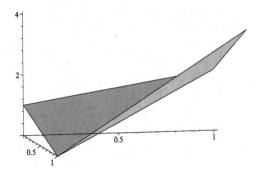

Figure 7.19: *A grid of four points in $\mathbb{R}^3$ does not constitute a rectangle, but rather is a triangular complex of two contiguous triangles.*

You can clearly see that these triangles do not lie in the same plane.

In summary, the general form of the GRID object is

$$\texttt{GRID(a..b,c..d,[[} z_{00}, \ldots, z_{0M} \texttt{],[} z_{10}, \ldots, z_{1N} \texttt{],} \ldots \texttt{,[} z_{N0}, \ldots, z_{NM} \texttt{]])}$$

This is gives a triangular mesh in $\mathbb{R}^3$, as described above and in Example 7.6, with vertices (x_i, y_j, z_{ij}), $i = 0, \ldots, N$, $j = 0, \ldots, M$, where

$$x_i = a + i(b-a)/N, \qquad y_j = c + j(d-c)/M.$$

Note that there are $(N+1) \times (M+1)$ points in the grid, and that the grid consists of $N \times M$ subrectangles.

7.4.2   The MESH Object

The MESH object has object data which are of a different form than those for the GRID object, but the resulting graphic is the same — a triangular complex representing some surface. The MESH object is generated when you use plot3d to plot a surface that is given parametrically.

For example, suppose $F : [0, 2\pi] \times [0, 1] \rightarrow \mathbb{R}^3$ is the vector-valued function:

$$F(\theta, z) = (\cos\theta, \sin\theta, z).$$

Then F parametrizes a cylinder of radius 1, height 1, and axis coinciding with the z-axis. The coordinate curves (cf. the Maple/Calculus Notes) are straight lines and circles. Specifically, for a fixed θ, the corresponding coordinate curve $z \rightarrow F(\theta, z)$ is a straightline on the cylinder (a generator parallel to the z-axis). On the other hand for a fixed z, the coordinate curve $\theta \rightarrow F(\theta, z)$ is a circle, at height z, on the cylinder. A standard plot3d command applied to F will produce polygonal approximations to these curves, with each polygon having 24 sides and 25 vertices. Since straight lines are perfectly approximated by a one-sided polygonal curve (a line segment), we can use a grid size of grid=[N,2]. We will use N=8 in the grid, which will not give a good approximation to the corresponding circles, but will keep the output small enough to examine and easily understand. Thus, we define F in Maple and use plot3d with grid=[8,2] as follows:

```
F:=(theta,z)->[cos(theta),sin(theta),z];
ps:=plot3d(F(theta,z),theta=0..2*Pi,z=0..1,grid=[8,2],
        axes=normal,scaling=constrained,
        gridstyle=triangular);
```

The output from the second command is

Maple V Output:

$$
\begin{aligned}
ps := \mathrm{PLOT3D}(\mathrm{MESH}([[[1,0,0],[1,0,1]], \\
[[0.6234, 0.7818, 0], [0.6234, 0.7818, 1]], \\
[[-0.2225, 0.9749, 0], [-0.2225, 0.9749, 1]], \\
[[-0.9009, 0.4338, 0], [-0.9009, 0.4338, 1]], \\
[[-0.9009, -0.4338, 0], [-0.9009, -0.4338, 1]], \\
[[-0.2225, -0.9749, 0], [-0.2225, -0.9749, 1]], \\
[[0.6234, -0.7818, 0], [0.6234, -0.7818, 1]], \\
[[1, -6.4621e - 015, 0], [1, -6.4621e - 015, 1]]]])), \\
\mathrm{AXESSTYLE}(NORMAL), \mathrm{SCALING}(CONSTRAINED), \\
\mathrm{GRIDSTYLE}(TRIANGULAR))
\end{aligned}
$$

Maple 6 Output:

$$
\begin{aligned}
ps := \mathrm{PLOT3D}(\mathrm{MESH}(\mathrm{Array}(1..8, \ 1..2, \ 1..3, \\
[[[...],[...]],[[...],[...]],[[...],[...]],[[...], \\
[...]],[[...],[...]],[[...],[...]],[[...],[...]],[[...],[...]]]], \\
\mathrm{datatype} = "oat[8], \ \mathrm{storage} = \mathrm{rectangular}, \\
\mathrm{order} = \mathrm{C\_order})), \mathrm{GRIDSTYLE}(TRIANGULAR), \\
\mathrm{SCALING}(CONSTRAINED), \mathrm{AXESSTYLE}(NORMAL))
\end{aligned}
$$

As above with the GRID data structure, the MESH data structures are essentially the same in Maple V and Maple 6, but the displayed output appears differently. We will discuss the Maple V output, since it is more explicit.

As you can see from the output, the object data for MESH comprise a list of 8 items. Each item is a list of two points, which in this example are the endpoints of a line segment that lies on the cylinder (a generator).[7] Figure 7.20 shows the plot produced by the plot structure ps and clearly illustrates how the triangulation of the parameter domain for F gets transformed into a triangulation of the cylinder.

Note that besides plotting the vertical line segments (generators of the cylinder), the MESH procedure also includes plots of the diagonal and horizontal segments, as well as coloring of the facets. Only the vertical line segment data are shown in the plot structure data list.

In general, the form of the MESH object is

---

[7]For readability in the text, the numbers for the coordinates have been truncated to 5 digits.

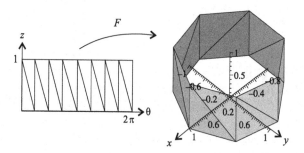

Figure 7.20: *A triangulation of the parameter domain* $[0, 2\pi] \times [0, 1]$ *of the parameter map* F *for the cylinder, and the corresponding triangulation of the cylinder in* $\mathbb{R}^3$.

MESH([L1,L2,...LN])

---

where `Li` is a list of data points of the form

$$L_i = [[x_i, y_1, z_{i1}], [x_i, y_2, z_{21}], \ldots, [x_i, y_M, z_{iM}]].$$

The vertices of the mesh are (x_i, y_j, z_{ij}), $i = 0, \ldots, N$, $j = 0, \ldots, M$, where

$$x_i = a + i(b - a)/N, \qquad y_j = c + j(d - c)/M.$$

Note that there are $(N + 1) \times (M + 1)$ points in the grid, and that the mesh consists of $N \times M$ subrectangles.

For the sake of comparison, the list L_i in the `MESH` data structure corresponds to the list `vert[i]` of vertical vertices in Example 7.6. See Figure 7.15. In that example, we also used lists of horizontal and diagonal vertices to create the three-dimensional plot. This was for convenience and clarity. We could have used only the vertical vertices, since all the necessary information is contained in them. This analogy should help you understand how the `MESH` data structure produces the plot.

As a last example, consider the surface which the following command produces:

```
PLOT3D(MESH(
    [[[0,0,0],[0,1,0]],[[1,0,0],[1,0,1]],[[1,1,0],[2,1,0]]]),
    AXESSTYLE(NORMAL),SCALING(CONSTRAINED),
    GRIDSTYLE(TRIANGULAR));
```

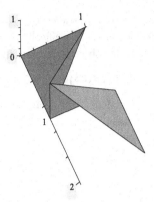

Figure 7.21: *A surface produced by the* MESH *data structure.*

As you can see in Figure 7.21, the squares (with diagonals) determined by
the list of vertical vertices:

$$[[[0,0,0],[0,1,0]],[[1,0,0],[1,0,1]],[[1,1,0],[2,1,0]]],$$

do not result in corresponding rectangular surfaces in the three-dimensional
plot.

7.5 Animations

You can create animations in several ways using high-level commands. For
example, you can use the basic **animate** command, or you can use the
display command with the option **insequence=true**. Before explaining
how to use these commands, we discuss what animations are and the use of
ANIMATE as an object in the **PLOT** data structure. This will give you a better understanding of the high-level commands for producing animations and
also help you build appropriate data structures for elaborate animations.

   An animation is just a sequence of pictures displayed in order, one after
another, on a screen (your monitor) with equal time intervals between the
respective displays. An animation is also commonly known as a movie.
Each picture is called a frame, and the interval of time between each frame
is governed by the frame rate so many frames per second. Typically the
images in the frames are related to each other and arise by filming a
scene in which something is moving. Each frame records the scene at a
particular instant, and when the frame rate is high enough, the playback of
the sequence of frames will simulate the motion observed in the scene.

In Maple each frame in an animation is a plot of something along with axes, labels, text, etc., to be included in the picture. Specifically, a frame is a list of CURVES objects, each with its particular object information:

```
frame = [CURVES(object-info),..., CURVES(object-info)]
```

A sequence of such frames constitutes a sequence of arguments for the ANIMATE object. The general form is as follows:

```
ANIMATE(frame1, frame2,..., frameN)
```

Example 7.7 As a simple example, consider creating an animation with five frames and each frame consisting of a polygonal curve with two sides. We first define the frames as

```
frame1:=[CURVES([[0,0],[2,2],[4,0]],COLOUR(RGB,0,0,0))];
frame2:=[CURVES([[0,0],[2,1],[4,0]],COLOUR(RGB,0,0,0))];
frame3:=[CURVES([[0,0],[2,0],[4,0]],COLOUR(RGB,0,0,0))];
frame4:=[CURVES([[0,0],[2,-1],[4,0]],COLOUR(RGB,0,0,0))];
frame5:=[CURVES([[0,0],[2,-2],[4,0]],COLOUR(RGB,0,0,0))];
```

The object information, in each case, consists of a list of the three vertices for the polygon and a specification that its color be black.

For convenience, we make a sequence of these frames and assign it to the name movie1:

```
movie1:=frame1,frame2,frame3,frame4,frame5;
```

To create the animation and have the axis style be a box and the scaling be constrained, we can use the following command:

```
PLOT(ANIMATE(movie1),AXESSTYLE(BOX),SCALING(CONSTRAINED));
```

Figure 7.22 shows each of the five frames all in the same figure. You will have to imagine the actual motion portrayed in the animation from this. If each polygonal curve represents the position of a guitar string, then the animation roughly simulates a vibrating string.

To illustrate other aspects that can comprise an animation, we extend the above animation as follows. We include another CURVES object in each frame. This will be a blue line segment that joins the middle of the string to the point $(1,0)$ on the x-axis. In addition, we create a green background for each frame by defining

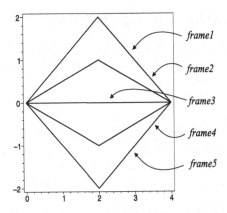

Figure 7.22: *A composite of all the frames in* movie1.

```
backgrd:=POLYGONS([[0,2],[0,-2],[4,-2],[4,2]],
                   COLOUR(RGB,0,1,0));
```

and then include this **backgrd** in each frame. The frames are created as before, but now with these new elements in them:

```
frame1:=[CURVES([[0,0],[2,2],[4,0]],COLOUR(RGB,0,0,0)),
         CURVES([[1,0],[2,2]],COLOUR(RGB,0,0,1)),backgrd];
frame2:=[CURVES([[0,0],[2,1],[4,0]],COLOUR(RGB,0,0,0)),
         CURVES([[1,0],[2,1]],COLOUR(RGB,0,0,1)),backgrd];
frame3:=[CURVES([[0,0],[2,0],[4,0]],COLOUR(RGB,0,0,0)),
         CURVES([[1,0],[2,0]],COLOUR(RGB,0,0,1)),backgrd];
frame4:=[CURVES([[0,0],[2,-1],[4,0]],COLOUR(RGB,0,0,0)),
         CURVES([[1,0],[2,-1]],COLOUR(RGB,0,0,1)),backgrd];
frame5:=[CURVES([[0,0],[2,-2],[4,0]],COLOUR(RGB,0,0,0)),
         CURVES([[1,0],[2,-2]],COLOUR(RGB,0,0,1)),backgrd];
```

Now we can create the movie and display it just as we did above:

```
movie2:=frame1,frame2,frame3,frame4,frame5;
PLOT(ANIMATE(movie2),AXESSTYLE(BOX),SCALING(CONSTRAINED));
```

These two movies are extremely simple, but should serve to show you the structure of animations built from the **ANIMATE** command. Once you understand these elementary movies, you can easily create much more elaborate ones, as the example and exercises below show.

7.5.1   The Display and Animate Commands

You have already used the `display` command to exhibit a sequence of different plots in the same picture. If the plots are named `p1,p2,...,pN`, then

    display({p1,p2,...,pN})

will produce a picture showing all the plots together. If on the other hand, the separate plots represent frames in a movie that you wish to create, then using the display command with the option `insequence=true` will produce an animation in which the plots `p1,p2,...,pN` are shown in succession:

    display([p1,p2,...,pN],insequence=true)

Note: You must use the list data structure in the above command if you want `p1,p2,..,pN` to occur in the specified order in the animation (`p1` in the first frame, `p2` in the second frame, etc.). Use of the set data structure will not guarantee this. As an example of this (one which is simple enough to display all the output here), consider the plots of the two straight lines $y = x$ and $y = x/2$, created by

    p1:=plot(x,x=0..1,color=black,adaptive=false,numpoints=2);
    p2:=plot(x/2,x=0..1,color=black,adaptive=false,numpoints=2);

These give the following outputs to the monitor screen:

$$p1 := \text{PLOT(CURVES}([0,0],[1.,1.]]),\text{COLOUR}(RGB,0,0,0),$$
$$\text{AXESLABELS}(\ x\ ,\ ),\text{VIEW}(0..1.,DEFAULT))$$
$$p2 := \text{PLOT(CURVES}([0,0],[1.,.5]]),\text{COLOUR}(RGB,0,0,0),$$
$$\text{AXESLABELS}(\ x\ ,\ ),\text{VIEW}(0..1.,DEFAULT))$$

Each `PLOT` structure has a single `CURVES` object (consisting of a list of two points determining the line) and three pieces of global information (`COLOUR`, `AXESLABELS`,`VIEW`). As usual, we can display these two plots in the same picture by using:

    A:=display({p1,p2});

The output from this command is

$$A := \text{PLOT(CURVES}([0,0],[1.,1.]],\text{COLOUR}(RGB,0,0,0)),$$
$$\text{CURVES}([0,0],[1.,.5]],\text{COLOUR}(RGB,0,0,0)),$$
$$\text{AXESLABELS}(\ x\ ,\ ),\text{VIEW}(0..1.,DEFAULT))$$

From this you can see how the display command produces a single `PLOT` structure with two `CURVES` objects from the plot structures `p1,p2`.

To produce an animation that displays `p1` in the first frame and `p2` in the second frame, we simply use the following command:

```
B:=display([p1,p2],insequence=true);
```

The output from this command is

$$B := \text{PLOT(ANIMATE(}$$
$$[\text{CURVES}([[0,0],[1.,1.]],\text{COLOUR}(RGB,0,0,0)),$$
$$\text{AXESLABELS}(\ x\ ,\ ),\text{VIEW}(0\,..\,1.,DEFAULT)],$$
$$[\text{CURVES}([[0,0],[1.,.5]],\text{COLOUR}(RGB,0,0,0)),$$
$$\text{AXESLABELS}(\ x\ ,\ ),\text{VIEW}(0\,..\,1.,DEFAULT)]))$$

From this you can see how the information in the plot structures `p1,p2` is stripped out and incorporated as information in the `ANIMATE` command. You can now view the animation by making the command `B;`. The movie is very simple, showing the plot of $y = x$ in the first frame and the plot of $y = x/2$ in the second frame. See Figure 7.23.

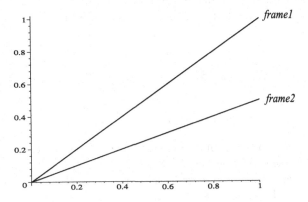

Figure 7.23: *Two frames in a very simple animation.*

Often it is necessary to have an object that appears in one frame of an animation also appear in each ensuing frame as well. This was the case in Example 7.7, which had a green background displayed in each frame. To show how to do this with the **display** command, we modify the animation in B as follows.

As it stands the animation consists of two frames with the first frame containing the plot from **p1** and the second the plot from **p2**. If we also want the plot from **p1** to appear in the second frame, we combine both plots with

```
p12:=display({p1,p2});
```

and then create the animation with

```
C:=display([p1,p12],insequence=true);
```

which gives the output

$$C := \text{PLOT}($$
$$\text{ANIMATE}([\text{CURVES}([[0,0],[1.,1.]],\text{COLOUR}(RGB,0,0,0)),$$
$$\text{AXESLABELS}(\ x\ ,\ ),\text{VIEW}(0\ ..\ 1.,DEFAULT)],$$
$$[\text{CURVES}([[0,0],[1.,1.]],\text{COLOUR}(RGB,0,0,0)),$$
$$\text{CURVES}([[0,0],[1.,.5]],\text{COLOUR}(RGB,0,0,0)),$$
$$\text{AXESLABELS}(\ x\ ,\ ),\text{VIEW}(0\ ..\ 1.,DEFAULT)]))$$

As you can see the arguments of the **ANIMATE** object consist of two frames **frame1**, **frame2** and the second frame has the desired two **CURVES** objects.

The **animate** command not only produces plot structures like the ones discussed above, but also it facilitates construction of the object information by generating all the points in the **CURVES** objects from a given function of two variables. It is convenient to think of the second variable of this function as the time t. For example, if

$$F(x,t) = \frac{x}{t},$$

then the command

```
animate(F(x,t),x=0..1,t=1..2,frames=2);
```

will produce an animation with two frames. The first frame will contain a plot of $F(x,1) = x$ and the second frame will contain a plot of $F(x,2) = x/2$. This will be same as the animation in B above (except that the graphs will have the default color red). To produce essentially the same output as B, we can use

```
B2:=animate(F(x,t),x=0..1,t=1..2,frames=2,color=black,
            numpoints=2);
```

which produces the output

$$B2 := \text{PLOT}($$
$$\text{ANIMATE}([\text{CURVES}([[0,0],[1.,1.]],\text{COLOUR}(RGB,0,0,0))],$$
$$[\text{CURVES}([[0,0],[1.,.5]],\text{COLOUR}(RGB,0,0,0))]),$$
$$\text{AXESLABELS}(x,\ ),\text{VIEW}(0\ ..\ 1.,DEFAULT))$$

Generally, if you specify the number of frames as **frames=n** and the time interval is **t=a..b**, then the animation consists of the plots of $F(x, t_i)$, $i = 1, \ldots, n$, where $t_i = a + (i-1)(b-a)/(n-1)$. The ith frame is the plot of $F(x, t_i)$. For example, the command

```
B3:=animate(F(x,t),x=0..1,t=1..2,frames=10,color=black,
            numpoints=2);
```

produces the output

$B3 :=$ PLOT(ANIMATE(
[CURVES([[0, 0], [1., 1.]],COLOUR(RGB,0,0,0))],
[CURVES([[0, 0], [1., .8999999999999999]],COLOUR(RGB,0,0,0))],
[CURVES([[0, 0], [1., .8181818181818181]],COLOUR(RGB,0,0,0))],
[CURVES([[0, 0], [1., .7499999999999999]],COLOUR(RGB,0,0,0))],
[CURVES([[0, 0], [1., .6923076923076922]],COLOUR(RGB,0,0,0))],
[CURVES([[0, 0], [1., .6428571428571428]],COLOUR(RGB,0,0,0))],
[CURVES([[0, 0], [1., .5999999999999999]],COLOUR(RGB,0,0,0))],
[CURVES([[0, 0], [1., .5624999999999999]],COLOUR(RGB,0,0,0))],
[CURVES([[0, 0], [1., .5294117647058822]],COLOUR(RGB,0,0,0))],
[CURVES([[0, 0], [1., .4999999999999999]],COLOUR(RGB,0,0,0))]),
AXESLABELS(x,),VIEW(0 .. 1.,$DEFAULT$))

which is a movie with ten frames. *Note:* Since for a fixed t, the graph of $y = x/t$ is a straight line, we have used **numpoints=2** so that there are only two calculated points on the graph. In general, for graphs of other functions, you will need more points than this.

Example 7.8 (Volumes by the Slice) The above examples are very simple since it is necessary to understand the basic aspects of animations before creating a more complex one. Here we look at a more complicated animation, one that will help us visualize how certain volumes are computed by iterated integrals.

Previously in the text, we have considered the double integral

$$V = \int_D f(x, y) \, dA,$$

of a function f over various types of regions D in $\mathbb{R}^2$. If f is nonnegative on D, i.e., $f(x, y) \geq 0$, for all (x, y) in D, then V is the volume of a certain solid S. This is the solid that is bounded on the top by the surface $z = f(x, y)$, on the bottom by the plane $z = 0$, and on the sides by the surface generated by

translating the boundary of D parallel to the z-axis. Visualizing the solid
S can be di cult. This is so even in the following special case.

Consider the case when the region D is bounded by the graphs of two
functions $y = g(x)$ and $y = h(x)$, with $g(x)$ $h(x)$ for x in $[a, b]$, and the
vertical lines $x = a$, $x = b$. See Figure 7.24.

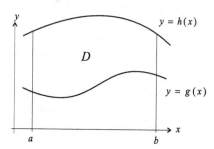

Figure 7.24: *A Type* I *region is bounded by the graphs of two functions*
$y = g(x)$, $y = h(x)$, *with* $g(x)$ $h(x)$, *and the vertical lines* $x = a$, $x = b$.

Such a region is known as a Type I region, and it is shown in calculus
that for such regions, the double integral can be computed as an iterated
integral, specifically:

$$\int_D f(x, y)\, dA = \int_a^b \left[\int_{g(x)}^{h(x)} f(x, y)\, dy \right] dx. \tag{7.4}$$

This result (or theorem) is motivated by showing how the iterated integral
measures the volume of S slice by slice. Namely, for a fixed x_0 $[a, b]$
consider the slice S_{x_0}, or section of S, which is the intersection of S and
the plane $x = x_0$. See Figure 7.25. This slice is a region in the plane
$x = x_0$ which is bounded by three straight lines and the graph of the function
y $f(x_0, y)$. Thus, the area of this region is given by

$$A(x_0) = \int_{g(x_0)}^{h(x_0)} f(x_0, y)\, dy.$$

Multiplying this by the infinitesimal dx_0 gives, heuristically, an infinitesimal
volume $dV = A(x_0)\, dx_0$, and adding all of these up (integrating) gives the
volume V. This heuristic argument motivates the iterated integral formula
(7.4) for measuring the volume.

We can make the argument precise and create graphics for the slices of
S by using Riemann sums as follows. We employ the standard partition of

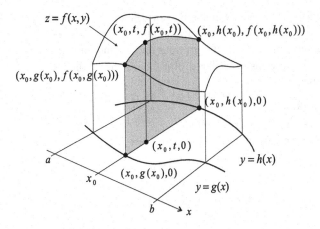

Figure 7.25: *A slice through the solid S by the plane $x = x_0$.*

$[a, b]$ into N equal subintervals: $x_i = a + i\Delta x$, where $\Delta x = (b - a)/N$. For each x_i, the slice S_{x_i} is bounded by the spacecurves:

$$
\begin{aligned}
G(t) &= (x_i, g(x_i), t), & (t \quad [0, f(x_i, g(x_i))]\,) \\
H(t) &= (x_i, h(x_i), t), & (t \quad [0, f(x_i, h(x_i))]\,) \\
L(t) &= (x_i, t, 0), & (t \quad [g(x_i), h(x_i)]\,) \\
F(t) &= (x_i, t, f(x_i, t)), & (t \quad [g(x_i), h(x_i)]\,).
\end{aligned}
$$

Here G, H, L are the straight-line segments forming the two sides and bottom of the region S_{x_i}, while F is the curve forming the top of the region. We plot each of these with the `spacecurve` command and display them together to give the image of the ith slice, which we store in the variable named `slice[i]`. The approximation to the volume V will be

$$
V_{approx} = \sum_{i=1}^{N} \left[\int_{g(x_i)}^{h(x_i)} f(x_i, y)\,dy \right] \Delta x.
$$

This will be calculated and returned through the parameter `approxvol` in the procedure. Maple's decimal approximation to the volume will be returned through the parameter `vol`.

The default output from the procedure will be a graphical figure that is one of two types, based on the input parameter `ct`. For `ct=0`, the output will be a static display of all $N + 1$ slices, which for N large will give a fairly good picture of the solid S. For `ct=1`, the output will be an animation showing the same sequence of slices and will exhibit how the solid S is built up from the slices S_{x_i}, $i = 0, 1, \ldots, N$. Each frame in the movie should display not only the ith slice, but also all the previous ones (a motion with tracks, or traces, of all the previously displayed objects). Thus, if the ith frame is named `fr[i]` then

```
fr[0]:=display3d(slice[0]);
fr[1]:=display3d({slice[0],slice[1]});
fr[2]:=display3d({slice[0],slice[1],slice[2]});
        .
        .
        .
```

The other input parameters will be the functions g, h, f and the endpoints a, b of the interval $[a, b]$.

Procedure for Animating Type I Slices

```
typeI:=proc(g,h,f,a,b,N,ct,approxvol::evaln,vol::evaln)
   local s,X,dX,slice,fr,i,j,bottom,side1,side2,top;
   with(plots,spacecurve,display3d):
   X:=array(0..N):slice:=array(0..N):fr:=array(0..N):
   s:=0;dX:=evalf((b-a)/N);
   for i from 0 to N do
     X[i]:=a+i*(b-a)/N;
     bottom:=spacecurve(
             [[X[i],g(X[i]),0],[X[i],h(X[i]),0]],color=red);
     side1:=spacecurve(
             [[X[i],g(X[i]),0],[X[i],g(X[i]),f(X[i],g(X[i]))]],
             color=black);
     side2:=spacecurve(
             [[X[i],h(X[i]),0],[X[i],h(X[i]),f(X[i],h(X[i]))]],
             color=black);
     if h(X[i])<=g(X[i]) then
       top:=spacecurve([[X[i],g(X[i]),f(X[i],g(X[i]))]])
     else
       top:=spacecurve([X[i],t,f(X[i],t)],t=g(X[i])..h(X[i]),
                     color=black);
     end if;
     slice[i]:=display3d({bottom,side1,side2,top});
     s:=s+evalf(int(f(X[i],t),t=g(X[i])..h(X[i])))*dX;
   end do:
```

```
approxvol:='approximate volume'=s;
vol:=Int(Int(f(x,y),y=g(x)..h(x)),x=a..b)=
     evalf(int(int(f(x,y),y=g(x)..h(x)),x=a..b));
if ct=0 then
   display3d({seq(slice[i],i=0..N)},axes=frame,
             orientation=[-12,80],projection=.8);
else
   for i from 0 to N do
      fr[i]:=display3d({seq(slice[j],j=0..i)})
   end do:
   display3d([seq(fr[i],i=0..N)],insequence=true,axes=frame,
             orientation=[-12,80],projection=.8);
end if;
end proc:
```

Note that in creating the top, curved part, of each slice, an `if-then-else` statement must be used to prevent errors. The possibility of errors arises in the statement

```
top:=spacecurve([X[i],t,f(X[i],t)],t=g(X[i])..h(X[i]),
                 color=black);
```

The specification of the parameter range `t=a..b` in such a plotting command must have `a<b`, otherwise you will get an error message saying that the range is empty. In the case here, we would get this message if `g(X[i])=h(X[i])` for some i, which typically can occur for $i = 0$ or $i = N$ when the curves $y = g(x)$, $y = h(x)$ intersect at $x = a$ or $x = b$.

We test the above procedure with the region D being the triangular region bounded by the graphs of $g(x) = -2 + x$ and $h(x) = 2 - x$ on the interval $[0, 2]$ and the function $f(x, y) = 4 - x^2 - y^2$, whose graph is a paraboloid, turned downwards. Figure 7.26 shows a representation of the solid S made up of $N = 20$ slices. This is the default graphic output returned by the call: `typeI(g,h,f,0,2,20,0,approxvol,vol)`. The other output can be viewed by entering `approxvol;vol;` at the prompt. This gives

$$approximate\ volume = 11.20000000$$

$$\int_{0}^{2} \left[\int_{-2+x}^{2-x} (4 - x^2 - y^2)\, dy \right] dx = 10.66666667$$

as output. The second result, which is what Maple calculates, is the exact answer $32/3$ converted to a 10-digit decimal approximation.

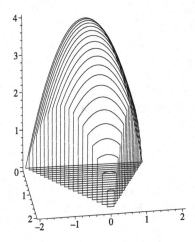

Figure 7.26: *The solid S represented by* 20 *equally spaced slices, or sections, through it.*

Exercise Set 7

1. (**Approximating Solids of Revolution**) The cylindrical disk approximations to solids of revolutions in Example 7.2 arise from step-function approximations to the function $y = f(x)$. This exercise examines a different
   approximation that arises from using polygonal approximations to f. Revolving a polygonal approximation about the x-axis creates a solid composed of
   frustra of cones, and so we call this solid a *conical frustrum approximation*
   to the solid of revolution S. Since polygonal approximations are better than
   step-function approximations, you would expect that the approximate volume of S measured by a conical frustrum approximation would be better
   than that measured by a cylindrical disk approximation. For example, Figure 7.27 shows such an approximation for the function $f(x) = e^{-x} \sin 2 \; x$ on
   the interval $[0, 2]$. (See Figure 7.1 also.) You should compare this with Figure
   7.5, which is the corresponding cylindrical disk approximation obtained from
   a step-function approximation to f.

   A *frustrum* of a cone is a solid obtained by cutting a cone with a plane parallel
   to its base and removing the portion toward its apex. See Figure 7.28.

   In this exercise you are to write a procedure like the `diskmethod` procedure in
   Example 7.2, naming it, say, `frustrummethod`. Specifically, do the following:

   (a) Figure 7.29 shows a line segment in the x-y plane joining the points
       (x_1, y_1), (x_2, y_2). When this line segment is revolved about the x-axis,
       it generates a surface that bounds a frustrum of a cone. Show that the

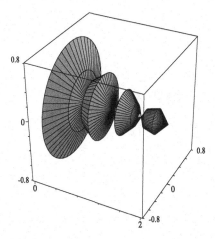

Figure 7.27: *A conical frustrum approximation to the solid of revolution S obtained from revolving $f(x) = e^{-x} \sin 2\ x$ about the x-axis.*

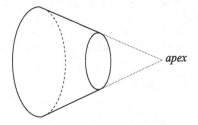

Figure 7.28: *A frustrum of a cone.*

line segment is the graph of the function

$$g(x) = \frac{y_2 - y_1}{x_2 - x_1}(x - x_1) + y_1,$$

on the interval $[x_1, x_2]$. Use this and formula (7.1) to show that the volume of the frustrum of a cone is

$$V = \frac{1}{3}(y_2^2 + y_2 y_1 + y_1^2)(x_2 - x_1). \tag{7.5}$$

Discuss how this gives the volume of a cone as a special case.

(b) For a function f on an interval $[a, b]$, approximate f by a polygonal function using a subdivision of $[a, b]$ into N equal subintervals, each of length $\Delta x = (b - a)/N$, and having subdivision points $x_i = a + i(b - a)/N$, $i = 0, \ldots, N$. The sides of the polygonal approximation

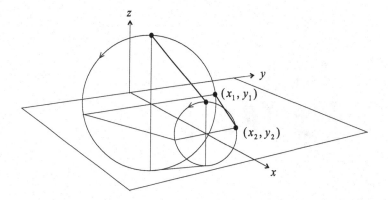

Figure 7.29: *The line segment joining* (x_1, y_1), (x_2, y_2) *generates a frustrum of a cone when revolved about the x-axis.*

have vertices (x_{i-1}, y_{i-1}) and (x_i, y_i), for $i = 1, \ldots, N$, where $y_i = f(x_i)$. The solid S obtained by revolving f about the x-axis is then approximated by a conical frustrum approximation, namely, the solid generated by revolving the polygonal approximation about the x-axis. Show that the canonical frustrum approximation has volume

$$V_N = \sum_{i=1}^{N} \frac{\pi}{3} \left[f(x_i)^2 + f(x_i)f(x_{i-1}) + f(x_{i-1})^2 \right] \Delta x.$$

Use this to argue that the volume of S is

$$V = \lim_{N \to \infty} V_N = \pi \int_{a}^{b} f(x)^2 \, dx,$$

which is the same as that given by the disk method.

(c) Write a procedure like **diskmethod** that computes the volume V_N of the conical frustrum approximation and plots the conical frustrum approximation. For the plot, note that the surface of the ith frustrum is parametrized by

$$F(\theta, x) = (x, g_i(x) \cos \theta, g_i(x) \sin \theta),$$

where $g_i(x)$ is the function whose graph is the ith line segment. See Figure 7.29.

(d) Test your procedure on the following functions. Use several values of N, both small and large. Compare the approximate volumes and figures with the corresponding ones produced by **diskmethod**, which is the cylindrical disk approximation. Also compute, by hand, the exact volume V of S and judge which approximation, cylindrical disks or conical frustra, gives the better approximation scheme.

(i) $f(x) = x$, on $[0, 2]$.

(ii) $f(x) = \cos x$, on $[0, \ /2]$.

(iii) $f(x) = e^{-x} \sin 2\ x$, on $[0, 2]$.

2. **(The Shell Method)** Consider the solid of revolution S obtained by revolving the graph of $y = f(x)$, for $x \quad [a, b]$, about the y-axis. For such a solid, the *shell method* is the standard method for calculating the volume of S. In cases where we can solve for x as a function of y, we can also use the disk method to compute the volume, but even in these cases it is often easier to use the shell method.

The volume formula used in the shell method is based on approximating S by *cylindrical shells*. Specifically, we first approximate f by a step function in one of the standard ways, say, with equal subintervals $[x_{i-1}, x_i]$, $\Delta x = (b-a)/N$, and right-hand endpoints $y_i = f(x_i)$, as in Example 7.2. Viewing the step-function approximation as consisting of rectangles as shown in Figure 7.30, we revolve each rectangle about the y-axis to create a cylindrical shell.

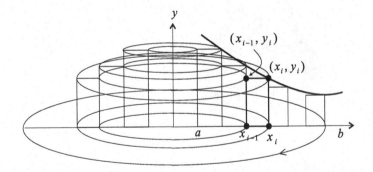

Figure 7.30: *The cylindrical shells obtained by revolving the rectangles in a step-function approximation about the y-axis.*

Each shell is bounded by the cylinders generated by the vertical sides of the rectangle. The collection of these cylindrical shells comprises a solid, as shown, that approximates the solid S, and we can approximate the volume of S by measuring the volume of the cylindrical shells and adding these up.

This exercise is to develop the cylindrical shell approximation scheme and produce graphics much as was done in Example 7.2 for the disk method. In fact, you can easily modify the code in the `diskmethod` procedure to handle the shell method here. Specifically:

(a) Argue that the ith cylindrical shell has volume $2\pi x_i f(x_i)\Delta x$ and thus that the cylindrical shell approximation to S has volume

$$V_N = \sum_{i=1}^{N} 2\pi\, x_i f(x_i)\,\Delta x.$$

Use this to argue that the volume of S is given by

$$V = \int_a^b 2\pi\, xf(x)\,dx.$$

(b) Modify the `diskmethod` procedure in Example 7.2, so that it calculates the volume of a cylindrical shell approximation and draws a graphic of this approximation. For the graphic, just plot each of the cylinders of radius x_i, height $y_i = f(x_i)$, for $i = 1,\dots,N$, and centered on the y-axis. See Figure 7.31.

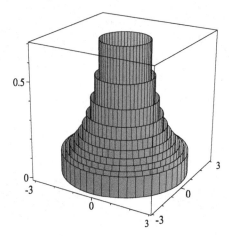

Figure 7.31: *The outer cylinders in a cylindrical shell approximation. Here* $f(x) = x^{-2}$ *and* $[a,b] = [1/2, 2]$.

This will only give an indication of what the approximating solid looks like. However, for N large, each cylindrical shell is very thin ($\Delta x \to 0$), and so the graphic will be more representative. *Note*: You will probably want to change the orientation used in `diskmethod` to something more appropriate for the shell method. Also it would be nice to add your own axes and labels to make the graphic read well.

(c) Test your procedure on the following functions, using several values of N, both large and small. Also calculate, by hand, the exact volume of S and compare this with the approximate values that your procedure returns.

(i) $f(x) = x + x^{-1}$, on $[0.1, 2]$.

(ii) $f(x) = \cos x$, on $[0, \ /2]$

(d) Discuss, as in Exercise 1, how the use of polygonal approximations to f can result in better approximations to the solid of revolution than the ones obtained by the shell method. Write a procedure to implement this. It will be entirely similar to the `frustrummethod` procedure you wrote in Exercise 1.

3. The graphic output produced by the code in Example 7.2 consists of cylindrical disks, with each disk represented by the lateral surface of the cylinder. Improve the graphics by adding a top and bottom to each cylindrical disk. One way to do this is to use the `spacecurve` command to generate the boundary circles for the tops and bottoms. Then pass this information (the points generated by `spacecurve`) to `polygonplot3d` which will render the interiors of the circles in color.

4. The procedure `surfplot` in Example 7.6 is based on a triangulation of R having the form shown in Figure 7.13. As shown in the figure, the triangles are produced by dividing each subrectangle in half using a diagonal going from upper left to lower right. Suppose we use the other choice of diagonal lower left to upper right to produce the triangulation. Based on this choice, alter the code in `surfplot` to produce the corresponding triangulation of the surface. This should be relatively easy with a little thought, however, be sure to describe the indexing scheme for the new diagonals and include diagrams.

5. **(Distance to a Curve)** Write a procedure to compute the approximate distance between a given point P and a curve C that is given parametrically by

$$F(t) = (\, g(t), \, h(t) \,),$$

for $t \quad [a, b]$. Note a special case of this is when C is the graph of a function f; then the parametrization is $F(x) = (x, f(x))$. Have your procedure produce output like the one in Example 7.5. Also do the following:

(a) Test your procedure on the curves

(i) $F(t) = (\, t^3 - 3t, \, t^2 - 1 \,)$, for $t \quad [-2, 2]$. Do two cases: $P = (1, 0)$ and $P = (0, 1)$.

(ii) The graph of $f(x) = x^2$, for $x \quad [-2, 2]$. Do two cases $P = (0, 1)$ and $P = (0, 1/2)$.

(b) Use calculus to derive the exact distances from the point to the curve in the situations in Part (a). Show your work.

6. **(Distance to a Set)** In advanced calculus and other math courses, you may encounter the notion of the distance $d(P, S)$ between a point P and a set S. This is the least of all the distances between P and points Q in S. When S is a finite set, $d(P, S)$ can be calculated easily on a computer. In this exercise do the following:

(a) Write a procedure that computes $d(P, S)$, where S is a finite set of points in $\mathbb{R}^2$. Use either a set or list data structure for S, whichever you prefer. Produce graphical output like that in Example 7.5.

(b) Write a procedure that computes the approximate distance between a plane curve C, and a point P. Design your procedure to work like this. Have one of the input parameters signal whether the curve is given parametrically or implicitly, and based on this compute a set S of vertices for a polygonal approximation to the curve. Then use the procedure you wrote in Part (a).

(c) Test the procedure in Parts (a) and (b) on suitable, and interesting, input data.

7. (**Surface Area**) This exercise is the two-dimensional analog of Exercise 4 in Chapter 3. That exercise developed an approximation for the length of a curve that is the graph of a function of one variable.

Write a procedure to compute the approximate surface area for a surface which is the graph of a function $f : R \quad \mathbb{R}$ defined on a rectangle $R = [a, b] \times [c, d]$. Base your approximation scheme on the method developed in Example 7.6, to approximate graphs of functions by triangular complexes. Specifically, let T_{ij}, T_{ij} be the two triangles in the complex as shown in Figure 7.14. Your procedure should compute the approximating area to be the sum of the areas of all these triangles, i.e.,

$$\mathcal{A}_{N,M}(f) = \sum_{i=1}^{N} \sum_{j=1}^{M} (A_{ij} + A_{ij}), \tag{7.6}$$

where A_{ij}, A_{ij} are the areas of T_{ij}, T_{ij}, respectively. Here are some further directions and parts to the exercise:

(a) Use vector methods, in particular the cross product, to show that the area A_{ij} of T_{ij} is

$$\frac{1}{2} \sqrt{1 + \left(\frac{f(x_i, y_{j-1}) - f(x_{i-1}, y_{j-1})}{\Delta x_i}\right)^2 + \left(\frac{f(x_i, y_j) - f(x_i, y_{j-1})}{\Delta y_j}\right)^2} \, \Delta x_i \, \Delta y_j, \tag{7.7}$$

where $\Delta x_i = x_i - x_{i-1}$ and $\Delta y_j = y_j - y_{j-1}$. Derive a similar formula for A_{ij}. *Note*: In your procedure, you can use these formulas or just code the vector methods by which they were derived.

(b) Extend the code in the procedure `surfplot` from Example 7.6, so that in addition to producing the plot of the triangular complex, it also calculates the surface area $\mathcal{A}_{N,M}(f)$ of this approximation.

(c) Test your procedure on $f(x, y) = 4x/3$ on $[0, 3] \times [0, 1]$, for which you know the exact surface area. Does your procedure give the exact answer? Why?

(d) Use your procedure to approximate the surface area of the hyperbolic paraboloid $f(x, y) = x^2 - y^2$ on $R = [-1, 1] \times [-1, 1]$. Use several appropriate choices of N, M, from small to large (be careful of memory requirements for large values). Show the approximate values of the surface area along with the approximating triangular meshes.

(e) Use your procedure to approximate the surface area of graph of the function $f(x, y) = \frac{2}{3}(x^{3/2} + y^{3/2})$ on $R = [0, 1] \times [0, 1]$. Use several appropriate choices of N, M, from small to large (be careful of memory requirements for large values). Show the approximate values of the surface area along with the approximating triangular meshes.

(f) Use formula (7.6) together with formula (7.7) (and the similar one for A_{ij}), to describe how, in the limit, the formula for the *exact* surface area is given by the double integral:

$$A(f) = \int_c^d \int_a^b \sqrt{1 + f_x(x, y)^2 + f_y(x, y)^2}\, dx\, dy. \qquad (7.8)$$

(g) Use Formula (7.8) to find the exact surface areas for the graphs of the functions in parts (d) and (e). For (d) you will have to use Maple (so the answer is only a good approximation), but for (e) do the work by hand (turn in your work). Along with the values of the areas, also plot the graphs of the functions.

8. **(Triangulating General Surfaces)** Extend the code in the procedure surfplot from Example 7.6, so that it applies to a surface given parametrically by a parameter map $F : R \to \mathbb{R}^3$:

$$F(s, t) = (\, f(s, t),\, g(s, t),\, h(s, t)\,),$$

where $(s, t) \in R = [a, b] \times [c, d]$. Apply the procedure to produce triangulations of the cone, cylinder, sphere, torus, and Möbius strip as given in the Maple/Calculus Notes at the end of this chapter. Use various choices of N, M from small to larger (5 to 25) and compare, qualitatively, the goodness of the approximations of the respective triangulations of these surfaces.

9. **(Surface Area in General)** Extend the procedure you wrote in Exercise 7 for approximating surface areas of surfaces that are graphs to the general case of surfaces that are given parametrically. Assume (cf. Exercise 8) that the surface is parametrized by a map $F : R \to \mathbb{R}^3$:

$$F(s, t) = (\, f(s, t),\, g(s, t),\, h(s, t)\,),$$

where $(s, t) \in R = [a, b] \times [c, d]$. As in Exercise 8 and Example 7.6, the triangulation of the surface consists of triangles with vertices

$$Q_{ij} = F(s_i, t_j) = (\, f(s_i, t_j),\, g(s_i, tj),\, h(s_i, t_j)\,),$$

for $i = 0, 1, \ldots, N$, $j = 0, 1, \ldots, M$. Specifically, let T_{ij}, T_{ij}, denote the triangles which approximate the part of the surface S corresponding to the rectangle R_{ij} in the parameter domain. Then $Q_{i-1,j-1}, Q_{i,j-1}, Q_{i,j}$ are the vertices of the triangle T_{ij} and $Q_{i-1,j-1}, Q_{i-1,j}, Q_{i,j}$ are the vertices of T_{ij}. Your procedure should compute the approximating area to be the sum of the areas of all these triangles, i.e.,

$$\mathcal{A}_{N,M}(F) = \sum_{i=1}^{N} \sum_{j=1}^{M} (A_{ij} + A_{ij}), \tag{7.9}$$

where A_{ij}, A_{ij} are the areas of T_{ij}, T_{ij}, respectively. Here are some further directions and parts to the exercise:

(a) Use vector methods, in particular the cross product, to show that the area A_{ij} of T_{ij} is

$$\frac{1}{2} \sqrt{\left| \frac{F(s_i, t_{j-1}) - F(s_{i-1}, t_{j-1})}{\Delta s_i} \right|^2 + \left| \frac{F(s_i, t_j) - F(s_i, t_{j-1})}{\Delta t_j} \right|^2} \, \Delta s_i \, \Delta t_j, \tag{7.10}$$

where $\Delta s_i = s_i - s_{i-1}$ and $\Delta t_j = t_j - t_{j-1}$. Derive a similar formula for A_{ij}.

(b) Extend the code in Exercise 8, so that in addition to producing the plot of the triangular complex, it also calculates the surface area $\mathcal{A}_{N,M}(F)$ of this approximation.

(c) Test your procedure on the cylinder $F(\theta, z) = (r \cos \theta, r \sin \theta, z)$, for $(\theta, z) \in [0, 2\pi] \times [0, h]$. This is a cylinder with radius r and height h, for which you know the exact surface area is $2\pi rh$. Does your procedure give the exact answer? Why?

(d) Use your procedure to approximate the surface areas of the Möbius strip as defined in the Maple/Calculus Notes. Use several appropriate choices of N, M, from small to large (be careful of memory requirements for large values). Show the approximate values of the surface area along with the approximating triangular meshes.

(e) Use your procedure to approximate the surface areas of the sphere and torus as given in the Maple/Calculus Notes. Use several appropriate choices of N, M, from small to large (be careful of memory requirements for large values). Show the approximate values of the surface area along with the approximating triangular meshes.

(f) Use formula (7.9) together with formula (7.10) (and the similar one for A_{ij}) to describe how, in the limit, the formula for the *exact* surface area is given by the double integral:

$$\mathcal{A}(F) = \int_c^d \int_a^b \left| \frac{\partial F}{\partial s}(s, t) \times \frac{\partial F}{\partial t}(s, t) \right| \, ds \, dt. \tag{7.11}$$

Here

$$\frac{F}{s}(s,t) = \frac{f}{s}(s,t),\ \frac{g}{s}(s,t),\ \frac{h}{s}(s,t)$$

$$\frac{F}{t}(s,t) = \frac{f}{t}(s,t),\ \frac{g}{t}(s,t),\ \frac{h}{t}(s,t)$$

are the vectors that result from the respective partial derivatives of the component functions of F, and the vertical bars denote the length of the cross product vector.

(g) Use Formula (7.11) to find the exact surface areas of the surfaces in parts (d) and (e). For (d) you will have to use Maple (so the answer is only a good approximation), but for (e) do the work by hand (turn in your work). Along with the values of the areas also plot the graphs of the functions.

10. Write a procedure that animates the plot of a function $y = f(x)$ on an interval $[a, b]$, i.e., shows a movie of it being drawn as x goes from a to b. Specifically for the standard partition points $x_i = a + i\Delta x$, $i = 0, \ldots, N$, let L_i be the line segment joining $(x_{i-1}, f(x_{i-1}))$ and $(x_i, f(x_i))$ (i.e., the ith side in the polygonal approximation). The first frame in the movie should show L_1. The second frame should show L_1 and L_2. Generally, the kth frame should show $L_1, L_2, \ldots, L_k$.

11. Write a procedure that animates the plot of a parametrized curve $F(t) = (f(t), g(t), h(t))$, t $[a, b]$, in $\mathbb{R}^3$, i.e., shows it being drawn dynamically, line segment by line segment. This will be like the procedure in the last exercise.

12. Alter the procedure `diskmethod` in Example 7.2 so that its output is one of three types based on an input parameter `k`. For `k=0`, the output is the static plot of the cylindrical disk approximation (the current output). For `k=1`, have the output be an animation showing each cylindrical disk, `disk[i]`, being rendered one after the other, for `i=1..N`. Each disk should be rendered by showing its cylindrical surface being swept out as   moves through the 25 equally spaced angles on $[0, 2\ ]$. For `k=2`, have the output be an animation like that for `k=1`, but now have all N of the disks be rendered simultaneously as they are being swept out. Be aware of memory usage that could occur from using large values of N.

13. Alter the `frustrummethod` procedure in Exercise 1 above so that it produces either a static picture or two types of animations like those described in the last exercise for the `diskmethod`.

14. Alter the `surfplot` procedure in Example 7.6 so that its output is an animation showing the surface being rendered, line segment by line segment. The movie should show the vertical segments being drawn in succession, then the horizontal segments, and finally the diagonal segments.

15. The double integral of a function f over a Type II region can be expressed as

$$\iint_L f(x,y)\,dA = \int_a^b \left[\int_{g(y)}^{h(y)} f(x,y)\,dx \right] dy. \qquad (7.12)$$

Alter the `typeI` procedure in Example 7.8 to obtain a procedure `typeII` that produces similar output for double integrals over Type II regions.

16. Consider a solid S bounded by the graphs of two functions f_1, f_2 over a Type I region D, where $f_1(x,y)$ $f_2(x,y)$, for all (x,y) D. In calculus you learned that the volume of such a solid is given by

$$V = \int_a^b \left[\int_{g(x)}^{h(x)} f_1(x,y) - f_2(x,y) \;\; dy \right] dx. \qquad (7.13)$$

Write a procedure, like `typeI`, to approximate the solid S by slices and give similar output as that in `typeI`.

17. Write a program that will take the object information in the `MESH` object of a `PLOT3D` data structure and draw in all the triangles comprising the approximating surface.

18. Plot the standard surfaces, a cone, a cylinder, a sphere, a torus, and a Möbius strip, as given by the parametrizations in the Maple/Calculus Notes. In each case you will have to choose particular values for the constants that occur, such as r, h, a, b. Use constrained scaling so that the plots are not distorted. Describe the nature of the coordinate curves. Label these on the printout of the surface. Define the coordinate curves and in a separate picture combine the plots of 10 to 15 of each type of coordinate curve. If you choose these judiciously, the picture should look like the surface.

19. For each of the following that you are assigned: (i) Plot the two implicitly defined surfaces using `implicitplot3d` and display them in that same figure. Use a constrained scaling and choose the ranges `x=a..b,y=c..d,z=e..f` judiciously so that you get a good view of the curve C that is the intersection of the two surfaces. Use the default and also the `patchcontour` styles. Print out several views and trace the curve C on the printout. (ii) Find a explicit parameterizations of the various pieces of the curve C, using z as the parameter. Determine the domains for each of these parameter maps and then plot, in the same figure, each of the pieces of the curve C using the `spacecurve` command. Combine this figure with the figure showing the two surfaces.

   (a) $z^2 = x^2 + y^2$ and $y^2 + z^2 = 1$.

   (b) $x^2 + y^2 = 1$ and $z^2 - x^2 - y^2 = 1$

7.6   Maple/Calculus Notes

We discuss two mathematical ways of representing curves and surfaces in
$\mathbb{R}^3$. The first is the parametric representation, which represents a curve or
surface by a vector-valued function of one or two variables, respectively. The
second is the implicit representation, which describes the curve or surface
as the solution set of a system of one or two equations, respectively.

The discussion here is similar to that for plane curves given in the Chapter 2 exercises and Maple/Calculus Notes.

7.6.1   Parametric Representations of Curves and Surfaces

The parametric representations of curves and surfaces in mathematics and
Maple have the same forms   they are parameter maps $F : U \quad \mathbb{R}^3$, of one
or two variables. The plots use the **spacecurve** and **plot3d** commands,
respectively.

Curves: A curve C in $\mathbb{R}^3$ is given parametrically if each point (x, y, z) on
the curve can be expressed by

$$x = f(t), \ y = g(t), \ z = h(t),$$

where $f, g, h : [a, b] \quad \mathbb{R}$ are three real-valued functions of a real variable
$t \quad [a, b]$, called the *parameter*. Corresponding to such a parametrization is
a vector-valued function $F : [a, b] \quad \mathbb{R}^3$, defined by

$$F(t) = (\, f(t), \, g(t), \, h(t)\,).$$

Thus, F associates to each value t of the parameter, a point $F(t)$ in $\mathbb{R}^3$ that
lies on the curve C. As t varies through $[a, b]$, the point $F(t)$ traces out the
curve C in $\mathbb{R}^3$. The function (or map) F is called a *parameter map*, or a
parametrization of the curve.

A curve that is given parametrically is easily defined and plotted in
Maple. The arrow notation can be used to define the parameter map,

```
F:=t->[f(t),g(t),h(t)];
```

Note that in Maple, the value `F(t)` of `F` is a list, not a 3-tuple as it is in mathematics. With `F` defined, the curve can be plotted using the **spacecurve**
command, which is part of the **plots** package:

```
with(plots,spacecurve);
spacecurve(F(t),t=a..b);
```

Alternatively, you can use the following to get the same plot:

```
with(plots,spacecurve);
spacecurve([f(t),g(t),h(t)],t=a..b);
```

Note that in either case the default number of points plotted is $N = 50$. This means that Maple plots the points $F(t_i)$, $i = 0, \ldots, N$, where $t_i = a + i(b-a)/N$, and joins these $N + 1$ points with line segments. Thus, the result is an N-sided polygonal approximation to the curve. To get an approximation that appears smoother, you can plot more points on the curve by using the option **numpoints=N**, where N is the number of desired points.

Surfaces: A surface S in $\mathbb{R}^3$ is given parametrically if each point (x, y, z) on the surface can be expressed by

$$x = f(s,t), \; y = g(s,t), \; z = h(s,t), \qquad (7.14)$$

where $f, g, h : [a, b] \times [c, d] \quad \mathbb{R}$ are three real-valued functions of two real variables $s \quad [a, b]$ and $t \quad [c, d]$. The variables s, t are called the *parameters* and $R = [a, b] \times [c, d]$ is called the *parameter domain.*

Generally, since a surface is two-dimensional, it takes two parameters to determine the points on the surface. On the other hand, as we have seen, the points on a curve can be described in terms of one parameter since a curve is one-dimensional.

Corresponding to the parametrization (7.14) is a vector-valued function $F : [a, b] \times [c, d] \quad \mathbb{R}^3$, defined by

$$F(s,t) = (\, f(s,t), \, g(s,t), \, h(s,t) \,).$$

Thus, F associates to an ordered pair (s, t) in the parameter domain, a point $F(s, t)$ in $\mathbb{R}^3$ that lies on the surface S. The function (or map) F is called a *parameter map*, or a *parametrization of the surface.*

It is important to view the parameter map $F : R \quad S$ as transforming the parameter domain R, which is a "at surface, into the surface S, which generally is curved. In particular, each horizontal or vertical line in the parameter domain gets transformed into a curve on the surface S. These curves are called the *coordinate curves.* Specifically, for (s_0, t_0) in R, we get two maps that parametrize coordinate curves on the surface:

$$G_1(s) \quad (\, f(s,t_0), \, g(s,t_0), \, h(s,t_0) \,)$$
$$G_2(t) \quad (\, f(s_0,t), \, g(s_0,t), \, h(s_0,t) \,),$$

for $s \in [a,b]$ and $t \in [c,d]$. The curves given by these maps are called coordinate curves because G_1 arises from F by fixing the second parameter, or coordinate, $t = t_0$, and allowing the first coordinate s to vary. Similarly, G_2 arises from F by fixing the first coordinate $s = s_0$ and varying the second.

The coordinate curves are images of horizontal and vertical lines in the parameter domain R, and thus if we divide R into a grid by horizontal and vertical lines, then the image of this grid will be a curved grid on the surface. This curved grid enables us to visualize the surface and is what underlies Maple's method for plotting surfaces.

Example 7.9 (Some Standard Surfaces) The following is a list of parameter maps $F : R \to \mathbb{R}^3$ for some of the standard surfaces in $\mathbb{R}^3$.

(a) **(A Cone)**

$$F(\theta,t) = (t\cos\theta,\, t\sin\theta,\, mt), \qquad (7.15)$$

for $(\theta,t) \in [0,2\pi] \times [0,r]$. Here $m = h/r$, where h is the height of the cone and r is the radius of the base. This cone has its apex at the origin and axis along the z-axis.

(b) **(A Cylinder)**

$$F(\theta,z) = (r\cos\theta,\, r\sin\theta,\, z), \qquad (7.16)$$

for $(\theta,z) \in [0,2\pi] \times [0,h]$. Here h is the height of the cylinder and r is the radius of its base.

(c) **(A Sphere)**

$$F(\phi,\theta) = (r\sin\phi\cos\theta,\, r\sin\phi\sin\theta,\, r\cos\phi), \qquad (7.17)$$

for $(\phi,\theta) \in [0,2\pi] \times [0,\pi]$. Here r is the radius of the sphere. Its center is at the origin.

(d) **(A Torus)**

$$F(\theta,\phi) = ((a+b\cos\phi)\cos\theta,\, (a+b\cos\phi)\sin\theta,\, b\sin\phi), \qquad (7.18)$$

for $(\theta,\phi) \in [0,2\pi] \times [0,2\pi]$. Here $a > b$ are constants.

(b) **(A Möbius Strip)**

$$F(\theta, t) = \left((a + t\cos(\theta/2))\cos\theta,\ (a + t\cos(\theta/2))\sin\theta,\ t\sin(\theta/2) \right),$$

(7.19)

for $(\theta, t) \in [0, 2\pi] \times [-b, b]$. Here $a > b$ and $2b$ is the width of the strip.

In each of these surfaces, the parameters θ, ϕ, and t have certain geometric meanings. There are exercises in this chapter that examine this and help you understand the nature of these parametrizations. For example, Figure 7.32 shows a plot of a sphere and how the coordinate curves are the images of certain horizontal and vertical lines in the parameter domain.

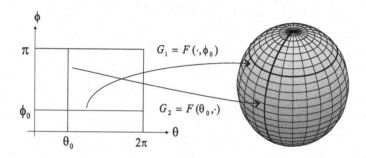

Figure 7.32: *Plot of a sphere showing the respective coordinate curves on the surface.*

In particular, each horizontal line gets mapped by the parameter map F onto a *latitudinal circle* on the sphere. Each vertical line gets mapped onto half of a *longitudinal circle* on the sphere. These are the coordinate curves in this case.

Plotting parametrically defined surfaces in Maple is analogous to what we did above for curves. The arrow notation can be used to define the parameter map,

```
F:=(s,t)->[f(s,t),g(s,t),h(s,t)];
```

With F defined, the surface can be plotted using the plot3d command as follows:

```
plot3d(F(s,t),s=a..b,t=c..d);
```

Alternatively, you can use the following to get the same plot:

```
plot3d([f(s,t),g(s,t),h(s,t)],s=a..b,t=c..d);
```

In each case, Maple uses a default grid of `grid=[25,25]`. This means that $[a,b]$ is divided into $N = 24$ subintervals of length $\Delta s = (b-a)/N$ and that $[c,d]$ is divided into $M = 24$ subintervals of length $\Delta t = (d-c)/M$. This gives subdivision points

$$s_i = a + i\Delta s, \qquad t_j = c + j\Delta t,$$

$i = 0,\ldots,N$, $j = 0,\ldots,M$, and a corresponding grid of 25 horizontal lines and 25 vertical lines. Thus, there is a corresponding curved grid on the surface consisting of 50 coordinate curves. In the above `plot3d` commands, the default `style=patch` will show the polygonal approximations to these coordinate curves. These are the polygons with vertices:

$$F(s_0,t_j), F(s_1,t_j),\ldots,F(s_{24},t_j)$$
$$F(s_i,t_0), F(s_i,t_1),\ldots,F(s_i,t_{24}),$$

for $i,j = 0,\ldots,24$. In addition to showing these polygons, Maple also fills in the grid elements with a colored   patch.

7.6.2 Implicit Representations of Curves and Surfaces

We discuss surfaces first, since their implicit representations are simpler.

Surfaces: The second standard way to represent a surface S in $\mathbb{R}^3$ is by a single equation

$$H(x,y,z) = 0,$$

where H is a real-valued function of three variables. More precisely, S is the following set of points:

$$S = \{(x,y,z) \quad \mathbb{R}^3 \mid H(x,y,z) = 0\}.$$

This is known as an *implicit* representation (or description) of the surface, since the points (x,y,z) on the surface are not given explicitly, but rather must be calculated from the defining equation $H(x,y,z) = 0$.

For example a sphere of radius 1 with center at the origin is given implicitly by

$$x^2 + y^2 + z^2 - 1 = 0.$$

Some standard points on the sphere are $(\pm 1,0,0), (0,\pm 1,0)$, and $(0,0,\pm 1)$, which is easily seen by verifying that each satisfies the defining equation.

Other points can be found in various ways. For example, $(x, 1/2, 1/2)$ will lie on the sphere if x satisfies $x^2 + 1/4 + 1/4 - 1 = 0$, i.e., if $x^2 - 1/2 = 0$. Hence, we find two points $(\pm 1/\sqrt{2}, 1/2, 1/2)$ on the sphere by this method.

As you would expect, Maple and other software packages have a tougher time rendering surfaces when they are given implicitly. The parametric representations are quicker and look better, primarily because no equations have to be solved to generate points on the surface.

To plot the surface $H(x, y, z) = 0$ in Maple, use an `implicitplot3d` such as

```
implicitplot3d(H(x,y,z)=0,x=a..b,y=c..d,z=e..f);
```

Generally, it helps to know something about the surface in advance so that the ranges `x=a..b,y=c..d,z=e..f` can be chosen judiciously. For example, using `x=-0.5..0.5,y=-0.5..0.5,z=-0.5..0.5` as ranges in plotting the sphere mentioned above, will give you nothing (the viewing box is too small). Also, choosing a finer grid size will give a more accurate approximation of the surface. (But this is at the expense of more time and memory taken to compute it. Additionally, for a very, very fine grid, the plots can become overrendered, being a dense mass of black lines.)

Curves: The implicit representation of a curve C in $\mathbb{R}^3$ involves two equations

$$H_1(x, y, z) = 0$$
$$H_2(x, y, z) = 0,$$

where H_1, H_2 are real-valued functions of three variables. More precisely, the curve C is the following set of points:

$$C = \{ (x, y, z) \in \mathbb{R}^3 \mid H_1(x, y, z) = 0, \ H_2(x, y, z) = 0 \}.$$

This is known as an *implicit* representation (or description) of the curve, since the points (x, y, z) on the curve are not given explicitly, but rather must be calculated from the defining equations $H_1(x, y, z) = 0$, $H_2(x, y, z) = 0$.

Geometrically, an implicit representation of a curve specifies it as the curve of intersection of two implicitly defined surfaces in $\mathbb{R}^3$, namely, the surfaces with equations $H_1(x, y, z) = 0$ and $H_2(x, y, z) = 0$, respectively.

A standard special case of this, discussed in the Chapter 5 exercises, is a straight line in $\mathbb{R}^3$, viewed as the intersection of two planes. For example, the solution set of the simultaneous linear system of equations

$$3x - 2y + z - 1 = 0$$
$$x + y - z - 2 = 0.$$

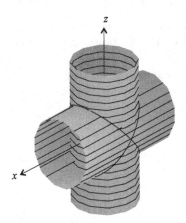

Figure 7.33: *Plot of two cylinders, $x^2 + y^2 - 1 = 0$ and $y^2 + z^2 - 1 = 0$, intersecting to give an implicitly defined curve C.*

is a certain line in $\mathbb{R}^3$. Using standard methods, you can express the solutions in terms of an arbitrary parameter, and this gives the parametric representation of the line that is the intersection of the two planes: $3x - 2y + z - 1 = 0$ and $x + y - z - 2 = 0$, respectively.

A similar thing happens in general. The nonlinear system $H_1(x, y, z) = 0$, $H_2(x, y, z) = 0$, has C for its solution set. If algebraic techniques permit, we can express C in parametric form. For example, consider the two cylinders given implicitly by the two equations

$$
\begin{aligned}
x^2 + y^2 - 1 &= 0 \\
y^2 + z^2 - 1 &= 0
\end{aligned}
$$

The solution set C of this nonlinear system of equations is the curve of intersection formed by the cylinders. Algebraically, we can write these equations as $x = \pm\sqrt{1 - y^2}$ and $z = \pm\sqrt{1 - y^2}$. Hence, with y as the parameter, we get four explicitly defined curves (one for each combination of $\pm$ signs) that make up the solution set C.

It is a simple matter to use Maple's `implicitplot3d` command to plot each of the cylinders and also the `spacecurve` command to plot the four curves comprising C. This is shown in Figure 7.33.

Generally you can always plot both surfaces in the same picture, and this will give you an indication of what the curve C of intersection looks like. But most often C is hard to visualize just from the two surface plots and so being able to parametrize C and then plot C will help. However, it is most often the case that you cannot solve the system $H_1(x, y, z) = 0$, $H_2(x, y, z) = 0$, to obtain an explicit representation of the solution set C. Thus, you are stuck. As a major programming project you could write your own code to generate and plot approximate solutions of the system $H_1(x, y, z) = 0$, $H_2(x, y, z) = 0$. This will not be easy and will require some research on your part.

Chapter 8

Recursion

In this chapter we discuss the concept of recursion as it occurs both in mathematics and in programming. There are many areas of mathematics that use recursion, or recurrence relations, as a fundamental element of the theory. We will look at a select few of these in the sections below.

In programming, recursion is often used to accomplish repetition, or looping, when the algorithm seems most clearly and easily expressed in this way. Many sorting algorithms are naturally constructed using recursion, and we will examine a some of these in the discussion and exercises.

On the other hand, recursive programs can be rewritten and expressed solely in terms of do loops, but often this is more cumbersome and unnatural. The examples below will show you how to accomplish certain tasks both with and without recursion.

Recursion is a powerful tool in programming, even though it can be confusing to master and ine cient if not used properly. After study and practice you will find it indispensable in many of your programming projects.

8.1 Recurrence Relations - Series Solutions

A prominent area of mathematics where recursion arises is in the theory of *recurrence relations*. A recurrence relation is a formula from which the terms in a sequence $\{a_n\}_{n=0}$ are determined, at each step, from one or more of the preceding terms. For example, suppose $a_0 = p$ and $a_1 = q$ are arbitrary, but all the other terms of the sequence are required to satisfy:

$$a_n = -\left[\frac{3(n-1)a_{n-1} + (n+1)a_{n-2}}{2(n-1)n}\right], \tag{8.1}$$

$n = 2, 3, \ldots$. This says that a_n is determined from the previous two terms, a_{n-1}, a_{n-2}, in the sequence. But each of these is likewise determined from

the two terms that immediately precede it, and so forth, until we run all
the way back (recur) to a_0, a_1. For example, to compute a_5 use recurrence
relation (8.1) with $n = 5$ to get

$$a_5 = -\frac{12a_4 + 6a_3}{40}.$$

To find a_4 and a_3 use (8.1) again:

$$a_4 = -\frac{9a_3 + 5a_2}{24}$$
$$a_3 = -\frac{6a_2 + 4a_1}{12}.$$

Finally we need a_2, which is

$$a_2 = -\frac{3a_1 + 3a_0}{4}.$$

Thus, if we know the values for a_0, a_1, then we can use the above to find a_5.

Equation (8.1) is a standard example of a recurrence relation and from
it the sequence $\{a_n\}_{n=0}$ is uniquely determined, up to its first two terms
$a_0 = p, a_1 = q$.

Often it is possible to solve the recurrence relation and express the
general nth term a_n explicitly as a function of p and q. However, in general,
one can view the specifications: $a_0 = p$ and $a_1 = q$, together with relation
(8.1) as *defining* the sequence a_n as a function of p and q.

In programming languages that allow procedures (subprograms, functions, etc.) to call themselves, the recursive definition of the sequence
$\{a_n\}_{n=0}$ above is very simple and natural. In Maple we could define the
sequence by the following procedure:

```
a:=proc(n)
  global p,q;
  if n=0 then p
  elif n=1 then q
  else  -(3*(n-1)*a(n-1)+(n+1)*a(n-2))/(2*(n-1)*n)
  end if;
end proc:
```

This is a recursive definition of a as a procedure since in the body of the
procedure a refers to itself twice through the expressions $a(n-1)$ and $a(n-2)$.
Having defined a this way, which makes it a function of n explicitly and

the global variables p, q implicitly, we can compute any particular value by functional evaluation: $a(n)$. For example, a(5) returns the value

$$-\frac{31q}{640} - \frac{39p}{640}.$$

It is instructive to think about all the recursive calls that are made when we ask for an evaluation such as $a(5)$. Figure 8.1 shows a tree diagram of the process.

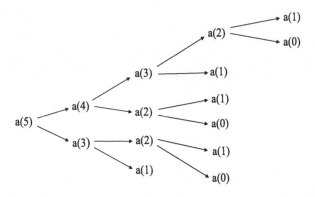

Figure 8.1: *The various recursive calls made to the procedure a when the evaluation of $a(5)$ is made.*

As you can see in the diagram, some evaluations (like $a(2)$) are done more than once. It is for this reason that recursive procedures *can be* ine - cient and memory consumptive. For example, evaluating $a(50)$ will take 20 minutes or more, depending on your computer. If you try to imagine the branching diagram for $a(50)$, which will be considerably larger than that for $a(5)$ in Figure 8.1, you can perhaps see why the evaluation of $a(50)$ takes so long   the evaluation of $a(2), a(3)$, etc., is done a large number of times. This is one of the drawbacks of writing recursive procedures. Even though they are often natural to write, they can consume a lot of time and memory when executing.

To make recursive procedures more e cient, Maple has an option that can be used in such procedures to circumvent multiple evaluations (such as $a(2)$ in the computation of $a(50)$). One simply includes the command **option remember** as the first statement in the procedure:

```
a:=proc(n)
```

No Remember Table		With Remember Table	
n	Time/Memory to Compute $a(n)$	Time/Memory to Compute $a(n)$	
5	$t = 0.06$, bytes $= 29926$	$t = 0.00$, bytes $= 8142$	
10	$t = 0.02$, bytes $= 82918$	$t = 0.00$, bytes $= 9802$	
15	$t = 0.17$, bytes $= 847818$	$t = 0.02$, bytes $= 10978$	
20	$t = 1.88$, bytes $= 9325198$	$t = 0.00$, bytes $= 12366$	
25	$t = 21.52$, bytes $= 103311426$	$t = 0.00$, bytes $= 13874$	
30	$t = 241.31$, bytes $= 1145023986$	$t = 0.00$, bytes $= 15158$	

Table 8.1: *Comparison of the times (in seconds) and memory used in computing $a(n)$, without and with a remember table.*

```
   option remember;
   global p,q;
   if n=0 then p
   elif n=1 then q
   else  -(3*(n-1)*a(n-1)+(n+1)*a(n-2))/(2*(n-1)*n)
   end if
end proc:
```

This causes Maple to create a *remember table* which holds all the values computed during execution of the procedure. Each time a new value is to be computed, the procedure first checks the remember table to see if this value has been computed before. If so, it uses the value in the table. Now, when you ask for the evaluation of $a(50)$, with the numbers in "oating-point, it takes less than 0.1 seconds for Maple to return the answer

$$.1120827468 \; 10^{-38} q - .1667198499 \; 10^{-38} p.$$

That's quite an improvement!

We can study the situation more precisely if we use Maple's `showtime` command. The results of a short study are shown in Table 8.1. As you can see, the times and memory usages are dramatically longer and bigger without the remember table. Computing $a(30)$ takes over four minutes without a remember table. Note that when a remember table is used, some of the results show a time of 0.00, which indicates that the actual time is less than the clock accuracy of the computer. Also note that you would expect the times to increase with n. However, we ran the results consecutively, using the remember table, and so $a(20)$ takes less time to compute than $a(15)$ (ostensibly because $a(20)$ uses some of the values from $a(15)$).

As another example of how the recursive procedure a above could be memory consumptive, consider the following use of it.

Example 8.1 (Series Solutions of Differential Equations) The sequence $\{a_n\}_{n=0}$ defined by the recurrence relation (8.1) arises in the study of series solutions of differential equations (which you may have studied, or soon will, in your upper-level courses). It can be shown that the power series

$$y(x) = \sum_{n=0} a_n(x-2)^n, \qquad (8.2)$$

is a solution of the differential equation

$$2y'' + (x+1)y' + 3y = 0,$$

provided the coefficients a_n are chosen to satisfy the recurrence relation (8.1). Since the a_n's determined by the recurrence relation depend on the choice of $a_0 = p$ and $a_1 = q$, we see that the series (8.2) involves these two arbitrary constants p, q. We denote by y_1 and y_2, the two particular series solutions that arise from the choices $p = 1, q = 0$ and $p = 0, q = 1$. These are fundamental solutions since all solutions can be expressed in terms of them.

Such series solutions have great theoretical and practical value, the latter coming from the fact that the sum of finitely many terms of the series will give a good approximation to the actual solution of the differential equation. For example, let $p = 1, q = 0$. Then

$$Y_1(x, N) = \sum_{n=0}^{N} a_n(x-2)^n,$$

is an approximation to the solution y_1 using $N+1$ terms of the series. We can employ the recursive procedure a, defined above, for the coefficients and set the global variables as p:=1;q:=0;. Then we can define the approximations to Y_1 in Maple by

```
Y1:=proc(x,N)
   local s,n;
   s:=0;
   for n from 0 to N do s:=s+a(n)*(x-2)^n; end do
end proc:
```

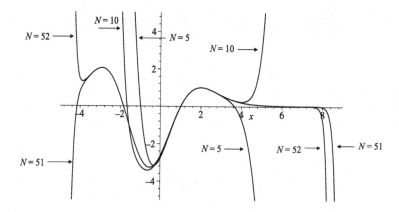

Figure 8.2: *Plots of $Y_1(x,5), Y_1(x,10), Y_1(x,51), Y_1(x,52)$ for the choice of $a_0 = 1, a_1 = 0$.*

Using this to produce the plots of the approximate solutions, as shown in Figure 8.2, can be a very, very slow process without the **remember** option included in the procedure a. Even with the remember option, the plots take approximately 0.7 seconds. As you can see, the approximations $Y_1(x,51), Y_1(x,52)$ on the interval $[-3.5, 7.5]$ appear to coincide with what would be the function $y_1(x)$ (not shown) obtained by using all the terms in the series. Plotting $Y_1(x,100)$ (not shown) takes about 0.4 seconds. Each of these times would be quicker were it not for the fact that in plotting a function f, Maple must compute the values $f(x)$ at su ciently many points x to give a smooth-looking polygonal approximation. In the case of $Y_1(x,100)$ it must, for each of the points x chosen, also use the recursive procedure to compute the coe cients $a(n)$, $n = 2, 3, \ldots, 100$. Imagine how slow this would be if we had not included the **option remember** command in the recursive procedure.

There is an alternative to the memory (and therefore time) consumptive problems with recursive procedures. One could, as some advocate, simply *not* write procedures that call themselves. In the power series example this philosophy can be implemented easily as follows.

Set up an array **a** that is large enough to hold the desired number of coe cients a_n of the power series and initialize it with the values $a_0 = 1, a_1 = 0$.

```
a:=array(0..100):a[0]:=1:a[1]:=0:
```

Then use a do loop to calculate and store the other values of the a_n's:

```
for n from 2 to 100 do
   a[n]:=-(3*(n-1)*a[n-1]+(n+1)*a[n-2])/(2*(n-1)*n)
end do;
```

This calculation is very quick, taking less than 0.1 seconds. Now simply redefine the approximate solutions by

```
Y1:=proc(x,N)
   local s,n;
   s:=0;
   for n from 0 to N do s:=s+a[n]*(x-2)^n end do
end proc:
```

Using this to produce the plot shown in Figure 8.2 takes less than 0.1 seconds. Thus, in this example the do loop code is better, and since it is just as simple to write and understand as the recursive code, it is recommended.

There is a general theory of recurrence relations that applies to general sequences, not just sequences of numbers (see, e.g., [Br], [Bo]). The next section discusses one such case, where the sequences are sequences of functions (Maple expressions).

8.2 Reduction Formulas for Integration

When you studied techniques of integration in calculus, you probably encountered the topic of reduction formulas. These formulas are nothing other than recurrence relations that allow you to compute indefinite integrals, such as $\int \cos^n x \, dx$, for a given power n, in terms of other integrals involving lesser powers.

The various reduction formulas are usually obtained by using integration by parts (perhaps several times) and algebra. For instance, integration by parts applied to $\int x^n e^{ax} dx$ gives (assuming $a \neq 0$):

$$\int x^n e^{ax} dx = \frac{x^n e^{ax}}{a} - \frac{n}{a} \int x^{n-1} e^{ax} \, dx.$$

Notice that this is a recursive formula (if $n > 1$) since the integral on the right-hand side is of the same type as the one we started with. To see this more clearly, let

$$E(n, a) \int x^n e^{ax} \, dx.$$

Then the reduction formula becomes

$$E(n,a) = \frac{x^n e^{ax}}{a} - \frac{n}{a}E(n-1,a),$$

which is a recurrence relation.

Writing a recursive procedure to obtain the value of the original integral is straightforward. We allow the more general case where a can be 0.

```
E:=proc(n::integer,a)
  local ans;
  if n<0 then RETURN("n should not be negative") end if;
  if n=0 then if a=0 then ans:=x
                    else ans:=exp(a*x)/a
             end if;
  end if;
  if n>0 then
   if a=0 then
     ans:= x^(n+1)/(n+1) else
     ans:= (x^n)*exp(a*x)/a-(n/a)*E(n-1,a);
   end if;
  end if;
  ans;
end proc:
```

The call E(4,3) to this procedure produces the output

$$\frac{1}{3}x^4 e^{(3x)} - \frac{4}{9}x^3 e^{(3x)} + \frac{4}{9}x^2 e^{(3x)} - \frac{8}{27}x e^{(3x)} + \frac{8}{81}e^{(3x)}$$

For $a \neq 0$ there will always be a common factor of $e^{(ax)}$, and we might want to write the result in factored form $p(n,x,a)e^{ax}$, for a polynomial $p(n,x,a)$. Using the above procedure this can be done easily as follows.

```
Ereduce:=proc(n::integer,a)
  local p;
  E(n,a);
  if a=0 then RETURN(E(n,a))
     else  p:=(n,x,a)->collect(E(n,a),exp(a*x));
  end if;
  p(n,x,a);
end proc:
```

Then the call $\texttt{Ereduce(4,3)}$ would give

$$\frac{1}{3}x^4 - \frac{4}{9}x^3 + \frac{4}{9}x^2 - \frac{8}{27}x + \frac{8}{81} \quad e^{(3x)}.$$

Another example of an integral reduction formula is the following.

$$x^n \sin ax \ dx = -\frac{x^n \cos ax}{a} + \frac{n}{a} \quad x^{n-1} \cos ax \ dx.$$

This is not a recursive formula since the integrals in the formula are not of the same type, one involving $\sin ax$ and the other $\cos ax$. We see, however, that if we apply integration by parts once more we will get a recursive formula.

$$
\begin{aligned}
x^n \sin ax \ dx \ &= \ -\frac{x^n \cos ax}{a} + \frac{n}{a}\ \frac{x^{n-1}\sin ax}{a} - \frac{(n-1)}{a} \quad x^{n-2}\sin ax \ dx \\
&= \ -\frac{x^n \cos ax}{a} + \frac{nx^{n-1}\sin ax}{a^2} - \frac{n(n-1)}{a^2} \quad x^{n-2}\sin ax \ dx.
\end{aligned}
$$

Letting $S(n,a)$ $x^n \sin ax \ dx$ and assuming that n 0, we get the recurrence relation

$$S(n,a) = -\frac{x^n \cos ax}{a} + \frac{nx^{n-1}\sin ax}{a^2} + \frac{n(n-1)}{a^2}S(n-2,a).$$

The corresponding Maple procedure is then

```
S:=proc(n::integer,a)
   local ans;
   if n<0 then ans:="n should be positive" end if;
   if n<=1 then ans:=(-x)*cos(a*x)/a+n*sin(a*x)/a^2 end if;
   if n>1 then
     if a=0
       then ans:=0
       else ans:=(-x^n)*cos(a*x)/a+n*x^(n-1)*sin(a*x)/a^2
                 -n*(n-1)/a^2*S(n-2,a)
     end if;
   end if;
   ans;
end proc:
```

The call S(4,3) results in

$$-\frac{1}{3}x^4\cos(3x) + \frac{4}{9}x^3\sin(3x) + \frac{4}{9}x^2\cos(3x) - \frac{8}{27}x\sin(3x) - \frac{8}{81}x\cos(3x),$$

which involves both $\sin(3x)$ and $\cos(3x)$. In general we can see that

$$S(n,a) = P(x,a)\sin ax + Q(x,a)\cos ax$$

for some polynomials P and Q.

As our last example in this section we will obtain a reduction formula for $(x^2 + a^2)^n\,dx$. Integrating by parts we get

$$(x^2 + a^2)^n\,dx = x(x^2 + a^2)^n - 2n \quad x^2(x^2 + a^2)^{n-1}\,dx. \qquad (8.3)$$

We do not see the recursion yet, but if we rewrite $x^2 = x^2 + a^2 - a^2$ we have

$$x^2(x^2 + a^2)^{n-1}\,dx = \quad (x^2 + a^2)^n\,dx - \quad a^2(x^2 + a^2)^{n-1}\,dx.$$

Now, substituting this in formula (8.3) and solving for $(x^2 + y^2)^n\,dx$ we arrive at the desired reduction formula:

$$(x^2 + a^2)^n\,dx = \frac{x(x^2 + a^2)^n}{2n + 1} + \frac{2na^2}{2n + 1} \quad (x^2 + a^2)^{n-1}\,dx.$$

8.3 Sorting

In some of the examples we have studied there has been a need to arrange a list of numbers in increasing order. For instance, we needed to do this to construct approximations to the Cantor function in Example 6.5. There we used Maple's **sort** command. The process of arranging a list of numbers in order is called *sorting* and in this section we will discuss some of the commonly occurring algorithms for sorting and how to implement them in Maple.

While these sorting algorithms are well known, with standard implementations readily available in most programming languages, and while translating them into Maple code is not di cult (or, is not necessary if you use the built-in **sort** command), our reasons for discussing them here are three-fold. First we want you to understand how these algorithms work in detail and be able to apply them to sort small lists by hand. Second, we want you to write modifications and extensions of the codes and produce animations of these algorithms. Third, we want you to see further examples where recursive procedures are natural and useful.

Although one can also write programs to sort lists of strings, we will only be concerned with sorting lists of numbers. Each sorting algorithm will take a list of numbers $a_1, a_2, \ldots, a_n$, and produce a list that is ordered,

$$a_1 \quad a_2 \quad \cdots \quad a_n,$$

from smallest to largest.[1] This is done by applying a basic *process* to the original list to produce one or more sublists. The basic process is then applied to the sublist(s) and the algorithm proceeds like this until the list is sorted.[2] A brief description of the four algorithms that we will study is:

Selection Sort: The process is: Select the largest element in the list and interchange it with the last element in the list. This produces a new list $b_1, b_2, \ldots, b_n$ in which b_n is the largest element (and thus in the correct position). Now apply the same process to the shorter sublist $b_1, b_2, \ldots, b_{n-1}$. Continue in this fashion until a sublist consisting of a single element is reached.

Bubble Sort: The process is: Scan through the list, from left to right, comparing each element a_i with its successor a_{i+1}, and interchanging them if they are out of order. This produces a new list $b_1, b_2, \ldots, b_n$ in which b_n is the largest (and thus in the correct position).[3] Now apply the same process to the shorter sublist $b_1, b_2, \ldots, b_{n-1}$. Continue in this fashion until a sublist consisting of a single element is reached.

Quick Sort: The process is: Rearrange the elements in the list so that the list is partitioned into two sublists

$$1, \ldots, \quad k, r_{k+1}, \ldots, r_n,$$

with $_i \quad r_j$, for all $i = 1, \ldots, k$ and $j = k + 1, \ldots, n$. Now apply this process to *each* of the sublists $1, \ldots, \quad _k$ and $r_{k+1}, \ldots, r_n$. Continue in this fashion until sublists consisting of single elements are reached.

---

[1]The   is used to indicate that the numbers in the original list have been relabeled as shown after they have been sorted.

[2]It is customary to refer to the numbers being sorted as a   list. For the Maple code however, we will use an array data structure to store the numbers.

[3]If this is not obvious to you, read the comments about bubble sort in Section 8.3.1.

Insertion Sort: The process is: Put the last two elements, a_{n-1}, a_n, of the list in order to get an ordered sublist: $a_{n-1} \quad a_n$. Then insert the element a_{n-2} into this ordered sublist to get a larger, ordered sublist: $a_{n-2} \quad a_{n-1} \quad a_n$. Continue in this fashion until all the elements in the original list have been inserted.

We will describe each of these algorithms separately and in more detail below the above is just an overview of them for the sake of comparison.

An important aspect that must be considered when implementing these algorithms is whether the sorting will be done *in-place* or *duplicate-place*. Sorting is called *in-place* if the elements $a_1, a_2, \ldots, a_n$ are stored in an array [A[1],A[2], ...,A[n]] in memory and all sorting is done by moving the a_i's around in these storage locations, without creating extra, duplicate storage in memory for the sorting. On the other hand, *duplicate-place* sorting refers to a sorting algorithm that creates a temporary array [B[1],B[2], ...,B[n]] in memory to aid in the sorting process. Duplicate-place sorting takes twice as much memory as in-place sorting, but some algorithms are easier to understand when written as duplicate-place algorithms.[4] The codes for in-place and duplicate-place sorting will generally have different logic and structure so it is important to understand the distinction when examining sorting algorithms from other books and sources from the Internet.

For the most part, we will construct in-place sorting algorithms and this inevitably uses the mechanism of interchanging two elements in the list $a_1, a_2, \ldots, a_n$. The interchanging of a_i and a_j requires that one of these be copied to a temporary location before overwriting its present value. The following short procedure accomplishes this.

Exchange Code

```
exchange:=proc(A::array,i,j)
  local temp;
    temp:=A[i];
    A[i]:=A[j];
    A[j]:=temp;
end proc:
```

Notice that the list to be sorted is input as a one-dimensional array and not as a list. This is because Maple will not permit the direct replacement of

---

[4]When you apply a sorting algorithm by hand, you inevitably create this duplicate storage usually it is a partially sorted list written right below the original list.

an element in a list by another element in the list when the list is a formal parameter in a procedure. One would have to make a copy of the list A and replace elements of the original list with elements from the copy. We can, however, directly replace an element of an array given by an input parameter with another element of the array. All of our sorting procedures repeatedly employ **exchange** to build up a sorted array. In fact the e ciency of an in-place sorting routine is judged in part by the number of exchanges necessary to completely sort the array.

8.3.1 Bubble Sort

To implement an in-place version of the bubble sort algorithm, we store the list to be sorted, $a_1, a_2, \ldots, a_n$, in an array $A:=[A[1],A[2],A[3],\ldots,A[n]]$. The first stage in the bubble sort procedure is

> Compare A[1],A[2] and exchange them if the second is less than the first.
> Compare A[2],A[3] and exchange them if the third is less than the second.
> $\vdots$
> Compare A[n-1],A[n] and exchange them if the nth is less than the $(n-1)$st.

After the last comparison and possible exchange, the largest element of the list $a_1, a_2, \ldots, a_n$ will be in the nth storage location A[n].[5] Note that the first stage above requires $n-1$ comparisons and at most $n-1$ exchanges.

For the second stage of the sorting, we repeat this process on the elements in the subarray $[A[1],A[2],\ldots,A[n-1]]$, and as a result the *second* largest element of the original list $a_1, a_2, \ldots, a_n$ will be placed in A[n-1]. Continuing this sequence of steps (stages) down to the subarray A[1],A[2] will result in an array completely sorted in increasing order. The code for this algorithm is straightforward and nonrecursive:

| Code for Bubble Sort |

```
bubblesort:=proc(A::array)
  local n,i,j;
  n:=nops(op(3,eval(A)));
  for i from 1 to n-1 do
    for j from 1 to n-i do
```

---

[5]Convince yourself of this by working a couple of examples by hand!

```
    if A[j]>A[j+1] then exchange(A,j,j+1) end if;
  end do;
  end do;
  eval(A);
end proc:
```

As an example of the use of this, consider the list $5, 4, 3, 2, 1$, which is in reverse order. The following command causes the array to be sorted in the sequence of steps shown.

```
bubblesort(array([5,4,3,2,1]));
```

$[5, 4, 3, 2, 1]$	*initial list*
$[4, 5, 3, 2, 1]$	
$[4, 3, 5, 2, 1]$	
$[4, 3, 2, 5, 1]$	
$[4, 3, 2, 1, 5]$	*end of the 1st stage*
$[3, 4, 2, 1, 5]$	
$[3, 2, 4, 1, 5]$	
$[3, 2, 1, 4, 5]$	*end of the 2nd stage*
$[2, 3, 1, 4, 5]$	
$[2, 1, 3, 4, 5]$	*end of the 3rd stage*
$[1, 2, 3, 4, 5]$	*end of 4th stage (final list)*

Notice that in this case, every comparison resulted in an exchange and so there were a total of $10 = 4 + 3 + 2 + 1$ comparisons and 10 exchanges. Thus, from an e ciency standpoint this example exhibits the worst possible case for sorting an array of 5 elements. Generalizing to an array of size n, we calculate the worst case would require $2((n - 1) + (n - 2) + \cdots + 1) = 2((n-1)n/2) = n^2 - n$ steps (comparisons and exchanges). Sorting an array in which some of the elements are already in order would require fewer steps (fewer exchanges).

The above example exhibits why the algorithm is called a bubble sort. During the first stage, the element 5 moves successively past the other elements until it reaches the right end of the array much like an air bubble in water would do in rising to the surface of the water. At the second stage the element 4 bubbles up until it reaches the larger element 5. And so on, until all the elements are bubbled into their correct positions.

This example is a rather special case, but the bubbling process works pretty much the same in general. At any stage, an element bubbles up until it encounters a larger element. Bumping into that larger element causes it to stop and the larger element to begin bubbling up. If you work some examples by hand, you will easily see how it goes.

One can write a recursive version of bubble sort (exercise), but the non-recursive version above is best for an introduction to sorting.

8.3.2   The Quick Sort Algorithm

Bubble sort is a simple algorithm but for long lists is not very e cient. Many sorting programs have been developed, some of which are more e cient than others. The idea for the quick sort algorithm was suggested in a paper by C.A. Hoare [Ho] and this algorithm is, on the average, faster than others, requiring on the order of $n \log n$ steps to sort a list of n numbers. Quick sort is a more complicated algorithm which uses recursion and the main idea is to divide the sorting task into two smaller sorting tasks. The basic part of the algorithm rearranges and partitions an unsorted list $a_1, a_2, \ldots, a_n$ into two smaller sublists so that each element in the first sublist is less than or equal to each element in the second sublist, i.e.,

$$\text{LOW} = {}_1, \ldots, {}_k, \quad \text{HIGH} = r_{k+1}, \ldots, r_n,$$

with ${}_i \quad r_j$, for all $i = 1, \ldots, k$ and $j = k + 1, \ldots, n$. If this step is repeated separately on each of the two sublists, ${}_1, \ldots, {}_k$ and $r_{k+1}, \ldots, r_n$, and then again on the subsequent partitions (i.e., recursively), then the original list will be sorted (as we shall see). The challenge then is to design the partitioning scheme.

It is important to note that the bubble sort algorithm has such a partitioning scheme. Namely, after the first stage of bubbling, the list is $b_1, b_2, \ldots, b_n$, with b_n the largest element in the list. Consequently, the sublists $\text{LOW} = b_1, \ldots, b_{n-1}$ and $\text{HIGH} = b_n$ are of the type needed in quick sort (every element in LOW is less than or equal to every element in HIGH). Thus, bubble sort could be considered as a special version of quick sort where the partitioning scheme always has the largest element on the high side and all the other elements on the low side. However, what makes quick sort better is that, on the average, it has partitions with sublists of nearly the same size (roughly $n/2$ elements in each), and so the recursion down to sublists of size two is much quicker. Similarly, the selection sort discussed in the exercises has a partitioning scheme exactly like bubble sort and so is generally slower than quick sort.

We examine two versions of quick sort the first one is duplicate-place sorting (and quite a bit easier to understand) and the second one is in-place

sorting. In each version, the unsorted list $a_1, a_2, \ldots, a_n$ is stored in an array
A=[A[1],A[2],...,A[n]].

Quick Sort (Duplicate-Place Version): This version creates an extra
array B=[B[1],B[2],...,B[n]] to temporarily store the elements of the
list as it is being divided into two parts. We explain the partitioning scheme
first by using examples.

Suppose the list has an odd number of elements (so there is a middle
element). Suppose the list is

$$2\ 5\ 4\ \boxed{8}\ 20\ 11\ 14$$

Then the middle element 8 partitions the list as required and we can tell
this by comparing the other elements *only* with 8. Namely, $2, 5, 4$ are each
less than 8, while $20, 11, 14$ are each bigger than 8. Thus we conclude,
without even comparing them, that each element in $\{2, 5, 4\}$ is less than
every element in $\{20, 11, 14\}$. We can include 8 in either the LOW or the
HIGH partition.[6] On the other hand, if the list is not already partitioned
as required by its middle element, say,

$$9\ 11\ 14\ \boxed{5}\ 20\ 4\ 2$$

then we can partition it as follows. We create extra (duplicate) workspace
below the list and use this space to do the partitioning.

$$9\ 11\ 14\ \boxed{5}\ 20\ 4\ 2$$
$$\textit{workspace}$$

As shown, we select the middle element 5 as a *pivot element*, i.e., an element
to which we will compare all the other elements. Now scan through the list,
from left to right comparing each element a_i with the pivot element 5. Be
sure to skip (or ignore) the pivot element when doing the scan. If a_i 5,
then copy a_i into the HIGH end of the workspace; otherwise copy it into the
LOW end. This produces the result

$$9\ 11\ 14\ \boxed{5}\ 20\ 4\ 2 \quad \textit{original list}$$
$$4\ 2\ \boxed{}\ 20\ 14\ 11\ 9 \quad \textit{workspace}$$

---

[6]This scheme reduces the number of comparisons necessary to form the low and high
sublists. In the example, we need only make six comparisons (the other elements compared
with 8) as opposed to twelve comparisons (every element of $\{2, 5, 4, 8\}$ compared with every
element of $\{20, 11, 14\}$).

The process will always leave one position vacant (a hole) in the duplicate list and we can now place the pivot element in that position and declare that the partition point is to the right of the pivot (i.e., include the pivot in the LOW sublist). This gives

$$9\ 11\ 14\ \boxed{5}\ 20\ 4\ 2 \quad \textit{original list}$$
$$4\ 2\ 5\ |\ 20\ 14\ 11\ 9 \quad \textit{workspace}$$

This completes the partitioning of the original list. Now we can apply the same process recursively to partition each of the sublists, the sublists of these, and so on until we reach a sublist that is either empty or contains a single element. We begin with the work on the LOW sublist:

$$4\ \boxed{2}\ 5\ |\ 20\ 14\ 11\ 9 \quad \textit{original list}$$
$$\square\ 5\ 4\ |\ 20\ 14\ 11\ 9 \quad \textit{workspace}$$
$$2\ |\ 5\ 4\ |\ 20\ 14\ 11\ 9 \quad \textit{workspace}$$

This shows the partition of the original low list into LOW = 2 and HIGH = 5, 4. (The elements in the original high list are carried along and will be worked on later.) Since LOW consists of a single element the work stops on it, and we proceed to partition HIGH = 5, 4. Since this list has two elements, there is no middle element to pick for the pivot. Our convention in this case will be: *for a list with an even number, $n = 2k$, of elements, the pivot will be the kth element.* Thus, for the example here, 5 is the pivot, and the work on the HIGH sublist is as follows:

$$2|\ \boxed{5}\ 4\ |\ 20\ 14\ 11\ 9 \quad \textit{original list}$$
$$2\ |\ 4\ \square\ |\ 20\ 14\ 11\ 9 \quad \textit{workspace}$$
$$2\ |\ 4\ 5\ ||\ 20\ 14\ 11\ 9 \quad \textit{workspace}$$

Now LOW = 4, 5 and HIGH = (the empty list, as indicated by the double partition bars ||. Processing of the high list terminates since it is empty. The low list is in order, but the computer will not know this until it is processed once more:

$$2|\ \boxed{4}\ 5\ |\ 20\ 14\ 11\ 9 \quad \textit{original list}$$
$$2\ |\ \square\ 5\ |\ 20\ 14\ 11\ 9 \quad \textit{workspace}$$
$$2\ |\ 4\ |\ 5\ |\ 20\ 14\ 11\ 9 \quad \textit{workspace}$$

The original low list, 4, 2, 5, is now ordered and the processing can begin on the original high sublist, 2, 14, 11, 9. You should try this yourself by hand to make sure you understand the process. Once you are familiar with how things work, it will be convenient not to separate all the steps in the processing, as we did above for clarity. Instead, let us record the quick sort processing as follows:

$$
\begin{array}{llllllll}
9 & 11 & 14 & \boxed{5} & 20 & 4 & 2 & \textit{list} \\
4 & \boxed{2} & 5 \mid 20 & 14 & 11 & 9 & & A \\
2 \mid & \boxed{5} & 4 \mid 20 & 14 & 11 & 9 & & B \\
2 \mid & \boxed{4} & 5 \parallel 20 & 14 & 11 & 9 & & C \\
2 \mid 4 \mid 5 \mid & 20 & \boxed{14} & 11 & 9 & & & D \\
2 \mid 4 \mid 5 \mid & 11 & \boxed{9} & 14 \mid 20 & & & & E \\
2 \mid 4 \mid 5 \mid 9 \mid & \boxed{14} & 11 \mid 20 & & & & & F \\
2 \mid 4 \mid 5 \mid 9 \mid & \boxed{11} & 14 \parallel 20 & & & & & G \\
2 \mid 4 \mid 5 \mid 9 \mid & 11 & \mid 14 \mid 20 & & & & & H \\
\end{array}
$$

This shows the original list on the first line. Each successive line shows an additional partition and the pivot element selected to partition this partition. This is the recommended way to do this type of sorting when using pencil and paper. Try it on this example and the ones in the exercises!

You should realize that each time a sublist is partitioned, there is a branching in the recursive process to further partition the sublists of the sublist. Figure 8.3 illustrates this.

Each sublist serves as a node in the tree and the branching from a node corresponds to the splitting into the low and high sublists. Branching out (splitting and growing) from a given node will continue until a node is empty or consists of a single element.

Designing the procedure, which we call **QS**, to implement this type of quick sort should now be quite simple. The initial list to be sorted is $a_1, a_2, \ldots, a_N$ and the duplicate list for intermediate sorting is $b_1, b_2, \ldots, b_N$. The procedure is designed to take a sublist $a_m, a_{m+1}, \ldots, a_n$, with $1 \quad m$ $n \quad N$ and sort it into the positions $b_m, b_{m+1}, \ldots, b_n$ as follows. If we let $r = [(m+n)/2]$, then we take $p \quad a_r$ as the pivot (middle element) of the a sublist.[7]

Next the sublist $a_m, a_{m+1}, \ldots, a_n$ is scanned, comparing each a_i with p, as $i = m, \ldots, n$. If $a_i \quad p$, then it is copied into the high end of the duplicate list. Otherwise it is copied into the low end. The first copy into the high end

---

[7]Here $[\cdot]$ denotes the greatest integer function: $[x]$ is the largest integer which is less than or equal to x. In Maple, the greatest integer function is `floor(x)`. Note that $(m+n)/2$ is the midpoint of the interval $[m, n]$ and is an integer when $n+m$ is even. You should verify the choice of the pivot as a_r, where $r = [(m+n)/2]$, agrees with our choices in the examples.

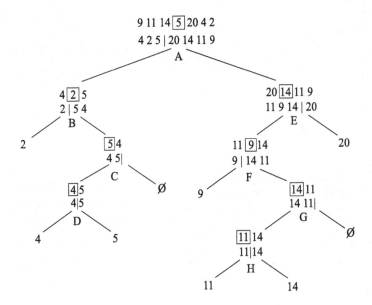

Figure 8.3: *A tree diagram for the branching process in the recursive calls to partition the sublists.*

goes into position n, and subsequent copies go into positions $n-1, n-2, \ldots$. The first copy into the low end goes into position 1, and subsequent copies go into positions $2, 3, \ldots$. To keep track of the positions in the high and low ends, we use two index variables called `high,low`. In the scanning process, `B[high]` holds the last element copied into the high end and `B[low]` holds the last element copied into the low end. When we find, `A[i]>=p`, we decrement, `high:=high-1`, and copy, `B[high]:=A[i]`. When the scanning is complete, the hole in B is to the left of the last element copied into the high end. So we copy the pivot there, `B[high-1]:=p`. Now the duplicate subarray `B[m],...,B[n]` holds all the elements, correctly partitioned into LOW and HIGH, and so we copy this back into the corresponding subarray of `A`. Then we can make the recursive calls

    QS(A,m,high-1); QS(A,high,n);

to apply the same process to LOW and HIGH. The code for this procedure is the following:

Quick Sort (Duplicate-Place)

```
QS:=proc(A,m,n)
  local r,p,i,low,high,B;
  if m<n then
    B:=array(m..n);
    r:=floor((m+n)/2);
    p:=A[r];
    low:=m-1;high:=n+1;
    for i from m to n do
      if i=r then i:=i+1 end if;
      if A[i]>=p
        then high:=high-1;B[high]:=A[i];
      else low:=low+1;B[low]:=A[i];
      end if;
    end do;
    B[high-1]:=p;
    for i from m to n do A[i]:=B[i] end do;
    print(A);
    QS(A,m,high-1); QS(A,high,n);
  end if;
end proc:
```

Note that the procedure returns without doing anything if $m = n$ (only one element in the subarray) or $m > n$ (which corresponds to an empty array). The statement **print(A)** is executed right before the branching to process the low and high subarrays. The entire array [A[1],...,A[n]] is printed, showing the state of the sorting at this point in the branching.

As an example, consider the list $9, 8, 7, 6, 5, 4, 3, 2, 1$, which is in reverse order. We define this, print it, and call the quick sort procedure with:

```
A:=array([9,8,7,6,5,4,3,2,1]);
print(A);QS(A,1,9);
```

This gives the output

$$[9, 8, 7, 6, 5, 4, 3, 2, 1]$$
$$[4, 3, 2, 1, 5, 6, 7, 8, 9]$$
$$[1, 2, 5, 3, 4, 6, 7, 8, 9]$$
$$[1, 2, 5, 3, 4, 6, 7, 8, 9]$$
$$[1, 2, 3, 4, 5, 6, 7, 8, 9]$$
$$[1, 2, 3, 4, 5, 6, 7, 8, 9]$$
$$[1, 2, 3, 4, 5, 6, 7, 9, 8]$$

$$[1, 2, 3, 4, 5, 6, 7, 9, 8]$$
$$[1, 2, 3, 4, 5, 6, 7, 8, 9]$$
$$[1, 2, 3, 4, 5, 6, 7, 8, 9]$$

You should try this example by hand and see if you get the same results (same number of lines and same order). Why do you think that the algorithm kept processing after line five, when clearly the list is in order at that time? Why did the algorithm switch 8 and 9 in line seven?

There are many ways to embellish the above output in order to make it more readable and to help us understand what is going on at each step in the processing. We reserve this for the in-place version of quick sort below, which is more di cult conceptually and harder to do by hand.

Quick Sort (In-Place Version): To write an in-place version of quick sort, we need a scheme for building the LOW and HIGH subarrays of `[A[m],A[m+1],...,A[n]]` that operates much like the duplicate-place version, i.e, this scheme should involve picking an element, called the pivot, roughly in the middle of the unsorted array, comparing the other elements to this pivot, and determining which elements go in LOW and which go in HIGH. However, now we cannot simply copy elements into selected positions of LOW and HIGH because that would overwrite the existing information in those positions. But, we can select *pairs* of elements, compare each to the pivot, and then exchange elements in each pair if necessary. Thus, we need a systematic way of successively selecting pairs and exchanging a way that guarantees a resulting partition into LOW and HIGH. Here is an example to help you see how the pair selection can be done (at least one way to do it).

Suppose the original list is

$$4 \ 17 \ 22 \ \boxed{8} \ 3 \ 9 \ 1$$

The pivot element 8 divides the list into two pieces: 4 17 22 and 3 9 1, but this is not a partition of the required type. However, scanning from *left to right*, we find that 4 is correctly positioned in the low end, but 17 needs to be copied to the high end. To find an element to exchange 17 with, we next scan the list *right to left*, starting at the extreme right of the high end, and find the element 1 which needs to be copied to the low end:

$$4 \ \boxed{17} \ 22 \ \boxed{8} \ 3 \ 9 \ \boxed{1}$$

Thus, we exchange the pair 17,1:

$$4 \boxed{1} \; 22 \; \boxed{8} \; 3 \; 9 \; \boxed{17}$$

After this exchange, we continue the scan from left to right, stopping when we find that 22 needs to go high. Then switch to a right to left scan in the high end, stopping when we find that 3 needs to go low:

$$4 \; 1 \; \boxed{22} \; \boxed{8} \; \boxed{3} \; 9 \; 17$$

Thus we exchange 22 and 3:

$$4 \; 1 \; \boxed{3} \; \boxed{8} \; \boxed{22} \; 9 \; 17$$

At the next step, the low-end scan meets the high-end scan:

$$4 \; 1 \; 3 \; \boxed{\boxed{8}} \; 22 \; 9 \; 17$$

This will achieve a partition of the desired type if we decide, say, to have the pivot element in LOW:

$$4 \; 1 \; 3 \; 8 \mid 22 \; 9 \; 17$$

This example illustrates the *general idea* for partitioning the array: First choose the pivot to be the element in the middle of the unsorted array. Then starting from opposite ends of the array, successively compare each element to the pivot. To the left of the pivot, if the first element is *less than* the pivot, leave it alone and check the next element in the array. Stop checking if you reach an element, say the ith one, greater than or equal to the pivot. Now go through the similar steps for the elements to the right of the pivot. If the last element in the array is *greater than* the pivot, leave it alone and check the next one down. Stop if you find an element, say the jth one, which is less than or equal to the pivot. This process will end since the pivot is neither greater than itself nor less than itself. If the process stops on the ith and jth elements in the array and $i < j$ then these two elements are exchanged, i is increased by one, and j is decreased by one. If then $i < j$, the checking process just described is continued. However, if $i > j$, then all the elements in the list have been compared to the pivot and the partitioning is complete. On the other hand, if $i = j$, then it may happen the ith element has not been compared to the pivot. If this is the case, then once this comparison is made the partitioning will be complete.

We look at another example to further explain the general idea and get a better understanding of what happens as the two scanning processes get close and either coincide ($i = j$) or cross ($i > j$). For this, note that the scanning processes, which search for pairs of elements to exchange, can be visualized as a pinching motion of two *scan-boxes*: Move the i scan-box to the right through the low end and stop when an element not less than the pivot is encountered. Then move the j scan-box to the left through the high end and stop when an element not greater than the pivot is encountered. Now, the elements in the scan-boxes can be exchanged and the scanning continued. In the last example above, the partition was achieved when the two scan-boxes coincided (the $i = j$ case).[8] Here is another example which illustrates how the scan-boxes can pass each other (the $i > j$ case).

Suppose the initial list is

$$15\ 3\ 10\ \boxed{9}\ 8\ 16\ 10$$

The i scan-box starts at 15 and stays there since 15 is not less than the pivot. So $i = 1$. Then the j scan box slides past 10, 16 (they are greater than the pivot) stopping at 8 (it is not greater than the pivot). So $j = 5$ and the picture is:

$$\boxed{15}\ 3\ 10\ \boxed{9}\ \boxed{8}\ 16\ 10$$

Exchanging the scan-box elements 15, 8 gives

$$\boxed{8}\ 3\ 10\ \boxed{9}\ \boxed{15}\ 16\ 10$$

Now move the scan-boxes one unit in their respective directions, $i := i + 1, j := j - 1$, to get

$$8\ \boxed{3}\ 10\ \boxed{\boxed{9}}\ 15\ 16\ 10$$

Since the scan boxes have not met or crossed (i.e., $i < j$ is still true), we continue the processing. Moving the i scan-box first, we find it stops at 10. The j scan-box is presently at 9 which is the pivot element and is not greater than itself, and so the j scan-box stops there. Thus at this time $i = 3, j = 4$ and the picture is

$$8\ 3\ \boxed{10}\ \boxed{9}\ 15\ 16\ 10$$

Exchanging 10 and 9 gives

$$8\ 3\ \boxed{9}\ \boxed{10}\ 15\ 16\ 10$$

---

[8] Just coincidentally, the scan-boxes coincided with the box around the pivot element. This is not always the case.

Now when we move the scan-boxes one increment in their respective directions ($i := i + 1 = 4$, $j := j - 1 = 3$), they pass each other and the picture is the same as the last one above, except that the boxes represent different things. We stop the process and achieve the desired partition by putting the partition mark between the two scan-boxes:

$$8\ 3\ 9\ |\ 10\ 15\ 16\ 10$$

It marks where they crossed.

With these two examples and a general description of the idea, we now turn to writing the code for the procedure. There are several ways to implement the general idea, but we must be careful that the algorithm we choose accurately handles all the possible cases of $i = j$ and $i > j$.[9]

It is probably best to present the code first and then analyze how it is structured and how it handles all the cases.

| **Quick Sort (In-Place Version)** |

```
quicksort:=proc(A::array,m,n)
  local p,i,j;
  if m<n then
    p:=A[floor((m+n)/2)];
    i:=m;j:=n;
    while i<j do
      while A[i]<p do i:=i+1 end do;
      while A[j]>p do j:=j-1 end do;
      if i<j then
      exchange(A,i,j);
      j:=j-1;i:=i+1;
      end if;
    end do;
    if i=j and A[i]<=p then
      i:=i+1
    end if;
    quicksort(A,m,i-1);
    quicksort(A,i,n);
  end if;
end proc:
```

---

[t]Beware. There are some algorithms in the literature that do *not* work on all the cases because the program logic and structure fail to account for some of the cases.

This is a deceptively simple-looking procedure, but unless you have had some experience with these things, understanding that it works correctly (maybe even proving that it does) can be surprisingly complex. To help the analysis we construct a "ow chart as shown in Figure 8.4.

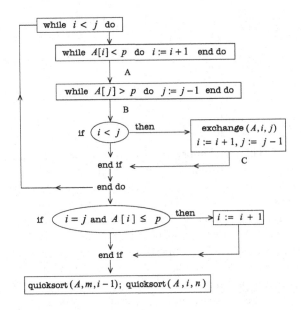

Figure 8.4: *A ow chart for the in-place version of quick sort.*

The scanning processes are controlled by the `while i<j do` loop. The condition $i < j$ is checked each time before entering the loop and when this condition is false (the $i = j$ or $i > j$ cases) control passes to the next statement after the loop. This statement is

```
if i=j and A[i]<=p then
        i:=i+1
end if;
```

and is used to handle the case where the i scan-box and j scan-boxes come together after the exchange-increment-decrement step, labeled C in the "ow chart. This is one of the $i = j$ cases, and since this occurs at the point C, passage back through the `while i<j do` loop does not occur. Consequently, the element x, which both scan-boxes are centered on, never gets compared to the pivot p. With a little thought, we can conclude that the element to

the left of p must be $\leq p$ and the one to the right of p must be $\geq p$. The local picture looks like this

$$p \;\; \boxed{\boxed{x}}^{\,i=j} \;\; p$$

Thus, we can put the partition mark to the right of p if $x \leq p$, but otherwise we *must* put the mark to the left of p. This is what the `if-then` statement accomplishes. It correctly positions the partition in this case. *Note*: Here we have used $\leq p$ and $\geq p$ to stand as symbols for *single* elements on either side of x. That is, rather than writing

$$y \;\; \boxed{\boxed{x}}^{\,i=j} \;\; z$$

we use $\leq p$ for y and $\geq p$ for z. This seems appropriate because, while we do not know what these elements are, we do know from the reasoning that they must be *less than or equal to p* and *greater than or equal to p*, respectively.

There are five other cases that can occur. The reasoning is that when control passes out of the `while i<j` do loop, we definitely know that $i \geq j$. We also know that this can only happen after an increment of i, a decrement of j, or both an increment and decrement. There are three places in the code, marked A,B,C in the flow chart, where we can analyze what happens right after a *possible* incrementing/decrementing. Before doing so it is instructive to note: When the scan-boxes are far apart during the scanning process, right after an exchange-increment-decrement step, the partially partitioned array must look like

$$\leq p \; \ldots \;\; \leq p \; \boxed{x}^{\,i} \; \ldots\ldots \; \boxed{z}^{\,j} \;\; \geq p \; \ldots \;\; \geq p$$

The reason that the inequalties shown here are not strict inequalities is that the array may contain many elements equal to p. The i scan and j scan both stop on encountering an element equal to p, these elements (a pair of p's) are exchanged and the processing continues.

Now try to visualize when the scan-boxes come together or cross. We examine this at places A,B,C in the flow of control.

Suppose that $i \geq j$ occurs at place A in the flow of control. If $i = j$, then i must have received at least one increment from the preceding `while A[i]<p` do loop and so the local picture must be

$$< p \;\; \boxed{\boxed{p}}^{\,i=j} \;\; \geq p$$

The element to the left of the i scan-box must be less than p (otherwise i would have stopped there in the comparison-increment process). The element in the i scan-box is p. If it is equal to p, then the next block of code will not decrement j and control will pass out of the processing loop and the partition mark will be put to the right of the overlapping scan-boxes. If, however, the element in the i scan-box is $> p$, then the next block of code will decrement j, but only by 1, and since then $i > j$, the processing stops. The local picture then is

$$\overset{j}{\boxed{<p}} \quad \overset{i}{\boxed{p}} \quad p$$

In this case, the code now makes the partition with the partition mark located between the scan-boxes (this is always so in the $i > j$ cases).

Next consider what happens if i j occurs at the place B in the "ow chart, but not before, in place A. Then the `while A[j]>p` do loop above B must have been executed at least once and so j has just been decremented from the value it had at place A. Thus, if $i = j$ and if i has been incremented in the prior `while A[i]<p` loop, then the local picture must look like

$$<p \quad \overset{i=j}{\boxed{\boxed{p}}} \quad >p$$

But since j stopped where it did, the element in the scan-boxes cannot be $> p$. Hence, it must be p. On passing out of the processing loop, the condition `if i=j and A[i]<=p` is true and the partition mark is put to the right of the overlapping scan-boxes. Next, if $i = j$ at place B, it could happen that i was not incremented in the prior `while A[i]<p` do loop. Then the local picture is

$$\overset{i=j}{\boxed{\boxed{p}}} \quad >p$$

Since j stopped where it is, the element in the overlapping scan-boxes must be p and again the logic statement puts the partition mark correctly to the right of the overlapping scan-boxes. Finally, in the case when $i > j$ occurs at place B, then the assumption that it did not occur at place A leads us to conclude that j was last decremented within the `while A[j]>p` loop and i was last incremented at place C (right after the exchange). Thus, the local picture is

$$\overset{j}{\boxed{p}} \quad \overset{i}{\boxed{>p}}$$

Thus, upon exiting the processing loop, the partition mark is correctly placed between the scan-boxes.

We discussed above what happens when $i = j$ at place C in the "ow chart. So all that is left to examine is the case when $i > j$ at place C. Then the local picture must look like this

Convince yourself of this! Thus, upon exiting the processing loop, the partition mark is correctly placed between the scan-boxes.

The exercises will give you some examples of lists to sort by hand using the in-place quick sort algorithm. It takes some practice to do this by hand, mainly because of the ambiguity of where to put the partition mark when the scan-boxes meet or cross. Perhaps the above, somewhat tedious, analysis will help you in quicksorting by hand.

We can write some code to graphically display the quick sorting, and this may also help you understand better how quick sort works. The graphics will be an animation showing the exchanging of elements as the various sublists are being processed into their partitions. Before describing the code, it is best to view Figure 8.5, which shows plots of eight frames in the animation that result from quick sorting the list $[2, 1, 5, 3, 6, 4]$. As you can see, each frame consists of six rectangular blocks with sizes representing the six numbers to be sorted. The original unsorted list is shown at the upper left of the figure. It is best to inspect the frames *in pairs*. The first frame in each pair shows two blocks (the shaded ones) which the algorithm selects to exchange, and the second frame in the pair shows these two blocks after the exchange. The first pair of frames in the figure (upper left) shows the exchange of 5 and 4 and the resulting partition into low and high sublists, L_1, H_1. The second and third pairs of frames show the algorithm working on the sublist L_1 and exhibit that it takes two partitions (L_2, H_2 and L_3, H_3) before the sublist L_1 is in order.[10]

Figure 8.5 actually shows more features than we want to produce with our code. We will simply draw the blocks before and after each exchange, coloring the exchanging blocks green, and put in a red circle as the partition mark after each sublist is partitioned.

---

[10]Technically there is a fourth partition L_4, H_4 of $L_{\bar{c}} = \{2, 3\}$, which is not shown in the "gure because there is no exchange. Remember, even though $\{2, 3\}$ is in order, the algorithm must check this before partitioning. When you run the animation on the CD-ROM, you will clearly see this fourth step.

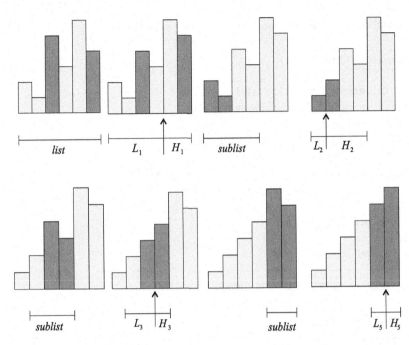

Figure 8.5: *Successive frames in the animation for the quick sort routine.*

We modify the code for **quicksort** to get the code, which we call **QSmovie**, for the animation. Two arrays called **block1,block2** will store the graphics for the block before and after each exchange, and the code for generating these data is inserted in the obvious places in **quicksort**. These will then be stored, in order, in an array called **pic**, which is output from the procedure. After **pic** is produced and returned, we can simply use the **insequence** option in a **display** command to get the animation. We use a local variable **r** to keep track of the number of elements in **pic** (i.e., the number of frames in the movie). It gets incremented by two right after we create the before/after pictures with **pic[r]:=block1;pic[r+1]:=block2**.

The partition marks (circles) are stored in an array called **pmark** and the local variable **s** is used to keep track of the number of partition marks.[11] This is created and another picture (element of **pic**) is produced right before we make the two recursive calls in the **quicksort** procedure.

Now we embed this modified version of **quicksort** in the procedure

---

[11]Do you think that we could predict the total number of partition marks in advance? What about the number of elements of **pic**?

`QSmovie`, which will act as the main procedure. The result is the following:

| Code for Animating Quick Sort |

```
QSmovie:=proc(A::array,m,n,pic::evaln)
  local k,r,s,N,quicksort,block1,block2,pmark;
  with(plottools,polygon,circle):
#--DRAW INITIAL PICTURE OF THE UNSORTED ARRAY--------------
  for k from 1 to n do
      block1[k]:=polygon([[k-.5,0],[k+.5,0],[k+.5,A[k]],
                          [k-.5,A[k]],[k-.5,0]]):
  end do;
  pic[0]:=seq(block1[k],k=1..n):
  r:=1;s:=1;
  N:=nops(op(3,eval(A)));
#------------------------------------------------------------
#          quicksort (WITH GRAPHICS) SUBPROCEDURE
#------------------------------------------------------------
  quicksort:=proc(A::array,m,n)
    local p,i,j,k;
      if m<n then
          p:=A[floor((m+n)/2)];
          i:=m;j:=n;
          while i<j do
              while A[i]<p do i:=i+1 end do;
              while A[j]>p do j:=j-1 end do;
              if i<j then
#--DRAW PICURE BEFORE EXCHANGE----------------------------
              for k from 1 to N do
                if (k=i or k=j)  then
                block1[k]:=polygon([[k-.5,0],[k+.5,0],[k+.5,A[k]],
                      [k-.5,A[k]],[k-.5,0]],color=green):
                else
                block1[k]:=polygon([[k-.5,0],[k+.5,0],[k+.5,A[k]],
                          [k-.5,A[k]],[k-.5,0]]):
                end if;
              end do;
#--EXCHANGE ELEMENTS---------------------------------------
          exchange(A,i,j);
#--DRAW PICTURE AFTER EXCHANGE-----------------------------
```

```
         for k from 1 to N do
         if (k=i or k=j)
          then
          block2[k]:=polygon([[k-.5,0],[k+.5,0],[k+.5,A[k]],
                     [k-.5,A[k]],[k-.5,0]],color=green):
         else
          block2[k]:=polygon([[k-.5,0],[k+.5,0],[k+.5,A[k]],
                     [k-.5,A[k]],[k-.5,0]]):
         end if;
         end do;
         pic[r]:=seq(block1[k],k=1..N);
         pic[r+1]:=seq(block2[k],k=1..N):
#--INCREMENT/DECREMENT------------------------------------
         r:=r+2;
         j:=j-1;i:=i+1;
      end if;
#--END OF EXCHANGING AND DRAWING. NOW LOOP BACK IF i<j--------
         end do;
#--ONE LAST CHECK AND THE PARTITIONING IS DONE:---------------
      if i=j and A[i]<=p then i:=i+1 end if;
#--PUT IN PARTITION MARK (pmark)-----------------------------
      pmark[s]:=circle([i-.5,0], .1, color=red,thickness=2) :
      for k from 1 to N do
       block2[k]:=polygon([[k-.5,0],[k+.5,0],[k+.5,A[k]],
                  [k-.5,A[k]],[k-.5,0]]):
      end do;
      pic[r]:=seq(block2[k],k=1..i-1),seq(pmark[k],k=1..s),
          seq(block2[k],k=i..N);
      r:=r+1;s:=s+1;
#--RECURSIVE CALLS-------------------------------------------
      quicksort(A,m,i-1);quicksort(A,i,n);
      end if;
    end proc:
#--END OF quicksort DEFINITION-------------------------------
#--INITIATE CALL TO quicksort--------------------------------
    quicksort(A,m,n);
    end proc:
```

After studying the above code, you should look at the material on the CD-

ROM corresponding to it and try it out on several examples. Of course, the code is only for tutorial purposes. You would not want to use it on extremely long lists or lists with numbers widely varying in magnitude.

8.4 Numbers

Recursion can be effectively used to determine the decimal representations for rational numbers $\frac{c}{d}$. Indeed, this is nothing other than the long division algorithm you learned in grade school, even though you probably never thought of it as a recursive algorithm.

While this topic may seem rather elementary, it requires the concept of infinite series to handle the case when the decimal representation is repeating and it also leads to many interesting related topics that are important in mathematics and computer programming. Therefore, we devote this section to a brief discussion of numbers (integers, rationals, and reals) in mathematics and in their implementations on a computer.

8.4.1 Numbers in Mathematics

Even though you have been using and calculating with numbers since grade school, you have perhaps never had a course that *defined* these numbers axiomatically, especially the set of real numbers $\mathbb{R}$. Generally this theory occurs in advanced mathematics courses at the senior or graduate level. We do not want to develop this theory here, but rather we only want to point out a few aspects of it which may be helpful to you.

Traditionally the set of natural numbers is denoted by $\mathbb{N}$ and can be defined rigorously by Peano's axioms (see [MM]). Using decimal notation for its elements, this set is descriptively written as

$$\mathbb{N} = \{\, 1, 2, 3, 4, \ldots \,\}.$$

From $\mathbb{N}$, one constructs the set $\mathbb{Z}$ of integers and endows this set with the operations of addition, subtraction, and multiplication. From the integers $\mathbb{Z}$, one constructs the set of rational numbers $\mathbb{Q}$ as the set consisting of ordered pairs (m, n) of integers m, n. Traditionally one writes $\frac{m}{n}$, or m/n for such an ordered pair and declares

$$\frac{a}{b} = \frac{c}{d},$$

if and only if $ad = bc$. This construction of the rationals gives a set $\mathbb{Q}$ that contains the integers as a subset, $\mathbb{Z} \quad \mathbb{Q}$, has extensions of the operations on

$\mathbb{Z}$ of addition, subtraction, and multiplication, and also permits the operation of division (except by zero). Mathematically, $\mathbb{Q}$ is an example of what is known as a *field*.

The set of real numbers $\mathbb{R}$ is also a field and it contains the rationals as a subfield $\mathbb{Q}$ $\mathbb{R}$. However, the construction of $\mathbb{R}$ from $\mathbb{Q}$ is quite a bit more complicated than the construction of $\mathbb{Q}$ from $\mathbb{Z}$. There are several methods of constructing (defining) the set $\mathbb{R}$ one can use *Dedekind cuts* (see [MM], [Sch]) or one can use *Cauchy sequences of rational numbers* (see [MM]). We mention here just a few things about the latter approach.

The need for a larger set of numbers than the rationals $\mathbb{Q}$ was realized in ancient times by the Pythagoreans. They considered it as somewhat shocking that the length d of the diagonal of any square was not commensurate with the length s of the side of the square. Otherwise put, the ratio $x = d/s$ is not a rational number. To see this, note that by the Pythagorean theorem, $d^2 = s^2 + s^2 = 2s^2$ and so $x^2 = d^2/s^2 = 2$. Thus the square of x is 2. Now if x were a rational number, then $x = p/q$ where p and q are positive integers. We can assume that p and q have no common factors (Otherwise, reduce the fraction p/q by eliminating common factors to get $p/q = p'/q'$ and p', q' have no common factors.) Now since

$$\frac{p^2}{q^2} = \left(\frac{p}{q}\right)^2 = x^2 = 2,$$

we get that $p^2 = 2q^2$. Hence, p^2 is an even integer and therefore p itself must be even (the squares of odd integers are odd). Thus p has the form $p = 2k$ for some integer k. But from this and $p^2 = 2q^2$, we get $4k^2 = 2q^2$. Dividing both sides by 2 gives $2k^2 = q^2$ and hence q^2 is even. Thus, as before, this says that q itself must be even. The reasoning has thus shown that *both* p and q are even. Hence, they have a common factor, namely 2, and this contradicts our initial assumptions that they had no common factors. To avoid the contradiction, we must admit that the assertion that $x^2 = 2$ has a rational solution x is indeed wrong.

With arguments like this the Pythagoreans went on to show that many other geometrical quantities x could not be represented by rational numbers and thus there was a need to extend $\mathbb{Q}$ to include non-rational numbers, i.e. *irrational numbers*. The larger set $\mathbb{R}$, comprised of the rational and irrational numbers and called the set of *real numbers*, was recognized and used in the ensuing centuries since Pythagoras. However, it is fair to say that only at the end of the 19th century did mathematicians finally get around to defining this set rigorously. Their definition is motivated by the following applied approach to understanding the solution of $x^2 = 2$.

While there is no rational number x that satisfies $x^2 - 2 = 0$, we can find many rational numbers that are *approximate* solutions of this equation. Indeed, Newton's method (see the case study in Chapter 3) gives us an algorithm for generating these rational approximations.

If we let $f(x) = x^2 - 2$, then Newton's method to approximate solutions of $f(x) = 0$ amounts to computing the map $T(x) = x - f(x)/f'(x)$, selecting a rational number r suitably close to the solution, and then iterating the map T:

$$T(r), \; T^2(r), \; T^3(r), \ldots, T^k(r), \ldots.$$

This is motivated in Chapter 3. In theory, $\lim_{k\to} T^k(r) = x$, where x is a root of the equation $f(x) = 0$. In the case at hand, $f'(x) = 2x$ and so the formula for the map T is

$$T(x) = \frac{x}{2} + \frac{1}{x}.$$

This is a particularly easy map to iterate. For instance, if we choose $r = 2$ as the initial value, then it is simple to calculate the first five iterates:

$$
\begin{aligned}
r_1 &= T(2) = 1 + \tfrac{1}{2} = \tfrac{3}{2} \\
r_2 &= T\,\tfrac{3}{2} = \tfrac{3}{4} + \tfrac{2}{3} = \tfrac{17}{12} \\
r_3 &= T\,\tfrac{17}{12} = \tfrac{17}{24} + \tfrac{12}{17} = \tfrac{577}{408} \\
r_4 &= T\,\tfrac{477}{408} = \tfrac{577}{816} + \tfrac{408}{577} = \tfrac{665857}{470832} \\
r_5 &= T\,\tfrac{665857}{470832} = \tfrac{665857}{941664} + \tfrac{470832}{665857} = \tfrac{886731088897}{627013566048}.
\end{aligned}
$$

It is easier to discern that the sequence $\{r_k\}_{k=1}$ is converging by converting the terms to their decimal representations:

$$
\begin{aligned}
r_1 &= 1.5 \\
r_2 &= 1.41666666666666666666666666666666 \cdots \\
r_3 &= 1.41421568627450980392156862745 \cdots \\
r_4 &= 1.41421356237468991062629557889 \cdots \\
r_5 &= 1.41421356237309504880168962350 \cdots
\end{aligned}
$$

Can we now use this sequence to define $x = \sqrt{2}$, i.e., a number whose square is 2? Why not just define $\sqrt{2}$ as the limit of this sequence of rationals:

$$\sqrt{2} = \lim_{k\to} r_k.$$

While this may appear to be a good definition of $\bar{2}$, it is not since it uses the concept of the limit, and the definition of the limit involves the limiting value, in this case $\bar{2}$. (Check the - definition of the limit of a sequence in your calculus book.) Of course, with a prior definition of the set of real numbers, the above limit is certainly a true statement. We just cannot use this statement to define $\bar{2}$. The way out of this dilemma is to *define* $\bar{2}$ by

$$\bar{2} \quad \{r_k\}_{k=1},$$

i.e., think of $\bar{2}$ as being given by the whole sequence $\{r_k\}_{k=1}$.[12]

These calculations help motivate the modern definition of a real number. The definition is that a real number x is a *Cauchy sequence of rational numbers*. Saying that a sequence is *Cauchy* means that differences, $r_k - r_\ell$, between the terms become as small as we please for large enough k and . In the above example, the differences of the consecutive terms are

$$
\begin{aligned}
r_2 - r_1 &= \frac{1}{12} \\
r_3 - r_2 &= \frac{1}{408} \\
r_4 - r_3 &= \frac{1}{470832} \\
r_5 - r_4 &= \frac{1}{627013566048}.
\end{aligned}
$$

These are easily seen to be getting successively smaller. This concept (definition) of real numbers is discussed further in the Maple/Calculus Notes at the end of this chapter.

Defining the integers, rationals, and reals in an axiomatic way as just indicated gives us a rigorous construct for the numbers that are the foundation of analysis. This foundation is necessary for all further theoretical work. However, for the practical aspects of actually calculating with these numbers, we must employ a system of notation and representation for these abstract quantities. For doing calculations by hand, the decimal system has been the standard for centuries, and the next subsection discusses some recursive programming connected with this.

8.4.2 Base N Representations of Numbers

An interesting recursive algorithm occurs in the process of computing representations of integers, and more generally rational numbers, in terms of a

---

[12]Technically, the modern de"nition de"nes $\sqrt{2}$ as an equivalence class of sequences.

given base N, where $N > 1$ is a fixed integer. When $N = 2$, the representation is called the *binary* representation of the rational number. For the other common choices, $N = 3, 8, 10, 16$, the representations are called the *ternary*, *octal*, *decimal*, and *hexadecimal* representations, respectively. For base N, the corresponding representation is often called the *N-ary representation*.

In general, the base N representation of a positive integer m comes from the assertion (actually theorem) that there is a finite sequence of integers $\{a_j\}_{j=0}^{s}$, with $0 \quad a_j < N$, such that

$$m = a_0 + a_1 N + a_2 N^2 + \cdots + a_s N^s \tag{8.4}$$

$$= \sum_{j=0}^{s} a_j N^j. \tag{8.5}$$

The proof that such a representation is possible is left as an exercise. The proof leads to an algorithm for determining the numbers $a_0, a_1, \ldots, a_s$, which can be implemented in Maple. (This is also an exercise.)

Example 8.2 (Base N Expressions of Integers) We look at the ternary, octal, and hexadecimal representations of a given integer, say, $m = 215$. This may help in the exercise on writing the code to automate this process.

It is important to stress at the beginning that all of your experience and calculations with particular numbers have been with the *decimal* representations of them. Abstractly, numbers exist (are defined or constructed from primitives) independently of the system used to represent them. However, in practice one must choose a particular representation for numbers in order to perform calculations. It was only in the 18th century that the decimal system was adopted as the standard for representing integers and doing calculations.[13]

To find the ternary representation of 215, we consider the highest power of three, 3^i, that is less than 215. This is $3^4 = 81$, since $3^5 = 243$. Dividing 215 by 3^4 gives quotient 2 and remainder 53. Thus

$$215 = 2(3^4) + 53.$$

Now repeat the process with 53. The highest power of three that is less than it is 3^3, and dividing by this gives quotient 1 and remainder 26. Thus,

$$53 = 1(3^3) + 26.$$

---

[1]Lagrange, who was on the committee to study which base to use for the number system, argued that base ten would be better than base twelve. D Alembert, another French mathematician on the committee, recommended base twelve, presumably because of its connection with time and the seasons   twelve hours from noon to midnight, twelve months in the year   and also because it has more divisors than ten.

Using this in the previous equation gives

$$215 = 2(3^4) + 1(3^3) + 26.$$

Continuing in this fashion, we eventually reach a remainder that is less than 3 and then we are done. You can easily verify that the result is

$$215 = 2(3^4) + 1(3^3) + 2(3^2) + 2(3) + 2.$$

The same process can be used to determine the representations of 215 in base eight (octal) and base 16 (hexadecimal). The results are

$$215 = 3(8^2) + 2(8) + 7$$
$$215 = 13(16) + 7$$

The decimal representation of the integer $m = 215$ requires no work. It is

$$215 = 2(10^2) + 1(10) + 5.$$

The reason that this requires no work is that when we express m as 215, we are using the *decimal positional notation* for m. Generally, if m is given in base N by

$$m = a_s N^s + a_{s-1} N^{s-1} + \cdots + a_2 N^2 + a_1 N + a_0,$$

where a_i $\{0, 1, \ldots, N-1\}$, then the *N-ary positional notation* for m is

$$m = a_s a_{s-1} \cdots a_2 a_1 a_0 \qquad \text{(base } N\text{)}.$$

For example, the decimal, ternary, octal, and hexadecimal positional notations for 215 are

$$
\begin{aligned}
m &= 215 & \text{(base 10)} \\
&= 21222 & \text{(base 3)} \\
&= 327 & \text{(base 8)} \\
&= \text{D7} & \text{(base 16)}
\end{aligned}
$$

Notice that the hexadecimal notation for m is D7. The letter D is required because base N positional notational needs $N-1$ distinct symbols, or digits, to stand for the numbers a_i. For hexadecimal, the fifteen symbols are 0, 1, 2, 3, 4, 5, 6, 7, 8, 9, A, B, C, D, E, with A, B, C, D, E representing the tenth, eleventh, twelfth, thirteenth, and fourteenth ordinal numbers, respectively. For the number m here, we computed that its hexadecimal coe cients (expressed decimally) are: $a_1 = 13$ and $a_0 = 7$, and its correct positional notation is in terms of two symbols (digits), $m = $ D7 (base 16). We *cannot* write $m = 137$ (base 16), because that is misleading and confuses it with the base 16 number $311 = 1(16)^2 + 3(16) + 7$.

An essential tool used in the algorithm for computing base N expressions is the *Euclidean algorithm*. It simply says that if m and n are positive integers, then we can write $m = nq + r$, where q, r are nonnegative integers and $r < n$. In perhaps more familiar terms, this says that we can write

$$\frac{m}{n} = q + \frac{r}{n},$$

i.e., we can divide m by n to get quotient q and remainder n. For example, $23/4 = 5 + 3/4$. In the special case when $m < n$, we take $q = 0$ and $r = m$. So the asserted representation is only significant for the case when m is larger than n.

The Euclidean algorithm is the mental process you perform when doing this division by hand. We can write a procedure to automate the process of computing the quotient q and remainder r as follows. To understand the simple algorithm involved, it helps to visualize things geometrically.[14] Represent the numbers m, n by line segments as shown in Figure 8.6.

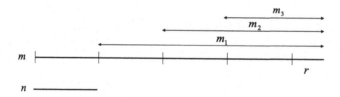

Figure 8.6: *Line segments whose lengths are m and n. The Euclidean algorithm gives, for this figure, $m = 4n + r$.*

Starting at one end of m, lay n along m and subtract to get a shorter line segment m_1 as the remainder. Now repeat the process, laying n out along m_1, from one end, subtract, and get a remaining line segment m_2. The sequence of lengths for $m, m_1, m_2, \ldots$ is

$$m, \quad m_1 = m - n, \quad m_2 = m_1 - n, \quad m_3 = m_2 - n, \ldots.$$

---

[14]The ancient Greeks did all their calculations geometrically with line segments representing the numbers.

The process continues until we get a remaining line segment that is less than n. This length of this last remainder is r and the number of subtractions in the process is q.

We can code this algorithm with a simple `while` loop, within which we continually subtract n from the previous remainder until we reach one that is strictly smaller than n.

| Code for the Euclidean Algorithm |

```
euclid:=proc(m,n,q::evaln,r::evaln)
  local M,Q;
  M:=m;Q:=0;
  while n <= M do
    M:=M-n;Q:=Q+1;
  end do;
  q:=Q;r:=M;
  RETURN()
end proc:
```

Note that it is necessary to introduce the local variables `M,Q` to represent the parameters `m,q`.

Example 8.3 (Base N Expressions for Rationals Less than 1)
Here we show how to determine the base N representation for any positive rational number c/d. The special case when the number is an integer ($d = 1$) was mentioned above and is assigned as an exercise. This special case is used in the general case as follows. By the Euclidean algorithm, we can write c/d as an integer plus a positive rational number less than 1:

$$\frac{c}{d} = q + \frac{m}{n}.$$

By your exercise, we can express q in base N form as

$$q = a_0 + a_1 N + a_2 N^2 + \cdots + a_s N^s,$$

where the a_i's are integers with $0 \le a_i < N - 1$. Thus, to obtain the base N expression of *any* positive rational, it suffices to show how to obtain such a representation for the special case of positive rationals $m/n < 1$ (i.e., $m < n$). For this we show

$$
\begin{aligned}
\frac{m}{n} &= b_1 N^{-1} + b_2 N^{-2} + b_3 N^{-3} + \cdots \\
&= \sum_{j=1} b_j N^{-j},
\end{aligned}
\tag{8.6}
$$

where $b_j \in \{0, 1, \ldots, N-1\}$ for all j, and one and only one of the following occurs:

(1) The series terminates, i.e., there exists an r such that $b_j = 0$ for all $j > r$.

(2) The series repeats, i.e., there exist r, p such that $b_{j+p} = b_j$ for all $j > r$. The number p is called the *period* of the repeating representation.

The series in equation (8.6) is called the base N representation of m/n and the corresponding positional notation for the two possible cases is

$$\frac{m}{n} = .b_1 b_2 \cdots b_r \qquad \text{(base } N\text{)} \qquad (8.7)$$

$$\frac{m}{n} = .b_1 b_2 \cdots b_r \overline{b_{r+1} \cdots b_{r+p}} \qquad \text{(base } N\text{)}. \qquad (8.8)$$

The overbar in the second representation indicates the digits that are repeating and is merely shorthand notation for what is otherwise expressed precisely by the series in equation (8.6).

To show the existence of the sequence $\{b_j\}_{j=1}$, we proceed as follows. Let i_1 be the first integer such that $mN^{i_1} > n$. By the Euclidean algorithm,

$$\frac{mN^{i_1}}{n} = q_1 + \frac{m_1}{n},$$

for integers q_1, m_1, with $0 \le q_1 < N$, $0 \le m_1 < n$. Note that our choice of i_1 guarantees that q_1 is less than N. Rearranging this gives

$$\frac{m}{n} = q_1 N^{-i_1} + \frac{m_1}{n} N^{-i_1}. \qquad (8.9)$$

Next, apply the same process to m_1/n, i.e., let i_2 be the first integer such that $m_1 N^{i_2} > n$. By the Euclidean algorithm:

$$\frac{m_1 N^{i_2}}{n} = q_2 + \frac{m_2}{n},$$

for integers q_2, m_2, with $0 \le q_2 < N$, $0 \le m_2 < n$. Rearranging this gives

$$\frac{m_1}{n} = q_2 N^{-i_2} + \frac{m_2}{n} N^{-i_2}. \qquad (8.10)$$

Substituting this into equation (8.9) gives

$$\frac{m}{n} = q_1 N^{-i_1} + q_2 N^{-(i_1+i_2)} + \frac{m_2}{n} N^{-(i_1+i_2)}. \qquad (8.11)$$

Next, apply the same process to m_2/n, to get i_3, q_3, m_3. Then apply the process to m_4/n, to get i_4, q_4, m_4, and so on, continuing this process inductively yields, after s steps,

$$\frac{m}{n} = \sum_{k=1}^{s} q_k N^{-j_k} + \frac{m_s}{n} N^{-j_s}. \tag{8.12}$$

Here for convenience we have let $j_k = i_1 + i_2 + \cdots + i_k$. Now note that while the process can be continued indefinitely, each of the remainders $m_1, m_2, \ldots$, is one of the finitely many numbers in $\{0, 1, 2, \ldots, n - 1\}$. Included among these numbers is the initial numerator m, which we label as m_0 m, for convenience. Thus, after at most n steps in this process, either (1) one of the remainders m_L is 0, in which case all the ensuing remainders are zero: $m_{L+1} = 0, m_{L+2} = 0, \ldots$, or (2) one of the remainders m_R coincides with a previous remainder, $m_R = m_L$, with $L < R$. These two possibilities lead to the desired representations in each case. (See the Maple/Calculus Notes for more details.)

Note that the process determines the nonzero b_j's, i.e., $b_{j_k} = q_k$, while $b_j = 0$ for j different than $j_1, j_2, j_3, \ldots$. If we let r j_L, then in the terminating case b_r is the last nonzero coe cient in the representation. In the repeating case p $j_R - j_L$ is the *period* of the representation, and the coe cients $b_{r+1}, b_{r+2}, \ldots, b_{r+p}$ form the repeating part of the representation. See Figure 8.7. For uniformity of notation, we take $p = 0$ in the case when the representation terminates.

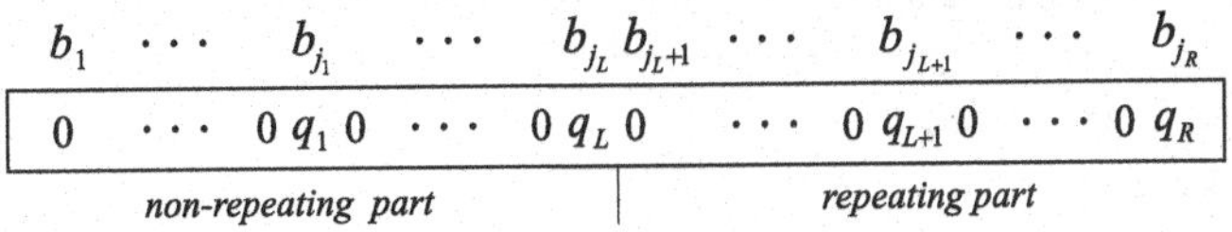

Figure 8.7: *In the repeating case, the indices $j_L + 1$ and j_R mark the first and last of the repeating coefficients, or digits.*

To write the code to determine the coe cients $\{b_j\}_{j=1}^{r}$ we first write a recursive procedure called **rep** that determines the nonzero b_j's for $j = j_1, j_2, \ldots$, exactly as in the above description. This procedure employs global variables for the remainders $m_1, m_2, \ldots$, the indices $j_1, j_2, \ldots$ (for which $b_{j_k} \neq 0$), and the counting index k. These variables are denoted by **M,J,K** respectively. A global variable **Jmax** is used to control the maximum number of coe cients b_j that you wish to consider. When **Jmax** is exceeded, the recursive calculation of the b_j ceases and the parameter **flag** is set to 1.

We now write a main procedure, called baseN, that includes the defining code for the procedure rep and euclid (see above). Thus these are sub-procedures to baseN, and baseN, after some definitions and initializations, makes a call to rep to begin the recursive calculation.

Code for Calculating the Base N Representation of $\dfrac{m}{n}$ $\quad$ 1

```
baseN:=proc(m,n,N,b::evaln,r::evaln,p::evaln)
  local M,J,K,Jmax,i,b1,euclid,rep;
  if m>=n then RETURN('m must be less than n')end if;
  # Define two sub-procedures:
  # --------------------------------
  # procedure for Euclidean algorithm
  # --------------------------------
  euclid:=proc(m,n,q::evaln,r::evaln)
    local M,Q;
    M:=m;Q:=0;
    while n <= M do
      M:=M-n;Q:=Q+1;
    end do;
    q:=Q;r:=M;
    RETURN()
  end proc:
  # ----------------------------------------------------
  # recursive procedure for finding the nonzero b[j] s
  # ----------------------------------------------------
  rep:=proc(m,n,b,flag::evaln)
    local i,q,m1,t;
    i:=0;
    while m*N^i < n do i:=i+1 end do;
    euclid(m*N^i,n,q,m1);
      K:=K+1;
      J[K]:=J[K-1]+i;
      if J[K]>Jmax
        then flag:=1;RETURN();
      end if;
      b[J[K]]:=q;
      M[K]:=m1;
      if M[K]=0 then
        r:=J[K];p:=0;RETURN();
```

```
        end if;
        for t from 0 to K-1 do
          if M[t]=M[K] then
            r:=J[t];p:=J[K]-J[t];RETURN();
          end if;
        end do;
        rep(m1,n,b,flag);
    end proc:
#  ----------------------------------------
#       Start main Program
#  ----------------------------------------
    Jmax:=200; #This is the max number of digits
    b1:=array(1..Jmax);M:=array(0..n);J:=array(0..n);
    K:=0;M[0]:=m;J[0]:=0;flag:=0;
    for i to Jmax do b1[i]:=0 end do:
    rep(m,n,b1,flag);
    b:=b1:
    if flag=1 then print('exceeded max digits') end if;
    RETURN()
end proc:
```

Here are several examples of the use of this procedure (see the CD-ROM also). All the output is returned through the parameters b,r,p, and from this we can use Maple's **cat** command to display the initial block of digits $b_1 \ldots b_r$ (if any) and then the block of repeating digits $b_{r+1} \ldots b_p$. For example, the commands

```
baseN(61,74,10,b,r,p);
cat(seq(b[k],k=1..r)),cat(seq(b[k],k=r+1..r+p));
```

produce the output

   8,243

which gives the decimal expression for 61/74, and indicates that the repeating digits are 243. The commands

```
baseN(11,127,10,b,r,p);
cat(seq(b[k],k=1..r)),cat(seq(b[k],k=r+1..r+p));
```

give the output

   ,086614173228346456692913385826771653543307

which indicates a block of 42 digits that form the repeating part. On the other hand, the command

```
baseN(300,431,10,b,r,p);
```

produces the output

exceeded max digits

which indicates that the decimal representation of 300/431 has not begun to repeat after 200 digits.

Summary: By combining the N-ary representations for positive integers with the N-ary representation for positive rational numbers less than 1, we get N-ary representations for all positive rational numbers. Including a minus sign gives representations for negative rationals as well. Then, in terms of the base N positional notation, we have that any $c/d \in \mathbb{Q}$ has one of the following representations:

$$\frac{c}{d} = \pm a_s \cdots a_1 a_0.b_1 b_2 \cdots b_r \qquad \text{(base } N\text{)} \qquad (8.13)$$

$$\frac{c}{d} = \pm a_s \cdots a_1 a_0.b_1 b_2 \cdots b_r \overline{b_{r+1} \cdots b_{r+p}} \qquad \text{(base } N\text{)} \qquad (8.14)$$

where $a_i, b_j \in \{0, 1, \ldots, N-1\}$. Here $a_s \cdots a_1 a_0$ is called the *integer part* and $b_1 b_2 \cdots$ is called the *fractional part* of c/d .

8.4.3 Numbers in Programming

We think it will be beneficial to include here a brief (and therefore simplistic) discussion of how computers store and operate on numbers. For the most part, the algorithms you have seen and written when studying this book do not require any knowledge of how computers and software (such as Maple) treat numbers. However, advanced projects in scientific computing inevitably require some concern with the accuracy and error problems that can arise from computer arithmetic and, while this is properly covered in courses in numerical analysis, the discussion here will give you an indication of what this concern is all about.

Any computing machine, whether it is Pascal's pascaline,[15] Charles Babbage's analytical engine, a Burrough's adding machine, a Texas Instruments calculator, or a modern Cray computer, can only store a finite set $\mathbb{M}$ of rational numbers, which we will call *machine numbers* (or hardware numbers).

---

[15]See [Aug] and [Asp] for an illustrated history of various calculating devices.

We describe three different schemes, *fixed-point*, *oating-point*, and *rational*, for storing and doing arithmetic on a computer. Each scheme leads to a different set $\mathbb{M}$ of machine numbers and has different problems/benefits with regard to arithmetical accuracy and errors. Using clever programming (as the Maple software designers did), one can extend the limitation of machine numbers and create special data structures for what could be called *software numbers* and provide more accurate operations on these numbers.

The fixed-point and "oating-point schemes for representing machine numbers could be discussed relative to any base N. Early calculating machines employed wheels and gears for which the decimal base was convenient, but present-day computers are built from electronic circuits for which the binary representation is most natural. Thus, for clarity the ideas and concepts are presented for base $N = 2$, even though they are independent of the base, and indeed, octal and hexadecimal representations of machine numbers are also important.

Computer Memory

The smallest memory element in a binary computer is called a *bit* and, regardless of the actual physical mechanism employed to construct it, this element is capable of being set in one of two states, say, *o* and *on*. The two states can be used to represent the numbers 0 and 1, which are the two binary digits in the binary system.[16] A string of eight bits constitutes a memory element known as a *byte*, and if we only needed to store nonnegative integers, then a byte is big enough to store any one of the numbers $0, 1, \ldots, 2^8 - 1 = 255$. Larger memory elements called *words* are composed of more bits, typically a multiple number of bytes. For example, present-day memory modules for personal computers have 32-bit words (four bytes per word). For the sake of our discussion, we will assume a fictitious computer memory with words w consisting of   bits. We will say that the length of w is   and picture w as shown in Figure 8.8.

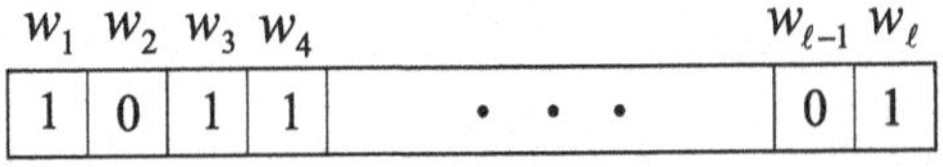

w_1	w_2	w_3	w_4			$w_{\ell-1}$	w_ℓ
1	0	1	1		$\cdots$	0	1

Figure 8.8: *A word w consists of   slots, or boxes, each capable of storing one or the other of the binary digits $0, 1$.*

---

[16] The word *bit* comes from a contraction of *bi*nary dig*it*.

Fixed-point Schemes

In the last subsection above, we saw that any rational number $x \in \mathbb{Q}$ has a base 2 positional representation of the form

$$x = \pm a_k \cdots a_1 a_0 . b_1 b_2 \cdots b_n \cdots \qquad \text{(base 2)}, \qquad (8.15)$$

where the $a_i, b_j \in \{0,1\}$. Irrational numbers x have a similar representation except that their fractional parts, $b_1 b_2 \cdots$, go on forever without terminating or repeating. This of course precludes storing an irrational number in a word composed of finitely many bits. The same is true for rational numbers with repeating fractional part — storing them exactly would require infinitely many bits. Even for rational numbers x with terminating fractional part, we must limit the lengths s and r of the integer part and the fractional part of x to store it in a word w of length ℓ.

A typical fixed-point scheme employs one bit, say the first one, of the word w to store the $\pm$ sign of x, and splits the remaining bits into two parts to store the integer and fractional parts of x (if they are not too long to fit). Figure 8.9 shows the allocation of the bits.

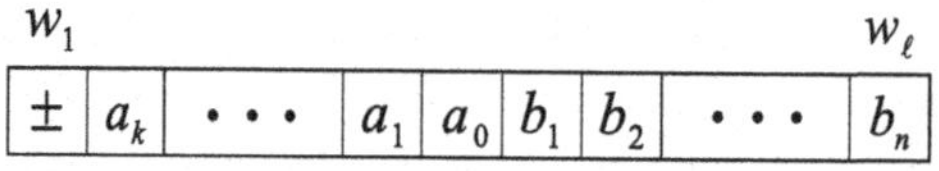

w_1									w_ℓ
$\pm$	a_k	$\cdots$	a_1	a_0	b_1	b_2	$\cdots$		b_n

Figure 8.9: *Allocation of the bits in a word w for fixed-point storage and arithmetic.*

Note that $k + 1$ of the bits are allocated to the integer part $a_k \cdots a_1 a_0$ and n bits are allocated to the fractional part $b_1 \cdots b_n$. Thus $\ell = k + n + 2$. The choice of k and n determines which rational numbers x get represented as machine numbers, and in particular what the largest positive number x_{max} and the smallest positive number Δx are.

To look at this in more detail, we let $\mathrm{M}^{k,n}_{fixed}$ denote the set of machine numbers that are possible using the fixed-point scheme in Figure 8.9. Thus by definition

$$\mathrm{M}^{k,n}_{fixed} = \left\{ x = \pm \sum_{i=0}^{k} a_i 2^i + \sum_{j=1}^{n} b_j 2^{-j} \; \middle| \; a_i, b_j \in \{0,1\} \right\}.$$

Clearly the smallest positive number in this set is 2^{-n} and we call this Δx:

$$\Delta x = \frac{1}{2^n},$$

for reasons that will be apparent in a moment. We can get a better description of the set of machine numbers if we rewrite the summation notation for
a typical machine number x as follows.

$$
\begin{aligned}
x &= \pm \left[\sum_{i=0}^{k} a_i 2^i + \sum_{j=1}^{n} b_j 2^{-j} \right] \\
&= \pm \left[\sum_{i=0}^{k} a_i 2^{n+i} + \sum_{j=1}^{n} b_j 2^{n-j} \right] \cdot 2^{-n} \\
&= \pm \left[\sum_{p=0}^{n-1} b_{n-p} 2^p + \sum_{i=0}^{k} a_i 2^{n+i} \right] \cdot 2^{-n} \\
&= \pm \left[\sum_{p=0}^{k+n} c_p 2^p \right] \cdot 2^{-n} \\
&= C \, \Delta x.
\end{aligned}
$$

Here $c_0, \ldots, c_{k+n}$ are just a relabeling of the a_i's and b_j's in the order indicated above. The number $C = \pm \sum_{p=0}^{k+n} c_p 2^p$ is an integer, and the largest
one of these is for the choice $c_p = 1$, for every p:

$$
C_{max} = \sum_{p=0}^{k+n} 2^p = 2^{k+n+1} - 1.
$$

All of this shows that every machine number is a multiple of $\Delta x = 2^{-n}$ (the
smallest machine number). More precisely

$$
\mathbb{M}_{fixed}^{k,n} = \left\{ \frac{C}{2^n} : C = 0, \pm 1, \ldots, \pm(2^{k+n+1} - 1) \right\}.
$$

From this it is easy to see that the maximum machine number in this fixed-
point scheme is

$$
x_{max} = \frac{2^{k+n+1} - 1}{2^n} = 2^{k+1} - \frac{1}{2^n} = 2^{k+1} - \Delta x.
$$

The machine numbers lie in the interval $[-x_{max}, x_{max}]$ and divide this interval into equal subintervals of length Δx.

For example, suppose $k = 1, n = 3$, so that the word length is $\ell = 6$ bits.
The fixed-point scheme is

$$
x = \pm a_1 a_0 . b_1 b_2 b_3,
$$

with two bits allocated to the integer part and 3 bits to the fractional part.
The machine numbers are

$$\mathrm{M}^{1,3}_{fixed} = \frac{C}{8} : C = 0, \pm 1, \ldots, \pm 31 \quad .$$

The smallest positive machine number is $\Delta x = \frac{1}{8}$ and the largest is $x_{max} = 31/8 = 4 - \Delta x = 3\frac{7}{8}$. Figure 8.10 shows the nonnegative machine numbers based on this scheme.

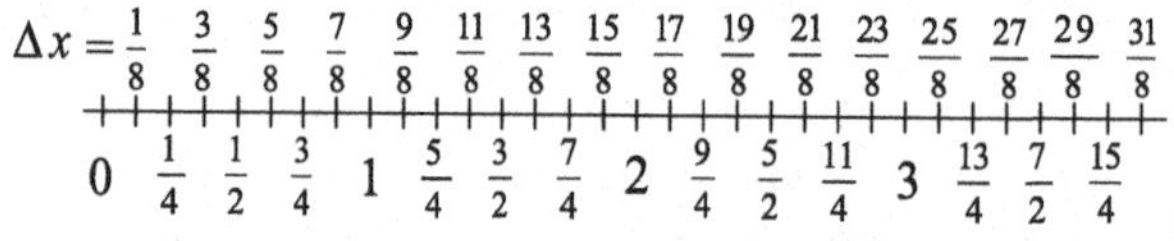

$$\Delta x = \frac{1}{8} \quad \frac{3}{8} \quad \frac{5}{8} \quad \frac{7}{8} \quad \frac{9}{8} \quad \frac{11}{8} \quad \frac{13}{8} \quad \frac{15}{8} \quad \frac{17}{8} \quad \frac{19}{8} \quad \frac{21}{8} \quad \frac{23}{8} \quad \frac{25}{8} \quad \frac{27}{8} \quad \frac{29}{8} \quad \frac{31}{8}$$

$$0 \quad \frac{1}{4} \quad \frac{1}{2} \quad \frac{3}{4} \quad 1 \quad \frac{5}{4} \quad \frac{3}{2} \quad \frac{7}{4} \quad 2 \quad \frac{9}{4} \quad \frac{5}{2} \quad \frac{11}{4} \quad 3 \quad \frac{13}{4} \quad \frac{7}{2} \quad \frac{15}{4}$$

Figure 8.10: *The nonnegative fixed-point numbers $\pm a_1 a_0.b_1 b_2 b_3$ for a computer with 6 bit words ($k = 1, n = 3$). These consist of all the rational numbers in the interval $[0, 3\frac{7}{8}]$ of the form $\frac{C}{8}$, with $C = 0, 1, \ldots, 31$.*

It is important to note that regardless of the word size   and the choice of k and n, the set $\mathrm{M}^{k,n}_{fixed}$ is *not* closed under any of the four arithmetical operations. That is, if x, y $\mathrm{M}^{k,n}_{fixed}$, then $x + y, x - y, xy, x/y$ need not be in $\mathrm{M}^{k,n}_{fixed}$. This is easy to see for the example $k = 1, n = 3$ above:

$$\frac{31}{8} + \frac{1}{8} = 4 \quad (\text{over"ow}), \qquad \frac{1}{8} \cdot \frac{1}{8} = \frac{1}{16} \ (\text{under"ow}).$$

In general, when an arithmetical operation results in a number that is larger in magnitude than x_{max}, then *over ow* is said to occur. When the result is smaller in magnitude than Δx, then *under ow* occurs. Over"ow generally halts the computation, but under"ow can be handled by setting the result to zero. Similarly, the operating system could be designed to handle the result $\frac{31}{8} \cdot \frac{1}{8} = \frac{31}{16}$ by rounding to $\frac{30}{16} = \frac{15}{8}$.

The nonclosure of the arithmetical operations on any set of machine numbers is the source of the problems with accuracy errors when using any computer. In general, for

$$x = \frac{C}{2^n}, \quad y = \frac{D}{2^n},$$

the exact arithmetical results are

$$x \pm y = \frac{C \pm D}{2^n}, \qquad xy = \frac{CD}{2^{2n}}, \qquad \frac{x}{y} = \frac{C}{D}.$$

In practice the numbers x, y are of reasonable magnitudes (neither too large nor too small), which means that C and D are quite a bit less than C_{max}. Thus $|C \pm D|$ is also less than C_{max} and so $x \pm y$ are machine numbers.

Floating-point Schemes

A "oating-point scheme for storing numbers generally results in a larger set of machine numbers and more accuracy than the corresponding fixed-point scheme (or schemes, if you vary k and n). A "oating-point scheme stores numbers x that can be expressed as a fractional part times a power of 2:

$$x = \pm .b \, 2^{\pm e} = \pm .b_1 b_2 \cdots b_n \, 2^{\pm e_{k-1} \cdots e_1 e_\ell}.$$

Here $e_i, b_j \quad \{0,1\}$ are binary digits. The corresponding picture of the way bits of a word w are allocated to store such numbers is shown in Figure 8.11.

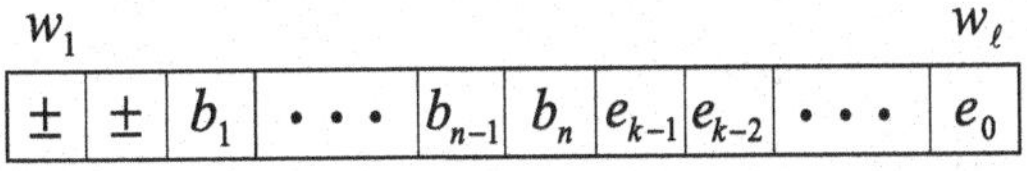

$$
\begin{array}{cc}
w_1 & \hspace{10em} w_\ell \\
\boxed{\pm \mid \pm \mid b_1 \mid \cdots \mid b_{n-1} \mid b_n \mid e_{k-1} \mid e_{k-2} \mid \cdots \mid e_0}
\end{array}
$$

Figure 8.11: *Allocation of the bits in a word w for "oating-point storage and arithmetic.*

The first bit stores the sign of x and the second bit stores the sign of the exponent (e is always a nonnegative integer). The next n bits store the fractional part $b = .b_1 b_2 \cdots b_n$, and the last k bits store the exponent $e = e_{k-1} \cdots e_1 e_0$. Thus $= k + n + 2$, as before. The exact set of machine numbers for this "oating-point scheme is

$$\mathbb{M}^{k,n}_{float} = \quad x = \pm \quad \sum_{j=1}^{n} b_j 2^{-j} \quad 2^{\pm \sum_{i=1}^{k-1} e_i 2^i} \quad e_i, b_j \quad \{0,1\} \quad .$$

This set of machine numbers is different from the ones that result from fixed-point schemes, and you may be surprised at the differences.

To better understand $\mathbb{M}^{k,n}_{float}$, we derive an alternative description of it as follows. First note that the maximum positive exponent e_{max} arises from having $e_i = 1$, for every i. The value of this integer is

$$e_{max} = \sum_{i=0}^{k-1} 2^i = 2^k - 1.$$

The smallest positive machine number Δx occurs when $b_n = 1$, $b_j = 0$, $j = 1, \ldots, n-1$, and the exponent is $-e_{max}$. Thus

$$\Delta x = \frac{1}{2^{2^k-1+n}}.$$

Next note that each $x \in \mathbb{M}^{k,n}_{float}$ can be rewritten as follows:

$$x = \pm \sum_{j=1}^{n} b_j 2^{-j} \cdot 2^{\pm e}$$

$$= \pm \frac{b_1 2^{n-1} + \cdots + b_n}{2^n} \cdot 2^{\pm e}$$

$$= \pm C \, 2^{-n \pm e}.$$

Here $C = \left(\sum_{p=0}^{n-1} b_{n-p} 2^p \right)$ is a nonnegative integer, the largest of which is $C_{max} = \sum_{p=0}^{n-1} 2^p = 2^n - 1$. From this it is easy to see that the largest positive machine number is $x_{max} = C_{max} \, 2^{-n+e_{max}}$, i.e.,

$$x_{max} = (2^n - 1)2^{(2^k-1-n)} = (1 - 2^{-n})2^{2^k-1}. \tag{8.16}$$

As in the fixed-point scheme, it will be convenient to note that each machine number is an integer multiple of the smallest machine number Δx. To see this, we take the above expression for x and rewrite it as

$$x = \pm C \, 2^{-n \pm e} = \pm C \, 2^{(2^k-1 \pm e)} \, 2^{-(2^k-1+n)} = \pm C \, 2^{(2^k-1 \pm e)} \Delta x.$$

Now note that $s = 2^k - 1 \pm e$ ranges from $s = 2^k - 1 - e_{max} = 0$ to $s = 2^k - 1 + e_{max} = 2(2^k - 1)$. Thus, the set of machine numbers for the "oating-point scheme can be expressed as

$$\mathbb{M}^{k,n}_{float} = \{ \pm C \, 2^s \Delta x \mid C = 0, \ldots, 2^n - 1, \text{ and } s = 0, \ldots, 2(2^k - 1) \}. \tag{8.17}$$

As in the fixed-point scheme, the "oating-point machine numbers all lie in the interval $[-x_{max}, x_{max}]$ and are multiples of the smallest machine number Δx. However, now the machine numbers are *not* equally spaced at a distance Δx apart. To analyze this, let B be the following subset of machine numbers:

$$B \quad \{ \pm C \, \Delta x \mid C = 0, \ldots, 2^n - 1 \}.$$

This is what you get from the representation in (8.17) for $s = 0$. The numbers in B are equally spaced at a distance of Δx apart. For $s = 1$, we get the subset $2B$ with numbers spaced at $2\Delta x$, for $s = 2$, we get the subset 2^2B with numbers spaced at $4\Delta x$, and so forth.[17] Hence, we can write $\mathbb{M}_{float}^{k,n}$ as the union of these subsets:

$$\mathbb{M}_{float}^{k,n} = B \quad 2B \quad 2^2B \cdots \quad 2^{2(2^k-1)}B. \tag{8.18}$$

The numbers in 2^sB are equally spaced at a distance of $2^s\Delta x$ apart. The analysis may be roughly interpreted as saying that the machine numbers near 0 are very tightly packed (dense), while those far from 0 are less tightly packed (sparse).

For example, suppose $k = 1, n = 3$, so that the word length is $= n + k + 2 = 6$ bits. The "oating-point machine numbers are

$$x = \pm.b_1b_2b_3\, 2^{\pm e_1 e_1},$$

with $e_i, b_j \quad \{0,1\}$. Note that $2^k - 1 = 1$ and $2^n - 1 = 7$. The smallest positive machine number is $\Delta x = 2^{-(2^k-1+n)} = 2^{-4} = \frac{1}{16}$, and the total set of machine numbers is

$$\mathbb{M}_{float}^{1,3} = \{\pm C\, 2^s\Delta x \,|\, C = 0,\ldots,7,\ s = 0,1,2\}.$$

The set of nonnegative machine numbers is the union of the following subsets:

$$\begin{aligned}
B_+ &= \{0, \tfrac{1}{16}, \tfrac{2}{16}, \tfrac{3}{16}, \tfrac{4}{16}, \tfrac{5}{16}, \tfrac{6}{16}, \tfrac{7}{16}\} \\
2B_+ &= \{0, \tfrac{1}{8}, \tfrac{2}{8}, \tfrac{3}{8}, \tfrac{4}{8}, \tfrac{5}{8}, \tfrac{6}{8}, \tfrac{7}{8}\} \\
4B_+ &= \{0, \tfrac{1}{4}, \tfrac{2}{4}, \tfrac{3}{4}, \tfrac{4}{4}, \tfrac{5}{4}, \tfrac{6}{4}, \tfrac{7}{4}\}.
\end{aligned}$$

The numbers are illustrated in Figure 8.12.

Figure 8.12: *The nonnegative oating-point numbers $.b_1b_2b_3\, 2^{\pm e_1 e_1}$ for a computer with 6 bit words. These consist of all the rational numbers in the interval $[0, \frac{7}{4}]$ of the form $C2^s/16$, with $C = 0, 1, \ldots, 7$ and $s = 0, 1, 2$.*

---

[1]If $B = \{x_1, \ldots, x_l\}$ is any set of numbers, then $mB \equiv \{mx_1, \ldots, mx_p\}$.

Fixed Point		Floating Point	
n x_{max}	Δx	x_{max}	Δx
1 63.50000000	.5000000000	1073741824	$.2328306437 \times 10^{-9}$
2 31.75000000	.2500000000	24576	$.7629394531 \times 10^{-5}$
3 15.87500000	.1250000000	112	.0009765625000
4 7.937500000	.06250000000	7.500000000	.007812500000
5 3.968750000	.03125000000	1.937500000	.01562500000

Table 8.2: *Comparison of the largest and smallest positive numbers, x_{max} and Δx, in the fixed-point and oating-point schemes for a one byte word and various lengths of the fractional part $.b_1 b_2 \cdots b_n$.*

A comparison of $\mathbb{M}_{float}^{1,3}$ and $\mathbb{M}_{fixed}^{1,3}$ does not indicate whether one scheme is better than the other. The Δx is smaller in the "oating-point scheme, but the fixed-point scheme has larger numbers. Both schemes, for these values of k and n, are poor for doing calculations. This is because the word length is too small.

For larger word lengths , the "oating-point scheme can easily be seen to be superior to the fixed-point scheme. With given, we can use $k+n+2 = $ to eliminate $k = -n-2$, and compare the sets $\mathbb{M}_{fixed}^{\ell-n-2,\,n}$, $\mathbb{M}_{float}^{\ell-n-2,\,n}$ for various values of $n = 1, \ldots, -3$. Table 8.2 shows a comparison of x_{max} and Δx for each scheme for a word of length $= 8$, i.e., one byte. As you can see, the "oating-point scheme is better for most values of n. This is true for all larger word lengths as well (see the exercises).

Rational Schemes

These schemes represent rational numbers $x = a/b$ as pairs of integers (a, b) (much like they are defined axiomatically in mathematics), and perform exact arithmetic on these representations. Thus, if $y = c/d$ is represented by (c, d) then

$$x + y = (ad + bc, bd), \qquad x - y = (ad - bc, bd),$$

$$xy = (ac, bd), \qquad x/y = (ad, bc).$$

The arithmetic is exact only when the resulting integers have magnitudes not exceeding the word size of the computer. Otherwise, over"ow occurs. To aid in preventing over"ow, the integers a, b in the representation (a, b) of $x = a/b$ can be factored and common factors can be removed, giving a representation (a', b') with integers of smaller magnitudes. This would require storing long integers in multiple word structures before eliminating common factors.

Exercise Set 8

1. The recurrence relation

$$a_n = -\frac{1}{2n}\, a_{n-2},$$

defines a_n for $n \geq 2$. As in the text, $a_0 = p$ and $a_1 = q$ are arbitrary constants and the recurrence relation determines all the other a_n's in terms of p or q. Compute, by hand, $a_2, a_3, \ldots, a_8$. You should see that all the even a_n's depend only on p, while all the odd ones depend only on q. Can you determine the formulas for a_{2k} and a_{2k+1}? (This is known as solving the recurrence relation.)

If we let $y(x) = \sum_{n=0}^{\infty} a_n x^n$, then y involves the two arbitrary constants p and q, and is the general solution of the differential equation

$$2y'' + xy' + y = 0.$$

Show that y can be written as $y(x) = py_1(x) + qy_2(x)$, where $y_1(x)$ and $y_2(x)$ are the functions defined by the series with even and odd coefficients a_{2k} and a_{2k+1}, respectively. Show that y_1 is a well-known function (involving the natural exponential function). Use this to show directly that y_1 satisfies the differential equation.

Write a recursive procedure (with a remember table) that defines the coefficients a_n in terms of p and q. Draw branching diagrams (like the one in Figure 8.1) which show the recursive calls that are initiated by the calls a(6) and a(5). Explain why the diagrams are different from that in Figure 8.1. Let

$$Y(x, N) = \sum_{n=0}^{N} a_n x^n,$$

be the approximation to the general solution y. Write a procedure to define Y and use this to compute $Y(x, N)$, $N = 0, 1, \ldots, 8$, for arbitrary p, q, and compare with what you got by hand.

Plot, all in the same figure, $Y(x, N)$, for $N = 1, 2, 3, 4, 10, 100$, $p = 1$, and $q = 0$. *Note:* You will have to choose an appropriate horizontal range x=-a..a for the figure, and you will definitely have to limit the vertical range, say y=-3..3, so that the detail is not lost. Explain why the vertical range must be limited like this. Annotate the printout of your figure. Do a similar figure for the case when $p = 0$ and $q = 1$. Explain why there are not six graphs in each figure.

2. **(Legendre Series/Polynomials)** Suppose $r > 0$ is a given constant, and consider the recurrence relation

$$a_n = -\frac{(r+n-1)(r-n+2)}{n(n-1)}\, a_{n-2}, \tag{8.19}$$

for $n \geq 2$. With $a_0 = p$ and $a_1 = q$ arbitrary, and the other a_n's determined by the recurrence relation (8.19), we get a power series $y(x) = \sum_{n=0}^{\infty} a_n x^n$, which is the general solution of the differential equation

$$(1 - x^2)y'' - 2xy' + r(r + 1)y = 0.$$

This is known as *Legendre's equation* (of order r) and is important in quantum mechanics and other areas of physics and mathematics. Use a recursive procedure to define the coefficients and study the approximate solutions $Y(x, N)$, much like you did in the previous exercise. Specifically:

(a) Show that the general solution can be written as $y(x) = py_1(x) + qy_2(x)$, where y_1, y_2 are functions defined by the power series with even coefficients, $a_{2k}, k = 0, 1, \ldots,$, and odd coefficients, $a_{2k+1}, k = 0, 1, \ldots,$, respectively.

(b) Find formulas for a_{2k} and a_{2k+1} for an arbitrary k. What happens in the special case when r is a positive integer?

(c) Suppose $r = 4.5$. For $p = 1, q = 0$, plot, in the same figure, the approximations $Y(x, N), N = 0, 2, 4, 6, 50, 100, 200$, for x in $[-2, 2]$ (limit the vertical range to y=-3..3). Set the number of points, **numpoints**, appropriately so that the plots look smooth. Do a similar plot for the case $p = 0, q = 1$ and $N = 1, 3, 5, 7, 51, 101, 201$. Annotate your figures and comment on any special features in the figures.

3. Study the amounts of time required for some of the computations in Exercise 2 above, when the recursive procedure does *not* have the **option remember** in its definition. To do this, close Maple, restart it, and open the worksheet you were using in Exercise 2. Define the procedure for the coefficients (after deleting **option remember** from it), and set $r = 4.5$. Make the command **readlib(showtime);**, which reads the **showtime** procedure from the library. Turn on this procedure by using the command **on;**. The prompt > on the interface changes. At the new prompt enter **a(5)**, and you will see the result along with the time and memory used. Continue doing this to find the time and memory used in computing **a(n)** for n=50,100,200.

Next turn the **showtime** procedure off by using the command **off;**. Then define the procedure for the approximations **Y(x,N)**, and take $p = 1, q = 0$. Turn **showtime** on again, and make the command to plot **Y(x,N)**, for N=0,2,4,6, all in the same figure. Then make a similar command for the case 50,100,200.

Record all the times and memory usage. Close Maple and then restart it. Go through the same steps but this time use the **option remember** in the procedure for the coefficients. Compare the corresponding times and memory usages and make comments and comparisons with the data in Table 8.1.

4. Write a procedure to compute $\int x^n \sin ax \; dx$ in the form:

$$\int x^n \sin ax \; dx = P(x, a) \sin ax + Q(x, a) \cos ax,$$

for some polynomials P and Q. Have your procedure first call the procedure $S(n, a)$ (given in the text) to calculate the integral in unfactored form. Then use the `collect` and `coeff` commands to produce the desired output.

5. Write a recursive procedure to calculate $(x^2 + a^2)^n \, dx$, assuming that n is a nonnegative integer.

6. Derive, by hand, a reduction formula for $\tan^n x \, dx$. Show your work. Use this to write a recursive procedure to compute $\tan^n x \, dx$, for any positive integer n

7. Alter the procedure `bubblesort` to stop if there are no exchanges made for one complete pass through the unsorted part of the list.

8. Write a recursive version of `bubblesort`.

9. Suppose A is a $2 \times n$ array where the first row consists of n numbers:

   A[1,1],A[1,2],...,A[1,n]

   and the elements in the second row are considered paired with corresponding elements in the first row,

   A[2,1],A[2,2],...,A[2,n].

   So, the number in A[1,i] is paired with the quantity in A[2,i]. For example, A[1,i] might be the midterm grade for a student with name A[2,i] in a class with n students. Write a procedure `bubblesort2` to sort the first row of A in increasing order so that this pairing is maintained.

10. **(Selection Sort)**

    (a) Write a procedure `selectsort`, which is a nonrecursive version of the selection sort algorithm. The procedure should have only one parameter A, which is an array of numbers to be sorted in increasing order. Use a call to the `exchange` procedure to swap elements as needed. You must write your own code to find the maximum element in an array. Apply this code to find, the largest element in A[1],...,A[n], and exchange it with A[n]. Then find the largest element in A[1],...,A[n-1], and exchange it with A[n-1], etc. At the $(n-k)$th step in this process, the array A is in the state:

    $$\underbrace{A[1], A[2], \ldots, A[k]}_{\text{unsorted list}}, \underbrace{A[k+1], \ldots, A[n]}_{\text{sorted list}}$$

    Thus, a do loop with index k can implement the sorting.

    (b) Write a procedure `Rselectsort`, which is a recursive version of the selection sort algorithm. The procedure should have two parameters: A, which is an array of numbers to be sorted in increasing order, and k, which is a positive integer indicating that the elements A[1],...,A[k] are to be sorted. Use a call to the **exchange** procedure to swap elements as needed.

11. **(Insertion Sort)**

   (a) Write a procedure **insertsort**, which is a nonrecursive version of the insertion sort algorithm. The procedure should have only one parameter A, which is an array of numbers to be sorted in increasing order. Use a call to the **exchange** procedure to swap elements as needed. The algorithm should work like this. At the first step, compare A[n-1],A[n] and exchange them if they are out of order. At the second step, insert A[n-2] in the appropriate place (by a sequence of exchanges) so that A[n-2],A[n-1],A[n] are in order (sorted). At the $(n-k)$th step in this process, the array A is in the state:

$$\underbrace{A[1], A[2], \ldots, A[k]}_{\text{unsorted list}}, \underbrace{A[k+1], \ldots, A[n]}_{\text{sorted list}}$$

   Now A[k] can be inserted in the appropriate place in the sorted list (by a sequence of exchanges) so that the resulting list is in order. The sequence of exchanges is like that in bubble sort, except you should use a **for-from-while-do** loop with the **while** condition stopping the process as soon as A[k] has bubbled into the correct position.

   (b) Write a procedure **Rinsertsort**, which is a recursive version of the insertion sort algorithm. The procedure should have two parameters: A, which is an array of numbers to be sorted in increasing order, and k, which is a positive integer (less than n) indicating that A[k] is to be inserted in the correct place in the sorted list A[k+1],...,A[n]. Use a call to the **exchange** procedure to swap elements as needed.

12. Using the duplicate-place quick sort algorithm, sort the following lists *by hand*. Use the recommended format for recording the sorting line by line and make sure you keep going with the processing as the algorithm dictates, even though at some line you clearly see that the list is sorted. Draw a tree diagram like that in Figure 8.3. Check your work with QS.

   (a) $[2,1]$, $[3,4]$, and $[6,2,4]$. These are warm-ups, but emphasize that you should always check that an algorithm works on very special cases.

   (b) $[8,2,5,1,6]$ and $[8,2,2,5,1,6]$.

   (c) $[2,2,1,1,1]$ and $[2,2,1,1,2]$.

   (d) $[9,8,7,6,5,4,3,2,1]$. This is the list that we sorted in the text using QS. Compare your work with the output from QS and answer the questions posed after that output in the text.

13. Using the in-place quick sort algorithm, sort the following lists *by hand*. Use the recommended format for recording the sorting line by line and make sure you keep going with the processing as the algorithm dictates, even though at some line you clearly see that the list is sorted. Check your work with `quicksort` and also view the corresponding animation with `QSmovie`.

    (a) $[2, 1]$, $[3, 4]$, and $[6, 2, 4]$. These are warm-ups, but emphasize that you should always check that an algorithm works on very special cases.

    (b) $[8, 2, 5, 1, 6]$ and $[8, 2, 2, 5, 1, 6]$.

    (c) $[2, 2, 1, 1, 1]$ and $[2, 2, 1, 1, 2]$.

    (d) $[9, 11, 14, 5, 20, 4, 2]$. This is the list that we sorted by hand in the text using the duplicate-place version of quick sort. Compare the successive steps here with those in the text.

    (e) $[7, 6, 5, 4, 3, 2, 1]$.

14. Write procedures to animate the sorting processes for one or more of the following sorting algorithms: (a) bubble sort, (b) the duplicate-place version of quick sort, (c) selection sort, (d) insertion sort.

15. Do a study that compares the performances of bubble sort, quick sort, selection sort, and insertion sort (assuming you wrote the code for the latter two in the exercises above). For this, generate arrays A with $n = 100, 200, \ldots, 1000$ elements, respectively. Use Maple's random number generator to generate the elements in each array. The command `rand(1..100)()` will generate a single random integer between 1 and 100. Thus to generate 500 such integers and store in an array, use the command

    ```
    A5:=array([seq(rand(1..100)(),n=1..500)]):
    ```

    Apply each of the four sorting schemes to each of the arrays `A1,A2,...,A10` generated by this method, and record the times that it takes each scheme to sort each array. *Note*: To record the times use the `showtime` procedure as described in Exercise 3 above. Plot the data, viewing the time t as a function of the number of elements n. Each plot will be a polygonal function. Combine the four plots in the same figure. Compare these graphs with the plots of $f(n) = n^2$ and $g(n) = n \ln n$. What conclusions can you make?

16. Prove (8.4). Write a procedure, called `baseNint`, that computes the base N representation of a positive integer m as given by equation (8.4).

17. Write a procedure, called `baseNrat`, that computes the base N representation of a positive rational number c/d. For this use the procedure from the previous exercise and the procedure `baseN` in the text.

18. Prove that $x = 2^{1/3}$ is irrational. Use Newton's method to compute rational approximations, $\{r_n\}_{n=1}^N$, to x. Start with $r_0 = 2$ and take N large enough so that the "oating-point digits for r_N agree with what your calculator gives. Compute r_1, r_2, r_3 by hand (and exactly).

19. Use the long division algorithm you learned in grade school to compute the decimal (base 10) representation of 35/64. Show your work in the customary way. Determine the same decimal representation using, *by hand*, the algorithm in Example 8.3. Compare each step with the corresponding step in the long division algorithm and explain how they are the same.

20. Use a calculator and Maple to compute the decimal representation of 23/67. Is the decimal representation repeating or terminating? Can you tell this from the results the calculator or Maple gives you? Set Maple's `Digits` variable to larger values than the default `Digits=10` and investigate. Use the procedure `baseN` from Example 8.3 to determine the decimal representation. Would you want to use long division to find the decimal representation by hand?

21. Suppose c/d is a positive rational number. Then, as we have seen above (also see the Maple/Calculus Notes that follow), c/d has one of the following base N representations:

$$\frac{c}{d} = a_s \cdots a_2 a_1 a_0 . b_1 b_2 \cdots b_r \qquad \text{(base } N) \qquad (8.20)$$

$$\frac{c}{d} = a_s \cdots a_2 a_1 a_0 . b_1 b_2 \cdots b_r \overline{b_{r+1} \cdots b_{r+p}} \qquad \text{(base } N) \qquad (8.21)$$

Write a program that will convert such N-ary representations back into the rational number c/d. Have the sequences $\{a_i\}_{i=0}^s$, $\{b_i\}_{i=1}^{r+p}$ input as lists to the procedure and the equivalent rational number as output.

22. For positive integers N_1, N_2, write a program that will convert a base N_1 representation of a positive rational into a base N_2 representation.

23. Suppose a computer has 32 bit words ($= 32$), and consider the sets $M_{fixed}^{k,r}$ and $M_{float}^{k,r}$ of fixed-point and "oating-point numbers for various values of n, k, with $n + k + 2 = $. Do a study that compares the largest and smallest machine numbers, x_{max} and Δx, for the two schemes. Use various plots for the comparison as well as a table like Table 8.2.

24. Design a fixed-point calculator to add, subtract, and multiply integers in the base 2 format. For this set up an array `W:=array(1..mem)`, which will represent the memory in the calculator. Here `mem` is the memory size (number of words) and `W[i]` is the ith word in memory. Each word should have a common, fixed length: `W[i]:=array(1..L)`. Write procedures to simulate the workings of the calculator. Specifically:

    (a) Have procedures `CONVERT(a,A)` and `RESTORE(A,a)` which convert a number `a` into its binary representation `A` (stored as a word) and vice versa. Generate an error message if `a` is too large in magnitude to store.

    (b) Have procedures `ADD(A,B)`, `SUBTRACT(A,B)`, and `MULTIPLY(A,B)` which add, subtract, and multiply `A,B` in their binary format. Your procedures must work exclusively with the 0's and 1's in the words `A,B` and not cheat by restoring `A,B` to Maple numbers and having Maple do the arithmetic. Print error messages if over"ow occurs.

(c) Have procedures `addr(a,b)`, `subtr(a,b)`, and `mult(a,b)` which act like the calculator buttons, i.e, they convert the input `a,b` to binary representation `A,B`, call `ADD`, `SUBTRACT`, `MULTIPLY`, respectively to do the artihmetic, and then display the result.

25. As an extended project, design a fixed-point calculator that extends the capabilities of the calculator in Exercise 24 to rational numbers. Include the fourth operation of division of rational numbers by designing procedures `DIVIDE(A,B)` and `divr(a,b)`. Provide for three data types, integer, rational, and fixed-point, which will be stored differently in the words, `W[i]`, `i=1..mem`, of the calculator. Design all the arthimetic operations which your procedures perform so that they recognize these data types and operate accordingly.

8.5 Maple/Calculus Notes

8.5.1 Real Numbers

As indicated in Section 8.4 the construction of the set $\mathbb{R}$ of real numbers from the set $\mathbb{Q}$ of rational numbers can be accomplished using Cauchy sequences of rational numbers. We give a few more details of this construction. We also complete the proof of the assertion that any rational number has a base N representation involving either a terminating or repeating series (see Example 8.3).

The basic idea of constructing $\mathbb{R}$ from $\mathbb{Q}$ is to identify an irrational number x with a sequence $\{r_n\}_{n=1}$ of rational numbers that approximates it, $r_n \to x$, for large n, and converges to it. Thus, we consider all possible sequences of rational numbers which are converging. Each such sequence represents a real number. To make this definition precise, we have to do two things.

First, we have to say what it means for a sequence $\{r_n\}_{n=1}$ to be converging.[18] This means, by definition, that it is a *Cauchy sequence*, i.e., for each *rational* number $\epsilon > 0$, there is a positive integer M such that

$$|r_n - r_m| < \epsilon ,$$

for all $n, m \geq M$. This is loosely interpreted as saying that far out in the sequence, all the terms differ very little from each other.

Second, we need to identify two sequences $\{r_n\}_{n=1}$ and $\{s_n\}_{n=1}$ which are converging to the same thing. Precisely, we say these two sequences

---

[18] The concept of converging here cannot involve what the sequence is converging to, because that might be an irrational number the very concept we are trying to de"ne!

are *equivalent* if for each *rational* number $\epsilon > 0$, there is a positive integer M such that

$$|r_n - s_n| < \epsilon,$$

for all $n \geq M$. Otherwise said, $\lim_{n \to \infty} |r_n - s_n| = 0$.

With these definitions, we can now define the *set of real numbers* as

$$\mathbb{R} \equiv \{ [\{r_n\}_{n=1}] \mid \{r_n\}_{n=1} \text{ is a Cauchy sequence of rationals} \}.$$

The brackets, [ ], indicate the equivalence class of the sequence. We can consider $\mathbb{Q}$ as a subset of $\mathbb{R}$ by identifying each rational number $r \in \mathbb{Q}$, with the equivalence class of the constant sequence $\{r, r, r, \ldots\}$. The operations of addition, subtraction, multiplication, and division can be extended from $\mathbb{Q}$ to $\mathbb{R}$ and, with these operations, $\mathbb{R}$ becomes a field. Unlike the field of rational numbers, the field of real numbers is complete, i.e., every Cauchy sequence of real numbers converges to a real number. All of these assertions require proof, and the interested reader can pursue this topic by consulting [MM] or [Sch].

We return now to some unfinished details in Example 8.3. The example discussed the N-ary, or base N, representation of positive rationals c/d. By division $c/d = q + m/n$, where q is an integer and $m < n$. Then the assertion is: relative to base N, the rational number m/n can be represented by a series

$$\frac{m}{n} = \sum_{j=1}^{\infty} b_j N^{-j}, \tag{8.22}$$

with coefficients $b_j \in \{0, 1, \ldots, N-1\}$ that either (1) terminate ($b_j = 0$, for $j > r$), or (2) repeat ($b_{j+p} = b_j$, for $j > r$). The discussion in Example 8.3 developed an algorithm for generating sequences q_k, j_j, m_k, $k = 1, 2, \ldots$, such that

$$\frac{m}{n} = \sum_{k=1}^{s} q_k N^{-j_k} + \frac{m_s}{n} N^{-j_s}, \tag{8.23}$$

for all s. This identity leads to the series representation as follows. The key observation is that the integers m_s are in the set $\{0, 1, \ldots, n-1\}$, for all s. Thus, after at most n steps in this process, either (1) one of the remainders m_L is 0, in which case all the ensuing remainders are zero: $m_{L+1} = 0, m_{L+2} = 0, \ldots$, or (2) one of the remainders m_R coincides with a previous remainder, $m_R = m_L$, with $L < R$. We discuss each case separately:

(1) When $m_L = 0$, equation (8.23), with $s = L$, gives

$$\frac{m}{n} = \sum_{k=1}^{L} q_k N^{-j_k}.$$ (8.24)

In this case the base N representation of m/n only requires a finite sum and coefficients $\{b_j\}_{j=1}$ of m/n are defined as follows. Let $r = j_L$ and define b_j by

$$b_j = \begin{cases} q_k & \text{if } j = j_k \text{ for some } k = 1, \dots, L \\ 0 & \text{if } j \neq j_k \text{ for any } k = 1, \dots, L \end{cases}.$$

Then the base N representation is

$$\frac{m}{n} = \sum_{j=1}^{r} b_j N^{-j}.$$

(2) When $m_R = m_L$, with $L < R \leq n$, then equation (8.23) will generate an infinite series that is the base N representation of m/n. This is shown as follows. Let $p = j_R - j_L$. This is called the *period* of the representation. To get the representation, use equation (8.23) for $s = L$ and then $s = R$ (remembering that $m_R = m_L$) to get

$$\frac{m}{n} = \sum_{k=1}^{L} q_k N^{-j_k} + \frac{m_L}{n} N^{-j_L}$$ (8.25)

$$= \sum_{k=1}^{L} q_k N^{-j_k} + \sum_{k=L+1}^{R} q_k N^{-j_k} + \frac{m_L}{n} N^{-j_R}.$$ (8.26)

Comparing (8.25) and (8.26) leads to the following identity for m_L/n:

$$\frac{m_L}{n} = \sum_{k=L+1}^{R} q_k N^{-(j_k - j_L)} + \frac{m_L}{n} N^{-p}.$$ (8.27)

For convenience let $A = \sum_{k=1}^{L} q_k N^{-j_k}$. Then substitute m_L/n from (8.27) into equation (8.25) to get

$$\frac{m}{n} = A + \sum_{k=L+1}^{R} q_k N^{-j_k} + \sum_{k=L+1}^{R} q_k N^{-(j_k+p)} + \frac{m_L}{n} N^{-(j_R+p)}$$ (8.28)

Now substitute m_L/n from (8.25) into equation (8.28), and continue in this fashion indefinitely to get the following series representation of m/n:

$$\frac{m}{n} = A + \sum_{k=L+1}^{R} q_k N^{-j_k} + \sum_{k=L+1}^{R} q_k N^{-(j_k+p)} + \sum_{k=L+1}^{R} q_k N^{-(j_k+2p)} + \cdots$$

In this case define the base N coefficients $\{b_j\}_{j=1}$ of m/n as follows. First, let $r = j_L$. Then $r + p = j_R$. Define b_j for $j \le r + p$ by

$$b_j = \begin{cases} q_k & \text{if } j = j_k \text{ for some } k = 1, \ldots R \\ 0 & \text{if } j \ne j_k \text{ for any } k = 1, \ldots, R \end{cases}.$$

Then extend this periodically by letting

$$b_{j+ip} = b_j,$$

for $j = r+1, \ldots, r+p$ and $i = 1, 2, 3, \ldots$. This then gives the base N representation in the repeating case

$$\frac{m}{n} = \sum_{j=1}^{r} b_j N^{-j} + \sum_{i=0}^{\infty} \sum_{j=r+1}^{r+p} b_j N^{-j} N^{-ip}.$$

Summary: Any positive rational number c/d has a base N representation and the N-ary positional notation for this representation is one of the following forms:

$$\frac{c}{d} = a_s \cdots a_1 a_0 . b_1 b_2 \cdots b_r \qquad \text{(base } N \text{)} \qquad (8.29)$$

$$\frac{c}{d} = a_s \cdots a_1 a_0 . b_1 b_2 \cdots b_r \overline{b_{r+1} \cdots b_{r+p}} \qquad \text{(base } N \text{)}. \qquad (8.30)$$

The overbar denotes the repeating part and $a_i, b_j \in \{0, 1, \ldots, N-1\}$. Note that to explicitly express c/d in positional (place-value) notation requires N distinct, single-digit symbols to stand for each of the ordinal numbers $0, 1, \ldots, N-1$. *Note:* The base N representation of negative rational numbers is an extension of the above that only requires affixing a minus sign to the corresponding representation for the positive case.

An interesting aspect arises from the above base N representation of rational numbers. It comes from the observation that for *any* sequence of coefficients $\{b_j\}_{j=1}$, subject only to the condition that $b_j \in \{0, 1, \ldots, N-1\}$, the series $\sum_{j=1} b_j N^{-j}$ converges. This follows from the comparison test:

$$\sum_{j=1} b_j N^{-j} < \sum_{j=1} N N^{-j} = \sum_{j=1} \left(\frac{1}{N}\right)^{j-1} = \frac{N}{N-1} < \infty.$$

Hence, we get a real number $x = \sum_{i=0}^{s} a_i N^j + \sum_{j=1} b_j N^{-j}$, with positional notation

$$x = a_s \cdots a_1 a_0 . b_1 b_2 \cdots b_j \cdots \qquad \text{(base } N\text{)}. \qquad (8.31)$$

By the above work x will be a rational number if and only if the b_j's are either eventually all zero or eventually repeat. In every other case x will be irrational. For example, consider the number represented by

$$x = 2.1011011101111 \cdots \qquad \text{(base 10)}. \qquad (8.32)$$

The pattern of the b_j's is successively a single one, two ones, three ones, four ones, etc., each separated by a 0. This indicates a nonrepeating, nonterminating decimal representation, and thus x is irrational. You can easily imagine other examples of decimal representations that give irrational numbers. The theorem is: *Any positive real number x has a base N representation*:

$$x = \sum_{i=0}^{s} a_i N^j + \sum_{j=1} b_j N^{-j},$$

where $a_i, b_j \in \{0, \ldots, N-1\}$, and x is irrational if and only if the series is nonterminating and nonrepeating.

Chapter 9

Programming Projects

This chapter contains a number of programming projects that can be assigned at any time during the course (but only after some initial programming skills have been developed). They can also form the basis for individual studies or a second course on computing in Maple.

Each section contains one or more sample projects to guide students in their work on the projects, which are related to the sample projects, but are more complicated. The codes for the solutions of the sample projects are given electronically on the CD-ROM along with other explanatory material about the projects.

9.1 Projects on Crystal Growth

The projects in this section are related to the topics of fractals and crystal growth and, as programming tasks, are mainly devoted to producing graphic displays and animations.

The growth described in all these projects involves starting with a given geometric figure (the initial state of the crystal) and successively adding new figures to it in some systematic way. Each step in the succession is viewed as a time step, hence the term growth. This growth is purely fictitious and not necessarily related to any particular physical process by which actual crystals grow, however, it will be convenient pedagogically to refer to the geometric objects as crystals.

Geometrically, the growth often will approach, in the limit, a certain geometric shape (the final state) and you will be asked to calculate various quantities that appear as limits in the final state. Mathematically these involve summing certain series.

9.1.1   Sample Project: Squares on Squares

This sample project serves as a prelude to the projects that you will be assigned. It is very elementary and is meant to help you get started on your assigned projects, which generally will be more di cult.

Problem: Write some code to produce the graphics to display the growth of the following geometric figure. The initial figure is a square (of any size). At the first growth stage two new squares, each with $1/4$ the area of the initial square, are added to the top-left side and right-bottom side of the initial square as shown in Figure 9.1.

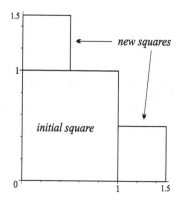

Figure 9.1: *The first stage of the growth whereby two new squares are added to the indicated sides of the initial square.*

At the second growth stage, two new squares are added, in the same way, to each of the squares produced at the first stage. The process is repeated indefinitely: At the nth growth stage, two new squares are added, in the same way, to each of the squares produced at the $(n-1)$st stage. The project is to do the following:

> Write Maple code that will draw the total figure at any growth stage and use it to produce the figures at stages $n = 2, 3, 4, 5, 6$.

> Produce an animation of the growth from the initial stage $n = 0$, out to the $n = 6$th stage.

> Assuming the process is repeated forever, determine the limiting (final state) figure to which the sequence of figures converges. Determine the area and pertinent lengths of the limiting figure.

Analysis of the Problem: It is useful to sketch by hand the growth process through several stages, at least out to where the squares get so tiny that they are hard to draw. This will give you an intuitive feel for the problem and help you readily guess that the limiting, final state of the crystal is a right triangle with sides of length 2 about the right angle (which will be confirmed mathematically below).

To produce the graphics, we need to choose an appropriate data structure to represent each square in the crystal and then develop an algorithm to produce all the new squares at each stage of the growth. Suppose we represent each square as a *list* data structure of the form

$$[[x_1, y_1], [x_2, y_2], [x_3, y_3], [x_4, y_4]],$$

where the (x_i, y_i), $i = 1, 2, 3, 4$, are the vertices of the square, enumerated in counter clockwise order starting from the lower left-hand vertex. See Figure 9.2.

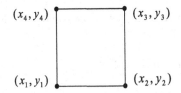

Figure 9.2: *Ordering of the four vertices of a square.*

If we name such a square sq, then sq[1], sq[2], sq[3], s[4] are the four vertices of the square and s[i][1], sq[i][2] are the x and y coordinates of the ith vertex.

Each stage of the growth produces a set of new squares appropriately placed on the sides of the squares produced at the previous stage. Abstractly Figure 9.3 shows the two new squares added to a given square from a previous growth stage and indicates how the coordinates of the vertices for the two new squares are determined. If we let S_n denote the set of squares produced at the nth stage of the growth, then at the next stage of growth S_{n+1} will contain twice as many squares and these are produced in pairs from each of the squares s S_n by the process outlined in Figure 9.3.

Mathematically, the area and side lengths of the limiting, steady state crystal are easy to compute by summing geometric series. Thus, if a denotes the area of the initial square, then the first stage of growth produces 2 squares each of area $a/4$, the second stage produces $2 \cdot 2 = 2^2$ squares each

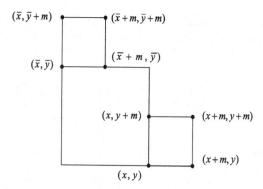

Figure 9.3: *Coordinates of the vertices for the two new squares added to a given square from the previous growth stage.*

of area $(a/4)/4 = a/4^2$, the third stage produces $2 \cdot 2^2 = 2^3$ squares each of area $(a/4^2)/4 = a/4^3$, etc. Inductively we establish that the nth stage of growth produces 2^n squares, each of area $a/4^n$. Thus, the total area added to the crystal at the nth stage is

$$\text{total area of all the squares in } S_n = 2^n \cdot \frac{a}{4^n} = \frac{a}{2^n}.$$

The sum of all the areas added to the crystal after it has been allowed to grow indefinitely is therefore

$$a + \frac{a}{2} + \frac{a}{2^2} + \frac{a}{2^3} + \cdots = \sum_{n=0} \frac{a}{2^n} = \frac{a}{1 - \frac{1}{2}} = 2a.$$

Thus, the crystal doubles in area after growing for an infinitely long time. This should be expected if we look ahead at Figure 9.4. From the figure it is apparent that the sides of the original square will double in length as the crystal converges to a right triangle shape. We can confirm this analytically by noting that at each stage the new squares have sides that are half the lengths of the ones at the previous stage. Thus, letting s denote the length of the sides of the original square, the sum of all the lengths of the sides is

$$s + \frac{s}{2} + \frac{s}{2^2} + \frac{s}{2^3} + \cdots = \sum_{n=0} \frac{s}{2^n} = 2s.$$

Thus, the right triangle is expected to have area $A = \frac{1}{2} \cdot (2s)(2s) = 2s^2 = 2a$, which is what we computed above.

Design of the Code: One element of our code should be a procedure, called `newsquares`, which takes a given square `sq` and produces the two new squares `nsq1,nsq2` as in Figure 9.3. Note that the x and y coordinates of the ith vertex of `sq` are `sq[i][1]` and `sq[i][2]`. Thus, the length of any of its sides is `sq[2][1]-sq[1][1]` (because of the ordering of the vertices). Half of this, which we call `m`, gives the length of the sides of the new squares and so by adding `m` in the appropriate way to the coordinates of two of the vertices of `sq`, we can produce the vertices of the two new squares. The code for the procedure is as follows:

```
newsquares:=proc(sq,nsq1::evaln,nsq2::evaln)
   local m,x,y;
   m:=(sq[2][1]-sq[1][1])/2.0;
   x:=sq[2][1];y:=sq[2][2];
   nsq1:=[[x,y],[x+m,y],[x+m,y+m],[x,y+m]];
   x:=sq[4][1];y:=sq[4][2];
   nsq2:=[[x,y],[x+m,y],[x+m,y+m],[x,y+m]];
   RETURN()
end proc:
```

Note that the assignment statements for `nsq1` and `nsq2` are the same, but the values of `x,y` are different in each.

Next we can write a short procedure, called `nextset`, that takes a set of squares `SS` and produces the next set of squares `NSS` in the growth process by making a call to `newsquares` for each square in `SS` and putting the resulting squares in `NSS` using the `union` command.

```
nextset:=proc(SS,NSS::evaln)
   local TSS,sq;
   TSS:={};
   for sq in SS do
     newsquares(sq,s1,s2);
     TSS:=TSS union {s1,s2}
   end do;
   NSS:=TSS;
   RETURN();
end proc:
```

Note that it is necessary to introduce a local variable `TSS`, which acts as a temporary set of squares in building the next set of squares `NSS`. There might be a tendency to try to use `NSS` in place of `TSS`, but you will get an error message about "too many levels of recursion" if you do so. This is because of the evaluation rules that are in effect when a procedure executes and the fact that NSS is an output parameter.

Now we can simply: (1) set up an array S to store the sets of squares at each stage (S[i] is the set of squares added at the ith stage), (2) assign the initial set of squares S[0] (which actually consists of only *one* square), and (3) use a do loop to calculate as many of the S[i]'s as we wish. (CAUTION: A large number will lock your computer when you try to display all the squares.) The code for this is:

```
S:=array(0..10)
S[0]:={[[0,0],[1,0],[1,1],[0,1]]};
for i to 6 do
  S[i]:= {};
  nextset(S[i-1],S[i])
end do:
```

To plot the resulting crystals for any stage, we first include several procedures from the **plots** package and then define an array to hold some colors.

```
with(plots,display,animate);
col:=array(0..15):
for i from 0 to 7 do
    col[i]:=COLOUR(RGB,.2+0.1*i,0,.8)
end do:
```

We next establish an array ps for the plot structures (ps[i] will be the graphic information for all the squares in S[i]), assign the plot structures with a do loop, and then display the result:

```
ps:=array(0..10):
for i from 0 to 6 do
  ps[i]:=seq(CURVES([seq(S[i][j][k],k=1..4),S[i][j][1]],
                          col[i]),j=1..nops(S[i]));
end do:
display({seq(ps[i],i=0..6)},scaling=constrained,tickmark=[3,3]);
```

Figure 9.4 shows the resulting geometric shapes of the crystal at the stages $n = 3$ and $n = 6$ of its growth. The animation for the growth out to the sixth stage can be accomplished by the following code:

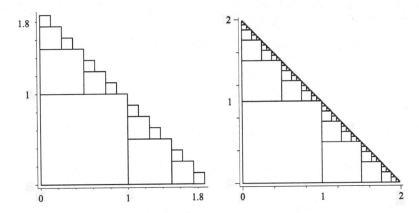

Figure 9.4: *Geometric shapes of the crystal at the stages $n = 3$ and $n = 6$ of its growth.*

```
FR:=array(0..10):
for q from 0 to 6 do
  FR[q]:=[seq(ps[i],i=0..q)]
end do:
PLOT(ANIMATE(seq(FR[q],q=0..6)),AXESSTYLE(NONE),
                        SCALING(CONSTRAINED));
```

All of this code is contained on the CD-ROM, which you should consult for the animations and related exercises and discussion.

9.1.2 Project: Crystal Growth (Squares)

Here is a project for you to work which is an extension of the above sample project. This project is more challenging and yields some interesting results.

Consider the geometric crystal constructed from squares by the following process. Start with a square of side x. Divide each side in half and, on the left-hand segment (going around clockwise) of each divided side, place a square of side $x/2$. Place these on the exterior of the original square. Each of the new squares has three *exposed* sides, i.e., not overlapping with a side of the initial square. Divide each exposed side in half and, on the left-hand segment (going around clockwise) of each divided side, place a square of side $x/4$. Continue this process whereby at each step, the three exposed sides of each of the squares added at the previous step are divided in half and on the left-hand segments (going around clockwise) of each divided side, a square of side one half the previous length is added. See Figure 9.5.

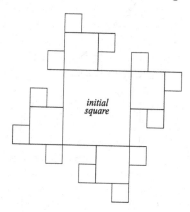

initial square

Figure 9.5: *The sets of squares S_0, S_1, and S_2 at the 2nd stage of the process.*

Let S_n be the set consisting of just the new squares which are added at the nth stage of the crystal growth. Thus, S_0 is the set consisting of the initial square, S_1 is the set consisting of the four squares added at the first stage, S_2 the set consisting of all the squares added (12 in all) at the second stage, etc. On the other hand let

$$E_n = S_0 \quad S_1 \quad \ldots \quad S_n,$$

be the total set of squares comprising the crystal at the nth stage of its growth.

Your problem is to write some code to graphically display the geometric figure at any stage and also to produce a moving picture of the way it is built up. Specifically:

(1) Represent each square as a *list* data structure of the form

$$[[x_1, y_1], [x_2, y_2], [x_3, y_3], [x_4, y_4], b],$$

where the (x_i, y_i), $i = 1, 2, 3, 4$ are the vertices of the square and $b = 1, 2, 3, 4$ is its type. *Note:* After steps zero and one, each added square is of one of four types depending on which of its sides overlaps a square from the prior step.

(2) Write a procedure **newsquares** to take a given square and, based on its type, produce three new squares.

(3) Write a procedure `nextset` which takes a set of squares `SS` and produces the next set of squares `NSS` based on the process outlined in the problem. This procedure will make calls to `newsquares` to accomplish its task.

(4) Initialize the initial set S_0 to consist of a single square. Use a do loop and the procedure `nextset` to produce the ensuing sets of squares: $S_1, S_2, \ldots, S_n$. *Note*: Going from S_0 to S_1 is different from the other steps in the construction. So either do this step separately or build the `nextset` procedure to recognize this.

For plotting, you will need to convert the squares to a different form, i.e., for each set of squares S_i, convert each square (as defined above) into a list of the form

$$[[x_1, y_1], [x_2, y_2], [x_3, y_3], [x_4, y_4], [x_1, y_1]].$$

This will be a square which has the right form for plotting its outline. On the other hand if you wish to plot the interior of this square with the `polygonplot` command, you will need to have it in the following form

$$[[x_1, y_1], [x_2, y_2], [x_3, y_3], [x_4, y_4]].$$

(5) Produce static plots of $E_1, \ldots, E_6$. (If you try this for E_n with $n > 6$, be sure you have enough memory for the plots.) Do colored outline plots as well as plots with the interiors of the squares displayed in various colors. Also produce an animation showing how the figure E_6 is built up successively from E_i, $i = 0, 1, \ldots, 6$.

(6) Do the following, by hand. Show your derivations and other work.

    (a) Let a_n be the total of all the areas of the squares in S_n and let p_n be the total of all the lengths of all the sides of all the squares in S_n. Find formulas for a_n and p_n.

    (b) If we let n in the geometric construction, then the totals for all the areas and lengths of sides of the squares are

$$A = \sum_{n=0}^{\infty} a_n \qquad P = \sum_{n=0}^{\infty} p_n,$$

    respectively. Find the values of A and P. Justify your answers.

(c) The geometric figure E_n is contained in a certain square S which it appears to approach in the limit as n . Indeed there is an S whose area is precisely A. Explain why this seems contradictory and then analyze how this is possible.

(7) As usual, write up how you analyzed the programming project and how you designed your code to solve it.

(8) (Extra Credit) The algorithm in this problem creates the set of squares S_{i+1} by adding three squares to each of the squares in S_i, for $i = 1, 2, \ldots$. What happens if four squares are added (following a similar pattern)? Redo the above parts to the problem in this case. Are you surprised by the values for the total area A ?

The following is a collection of crystal growth problems where the geometry is different from that in Project 9.1.2 above. Work each assigned project below following the general format of instructions laid down in Project 9.1.2.

9.1.3 Project: Crystal Growth (Isosceles, Right Triangles)

To an initial isosceles right triangle, two additional isosceles right triangles are added to *each* of the sides about the right angle as shown in Figure 9.6. The process is continued indefinitely.

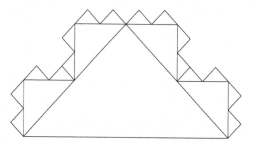

Figure 9.6: *In Project 9.1.3, a crystal grows by adding two new isosceles right triangles to each of the two sides about the right angle of a previous isosceles right triangle.*

9.1.4   Project: Crystal Growth (Equilateral Triangles)

Start with an initial equilateral triangle. On each of its three sides add an equilateral triangle with sides 1/2 the length of the original side and placed on half the original side as shown in Figure 9.7. At the second stage, and all ensuing stages, only two isosceles triangles are added to each triangle from the previous stage. These are added to the two sides that do not overlap a side from a prior stage. See Figure 9.7. The process is continued indefinitely.

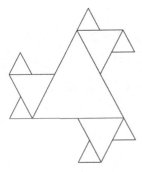

Figure 9.7: *The first and second stages of the growth of a crystal using equilateral triangles. See Project 9.1.4.*

9.1.5   Project: Crystal Growth (The Koch Snowflake)

Start with an initial equilateral triangle. On each of its three sides add an equilateral triangle with sides 1/3 the length of the original side and placed at the middle third of the original side. This yields a 12-sided polygon shaped somewhat like a star or snow"ake. On each of these 12 sides add an equilateral triangle with sides 1/3 the length of the original side and placed at the middle third of the original side. See Figure 9.8. The process is continued indefinitely.

9.1.6   Project: Crystal Growth (Squares and Triangles)

A crystal grows using squares and isosceles right triangles in alternating stages. Start with an initial square. On each of its four sides place an isosceles right triangle as shown in Figure 9.9. At the next stage place a square on half of the hypotenuse of each triangle as shown. At the third, and ensuing stages, when adding triangles to squares, only three triangles

Figure 9.8: *The Koch snow ake at the second stage of its growth. See Project 9.1.5.*   .

are added to the sides of each square. These are added to the sides of the square which do not touch a triangle from the previous stage. The process is continued indefinitely.

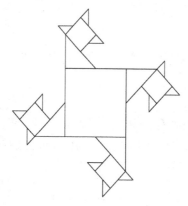

Figure 9.9: *The first, second, and third stages of the growth of a crystal using squares and isosceles right triangles at alternating stages in the growth. See Project 9.1.6.*

9.2   Projects on Inscribed Polygons

The case study in Chapter 3 described Newton's method for finding roots of an equation and how this method can be viewed as an application of the theory of iterated maps. Here we describe another application of the theory of iterated maps that arises from a particular method for obtaining a sequence of polygons inscribed in a circle. There are interesting graphics and mathematics associated with this problem (see [HZ]). We begin with the simplest case, inscribed triangles, for illustrating the ideas.

9.2.1 Sample Project: Sequences of Triangles

We produce a sequence of triangles inscribed in a given circle by the method described below. By animating the sequence of triangles, we will be able to clearly see that the sequence quickly converges to an alternating pair of equilateral triangles that together form a star of David figure.

The Method: Start with *any* initial triangle $T_0 = \Delta P_0 Q_0 R_0$ inscribed in a given circle. See Figure 9.10.

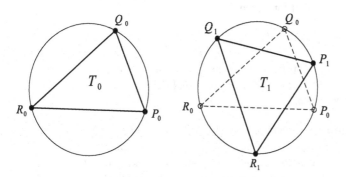

Figure 9.10: *An initial inscribed triangle T_0 and a new one T_1 obtained from it by bisecting arcs.*

Bisect each of the three arcs subtended by sides of T_0 to obtain three points P_1, Q_1, R_1 on these arcs. These points form the vertices of the next inscribed triangle $T_1 = \Delta P_1 Q_1 R_1$, as shown in the figure. Similarly, the second triangle T_2 in the sequence is obtained from T_1 in the same way, i.e., bisect the arcs subtended by the sides of T_1 to obtain the vertices of T_2. Continuing this process indefinitely, we get a sequence $\{T_n\}_{n=0}$ of inscribed triangles.

Problem: Write a procedure to produce several types of animations that can help in the analysis of the behavior of the sequence $\{T_n\}_{n=0}$. Use this procedure to investigate the following properties of the sequences that arise

from this process, and then prove these properties by mathematical arguments.

(i) If the initial triangle T_0 is equilateral, then the sequence $\{T_n\}_{n=0}$ is
an alternating sequence:

$$T_0, T_1, T_0, T_1, \ldots,$$

i.e., $T_{2k} = T_0$ and $T_{2k+1} = T_1$, for all k, and the figure consisting of
T_0, T_1 is a star of David. This sequence is also known as a *doubly
periodic* sequence of equilateral triangles.

(ii) More generally, for an arbitrary choice of initial triangle T_0, the sequence $\{T_n\}_{n=0}$ converges to an alternating sequence of equilateral
triangles. That is, there is a pair of equilateral triangles T, T , which
form a star of David, and for which

$$T_{2k} \quad T \qquad T_{2k+1} \quad T \ ,$$

for large k. Here the approximations mean that the vertices are approximately the same. One says the sequence $\{T_n\}_{n=0}$ is *eventually
periodic, with period two.*

(iii) Each of the sequences of vertices, $\{P_n\}_{n=0}, \{Q_n\}_{n=0}, \{R_n\}_{n=0}$, converges to a periodic sequence, of period six, that cycles counterclockwise through the six vertices of the limiting star of David figure. One
says that each of these sequences is *eventually periodic, with period
six.*

Analysis of the Problem: The key to understanding the behavior
of such a sequence of triangles $\{T_n\}_{n=0}$ is the observation of the central
angles n, n, n formed by joining the vertices of T_n to the center of the
circle. These angles will also be needed in writing the Maple code for the
animations. See Figure 9.11.

Inspection of the figure and a little thought reveal the central angles for
T_{n+1} are obtained from those for T_n via the formulas

$$n+1 \;=\; \tfrac{1}{2}\,n + \tfrac{1}{2}\,n \tag{9.1}$$

$$n+1 \;=\; \tfrac{1}{2}\,n + \tfrac{1}{2}\,n \tag{9.2}$$

$$n+1 \;=\; \tfrac{1}{2}\,n + \tfrac{1}{2}\,n. \tag{9.3}$$

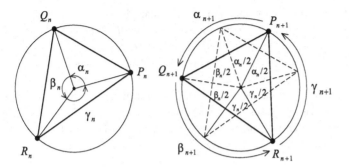

Figure 9.11: *The central angles* α_n, β_n, γ_n *for triangle* T_n *and the corresponding ones* α_{n+1}, β_{n+1}, γ_{n+1} *for* T_{n+1}.

In matrix form, these three equations can be written as one equation:

$$
\begin{pmatrix} \alpha_{n+1} \\ \beta_{n+1} \\ \gamma_{n+1} \end{pmatrix} = \begin{pmatrix} \frac{1}{2} & \frac{1}{2} & 0 \\ 0 & \frac{1}{2} & \frac{1}{2} \\ \frac{1}{2} & 0 & \frac{1}{2} \end{pmatrix} \begin{pmatrix} \alpha_n \\ \beta_n \\ \gamma_n \end{pmatrix}
\tag{9.4}
$$

or

$$
u_{n+1} = Au_n,
\tag{9.5}
$$

where $u_i = (\alpha_i, \beta_i, \gamma_i)$, $i = n, n+1$, are the 3×1 column matrices (vectors) and A is the 3×3 matrix shown. In this form, you can readily see that the vectors u_i, which have the central angles for components, arise by iterating the map A. Specifically, starting with $u_0 = (\alpha_0, \beta_0, \gamma_0)$, we get successively

$$
\begin{aligned}
u_1 &= Au_0 \\
u_2 &= Au_1 = A^2 u_0 \\
u_3 &= Au_2 = A^3 u_0 \\
&\vdots \\
u_n &= Au_{n-1} = A^n u_0.
\end{aligned}
\tag{9.6}
$$

Now it is easy to see that if T_0 is an equilateral triangle, then all of its central angles are equal

$$
u_0 = \left(\tfrac{2\pi}{3}, \tfrac{2\pi}{3}, \tfrac{2\pi}{3} \right),
$$

and thus from the form of the matrix A (or also from equations (9.1) (9.3)), we see that

$$
Au_0 = u_0.
$$

This gives that all the ensuing iterates of u_0 under the map A are the same $A^n u_0 = u_0$, for all n. Such a point u_0 is known as a *fixed point* of the map A. The point $u_0 = (\ _0, \ _0, \ _0)$ represents the central angles in the initial triangle, and since it is fixed under the iteration process, all the triangles T_n are equilateral. The triangles themselves, however, are not fixed, but rather alternate between T_0 and T_1, as can be easily seen from a drawing.

Next consider the more interesting case when the initial triangle T_0 is *not* equilateral. Then the central angles $u_n = (\ _n, \ _n, \ _n)$ are not fixed, but vary with n, and we would like to analyze how they vary and if they approach limiting values as n .

From equation (9.6) above, we see that $u_n = (\ _n, \ _n, \ _n)$ arises by applying the matrix A^n to the vector $u_0 = (\ _0, \ _0, \ _0)$, and so it is natural to analyze the nature of A^n as n . Experimentally, we can use Maple to investigate the various powers of the matrix A (see the CD-ROM) and determine a pattern. For example, one finds that

$$A^{10} = \begin{matrix} .3330078125 & .3330078125 & .3339843750 \\ .3339843750 & .3330078125 & .3330078125 \\ .3330078125 & .3339843750 & .3330078125 \end{matrix} . \qquad (9.7)$$

Investigations like this lead to the conjecture that the sequence of matrices $\{A^n\}_{n=1}$ converges to the matrix B which has $1/3$ as all of its entries:

$$\lim_{n \to} A^n = B \qquad \begin{matrix} \frac{1}{3} & \frac{1}{3} & \frac{1}{3} \\ \frac{1}{3} & \frac{1}{3} & \frac{1}{3} \\ \frac{1}{3} & \frac{1}{3} & \frac{1}{3} \end{matrix} .$$

It is left as an exercise to prove that this conjecture is correct. Accepting the truth of this, we find that the sequence of central angles has the following limit

$$\begin{aligned} \lim_{n \to} (\ _n, \ _n, \ _n) &= \lim_{n \to} u_n \\ &= \lim_{n \to} A^n u_0 \\ &= B u_0 \\ &= \tfrac{1}{3}(\ _0 + \ _0 + \ _0), \tfrac{1}{3}(\ _0 + \ _0 + \ _0), \tfrac{1}{3}(\ _0 + \ _0 + \ _0) \\ &= \left(\tfrac{2\pi}{3}, \tfrac{2\pi}{3}, \tfrac{2\pi}{3} \right). \end{aligned}$$

This is an important result. It says that regardless of the initial values of the central angles, in the limit each of these angles approaches $2 /3$. Thus,

T_n will be approximately equilateral for large n. Before analyzing how the sequence begins to alternate between two fixed triangles, we first write some Maple code to plot $T_0, T_1, \ldots, T_N$ in succession and also animate this.

Design of the Code: We use the standard circle $x^2 + y^2 = 1$, whose points are parametrized by $(\cos\theta, \sin\theta)$ for angles $\theta \in [0, 2\pi]$. The vertices P_n, Q_n, R_n of inscribed triangle T_n lie on this circle and are given by

$$P_n = (\cos p_n, \sin p_n) \tag{9.8}$$
$$Q_n = (\cos q_n, \sin q_n) \tag{9.9}$$
$$R_n = (\cos r_n, \sin r_n), \tag{9.10}$$

where p_n, q_n, r_n are the angles shown in Figure 9.12.

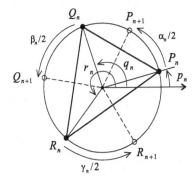

Figure 9.12: *The angles p_n, q_n, r_n that determine the vertices of T_n.*

Thus, once we know the angles p_n, q_n, r_n, it is a simple matter to plot the triangle T_n. The initial angles (angles of T_0) are known and the iteration scheme determines all the others from these. Specifically, from Figure 9.12, it is easy to see that

$$p_{n+1} = p_n + \tfrac{1}{2}\alpha_n \tag{9.11}$$
$$q_{n+1} = q_n + \tfrac{1}{2}\beta_n \tag{9.12}$$
$$r_{n+1} = r_n + \tfrac{1}{2}\gamma_n, \tag{9.13}$$

for all n. Thus, to determine and plot the sequence of triangles, we write the procedure to calculate, step by step, the sequence $u_n = (\alpha_n, \beta_n, \gamma_n)$ of central angles using the matrix multiplication indicated in equation (9.4), and at the same time use equations (9.11)–(9.13) to determine the sequence of angles that locate the vertices of the triangles.

We design the procedure to have the initial angles $_0$, $_0$ as input (note that $_0$ is not needed since $_0 + _0 + _0 = 2$). These angles are named a0,b0 and are required to be in radians.

Code for the Triangle Sequence

```
trisec:=proc(a0,b0,N,plotype)
   local u,T,Tplot,P,Q,R,FR,QR,A,tp,p,q,r,circle,n,i;
   with(plots,display);
   u:=array(0..N):T:=array(0..N):Tplot:=array(0..N):
   P:=array(0..N):Q:=array(0..N):R:=array(0..N):
   FR:=array(0..N):QR:=array(0..N):
   A:=matrix([[0.5,0.5,0],[0,0.5,0.5],[0.5,0,0.5]]);
   tp:=evalf(2.0*Pi);
   u[0]:=[a0,b0,tp-a0-b0];
   p:=0:q:=a0:r:=a0+b0:
   P[0]:=[cos(p),sin(p)];Q[0]:=[cos(q),sin(q)];
   R[0]:=[cos(r),sin(r)];
   T[0]:=[P[0],Q[0],R[0],P[0]];p:=0;
   for n to N do
     u[n]:=evaln(A&*u[n-1]);
     p:=p+u[n-1][1]/2;
     q:=p+u[n][1];
     r:=q+u[n][2];
     P[n]:=[cos(p),sin(p)];
     Q[n]:=[cos(q),sin(q)];
     R[n]:=[cos(r),sin(r)];
     T[n]:=[P[n],Q[n],R[n],P[n]];
   end do:
   circle:=plot([cos(t),sin(t),t=0..2*Pi]):
   for n from 0 to N do
     FR[n]:=seq(POLYGONS(T[i],
              COLOUR(RGB,n/N,(N-n)/N,.5)),i=n..n)
   end do :
   if plotype=0 then
       display([seq(FR[N-n],n=0..N),circle],scaling=constrained)
   elif plotype=1 then
       PLOT(ANIMATE(seq([FR[n],op(1,circle),
           TEXT(P[n]+.05*P[n],'P')],n=0..N)),
           SCALING(CONSTRAINED)));
```

```
   elif plotype=2 then
       for n from 0 to N do
       QR[n]:=seq(POLYGONS(T[n-i],COLOUR(RGB,(n-i)/N,
               (N-n+i)/N,.5)),i=0..n)
       end do :
       PLOT(ANIMATE(seq([QR[n],op(1,circle),
           TEXT(P[n]+.05*P[n],'P')],n=0..N)),
           SCALING(CONSTRAINED)));
   else RETURN('the fourth parameter must be 0,1, or 2');
   end if;
end proc:
```

We use this procedure to animate the sequence $\{T_n\}_{n=0}^N$ of triangles with T_0 having initial angles $_0 = /4$, $_0 = 3 /4$, $_0 =$. Figure 9.13 shows a composite picture of the first ten triangles in the sequence starting with T_0.

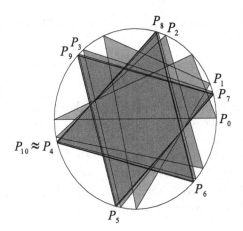

Figure 9.13: *The triangles* $T_0, T_1, \ldots, T_{10}$ *determined by an initial triangle* T_0 *with angles* $_0 = /4$, $_0 = 3 /4$, $_0 =$.

The figure also shows one of the sequences of vertices, $P_0, P_1, \ldots, P_{10}$, and exhibits how these begin to approximately repeat after ten steps. The CD-ROM has an animation of this and it is much easier to see the convergence of the sequence to an alternating sequence of equilateral triangles. In this case, $N = 10$ steps is all it takes for practical identity of the sequence $\{T_n\}_{n=0}$ with its limiting alternating sequence. The animation will also help you discern how a sequence of vertices, say, $\{P_n\}_{n=0}$, approaches its limiting, period six, sequence of points.

9.2.2 Project: Asymptotics of the Triangle Sequence

With the above experimental evidence as a guide, you can now prove properties (ii) and (iii) about the convergence of sequences of triangles that are generated by the method. Specifically, do the following:

(1) Show that if A is the matrix

$$A = \begin{array}{ccc} \frac{1}{2} & \frac{1}{2} & 0 \\[4pt] 0 & \frac{1}{2} & \frac{1}{2} \\[4pt] \frac{1}{2} & 0 & \frac{1}{2} \end{array},$$

then

$$\lim_{n \to} A^n = B \quad \begin{array}{ccc} \frac{1}{3} & \frac{1}{3} & \frac{1}{3} \\[4pt] \frac{1}{3} & \frac{1}{3} & \frac{1}{3} \\[4pt] \frac{1}{3} & \frac{1}{3} & \frac{1}{3} \end{array}. \tag{9.14}$$

Here are several suggested steps in the proof and hints for accomplishing them.

(a) **(Eigenvalues and Eigenvectors)** A key concept in the analysis of the limiting behavior of the sequence of powers A^n of a matrix is the concept of an *eigenvalue* of A. While this concept is usually covered in the first or second linear algebra course, we will explain some details here in case you have not yet encountered it.

By definition, an *eigenvalue* of A is a number , either real or complex, for which there is a nonzero vector v such that

$$Av = v.$$

The vector v is called an *eigenvector* of A corresponding to eigenvalue . While A has infinitely many eigenvectors, it has only finitely many eigenvalues (just three in the case here) and they are the roots, or solutions, of the polynomial equation

$$\det(A - I) = 0.$$

Here I is the identity matrix (a 3×3 matrix in our case) and det is the determinant operator. If necessary look up the definition of the determinant and then show that for the matrix A in our case:

$$\det(A - \lambda I) = \det \begin{array}{ccc} \tfrac{1}{2} - \lambda & \tfrac{1}{2} & 0 \\ 0 & \tfrac{1}{2} - \lambda & \tfrac{1}{2} \\ \tfrac{1}{2} & 0 & \tfrac{1}{2} - \lambda \end{array}$$

$$= -[(\lambda - \tfrac{1}{2})^3 - \tfrac{1}{8}]$$

Use this to find the three eigenvalues of A. Two of the eigenvalues are complex and all three are related to the *cube roots of unity*.

(b) **(Cube Roots of Unity)** The cube roots of unity are the three roots of the equation $x^3 = 1$. Use Euler's formula:

$$e^{i\theta} = \cos\theta + i\sin\theta ,$$

to show that the eigenvalues of A, labeled $\lambda_0, \lambda_1, \lambda_2$ are given by

$$\lambda_k = \tfrac{1}{2}(1 + \omega_k), \qquad (k = 0, 1, 2),$$

where

$$\omega_k = e^{2\pi k i/3}, \qquad (k = 0, 1, 2),$$

are the three cube roots of unity. Show that the angles involved in the cube roots of unity are $\theta = 0, 2\pi/3, 4\pi/3$ and that the explicit values for the roots are

$$\omega_0 = 1, \quad \omega_1 = -\tfrac{1}{2} + \tfrac{\sqrt{3}}{2}i, \quad \omega_2 = -\tfrac{1}{2} - \tfrac{\sqrt{3}}{2}i.$$

Show that if we identify a complex number $z = x + iy$ with the point (x, y) in $\mathbb{R}^2$ (thought of as the complex plane) then the cube roots of unity all lie on the unit circle $x^2 + y^2 = 1$ and are plotted as shown in Figure 9.14. Show that $\omega_1^2 = \omega_2$ and that ω_1, ω_2 are the two roots of the equation $\omega^2 + \omega + 1 = 0$.

(c) Show that in general, for any angle θ, the complex number $z = e^{i\theta}$ and all of its powers z^n, $n = 2, 3, 4, \ldots$, lie on the unit circle in the complex plane. Explain the geometrical relation of z^2, z^3 to z. Generalize to z^n.

(d) Show that the three vectors

$$v_k = (1, \omega_k, \omega_k^2), \qquad (k = 0, 1, 2),$$

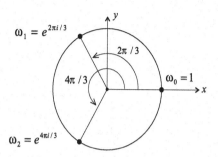

Figure 9.14: *The three cube roots of unity, plotted in the complex plane.*

are eigenvectors of A. Specifically, $Av_k = \omega_k v_k$. Write out these vectors explicitly. Let $M = [v_0, v_1, v_2]$ be the 3×3 matrix with v_0, v_1, v_2 as columns. Explicitly,

$$M = \begin{bmatrix} 1 & 1 & 1 \\ 1 & \omega_1 & \omega_1^2 \\ 1 & \omega_2 & \omega_2^2 \end{bmatrix}.$$

Show that $AM = MD$, where D is the diagonal matrix

$$D = \begin{bmatrix} \omega_0 & 0 & 0 \\ 0 & \omega_1 & 0 \\ 0 & 0 & \omega_2 \end{bmatrix} = \begin{bmatrix} 1 & 0 & 0 \\ 0 & \frac{1}{2}(1+\omega_1) & 0 \\ 0 & 0 & \frac{1}{2}(1+\omega_2) \end{bmatrix}.$$

The matrix M is invertible and so we can write the equation $AM = MD$ as $A = MDM^{-1}$. Show from this that

$$A^n = MD^n M^{-1}, \qquad \text{(for } n = 2, 3, 4, \ldots\text{)}.$$

Show that all the complex numbers $(1+\omega_1)^n, (1+\omega_2)^n$, $n = 1, 2, \ldots$, lie on the unit circle. Use this along with the fact that $\lim_{n \to \infty} (\frac{1}{2})^n = 0$, to complete the proof that $\lim_{n \to \infty} A^n = B$ where B is the matrix in equation (9.14).

(2) To analyze the asymptotic behavior of the sequences

$$\{P_n\}_{n=0}, \ \{Q_n\}_{n=0}, \ \{R_n\}_{n=0}$$

of vertices in the triangle sequence, consider the angles p_n, q_n, r_n that determine the vertices (see equations (9.11) (9.13)). These angle sequences will not converge, indeed they get bigger and bigger, but modulo multiples of 2π these sequences can be handled as follows.

(a) Use equation (9.11) to show that for any n 1

$$p_n = p_0 + \tfrac{1}{2}({}_0 + {}_1 + \cdots + {}_{n-1}).$$

Show that similar formulas hold for q_n and r_n. Use the divergence test to show that the series $\sum_{n=0} {}_n$ diverges. Similarly, the series $\sum_{n=0} {}_n,$ $\sum_{n=0} {}_n$ diverge.

(b) One can show, however, that each of the series

$$= \sum_{n=0} ({}_n - \tfrac{2\pi}{3}) \tag{9.15}$$

$$= \sum_{n=0} ({}_n - \tfrac{2\pi}{3}) \tag{9.16}$$

$$= \sum_{n=0} ({}_n - \tfrac{2\pi}{3}), \tag{9.17}$$

converges. (See Section 9.2.4 for the proof in the general case). Assuming that the series converge, find the formulas for , , and . To do this, you may find it easier to proceed by working the following parts:

(c) Let V be the vector $V = (\ ,\ ,\ )$. Show that the three series in equations (9.15) (9.17) can be combined into a single series in vector form:

$$V = (\ ,\ ,\ ) = \sum_{n=0} \left({}_n - \tfrac{2\pi}{3},\ {}_n - \tfrac{2\pi}{3},\ {}_n - \tfrac{2\pi}{3} \right) \tag{9.18}$$

$$= \sum_{n=0} A^n (u_0 - v), \tag{9.19}$$

where $u_0 = (\ {}_0,\ {}_0,\ {}_0)$ and $v = (\tfrac{2\pi}{3}, \tfrac{2\pi}{3}, \tfrac{2\pi}{3})$. To find a formula for the sum of the series, consider the sequence of partial sums

$$V_N \quad \sum_{n=0}^{N} A^n (u_0 - v),$$

for $N = 1, 2, 3, \ldots$. Show that for each N

$$(I - A)V_N = u_0 - A^{N+1} u_0, \tag{9.20}$$

and hence taking N gives the equation

$$(I - A)V = u_0 - Bu_0 \tag{9.21}$$

where B is the matrix in equation (9.14). Note that $Bu_0 = (\frac{2\pi}{3}, \frac{2\pi}{3}, \frac{2\pi}{3})$.

(d) Find the solution $V = (\ ,\ ,\ )$ of the system of equations (9.21), with , each expressed in terms of an arbitrary parameter. Note $(I - A)$ is not invertible, so there is not a unique solution. Show that the , defined by equations (9.15) (9.17) must satisfy $+ + = 0$. Use this to eliminate the arbitrary parameter to get the following formulas

$$= \tfrac{4}{3}\,{}_0 + \tfrac{2}{3}\,{}_0 - \tfrac{4\pi}{3} \tag{9.22}$$

$$= -\tfrac{2}{3}\,{}_0 + \tfrac{2}{3}\,{}_0 \tag{9.23}$$

$$= -\tfrac{2}{3}\,{}_0 - \tfrac{4}{3}\,{}_0 + \tfrac{4\pi}{3} \tag{9.24}$$

(e) Each sequence of vertices approaches a period six sequence of points. To show this, just consider the sequence $\{P_n\}_{n=0}$ of vertices. Vertex $P_n = (\cos p_n, \sin p_n)$, and we assume for convenience that the initial angle $p_0 = 0$. The nth angle is given by

$$p_n = \sum_{j=0}^{n-1} \tfrac{1}{2}\; j.$$

Let   be the angle given by equation (9.22), or equivalently equation (9.15), and let $\{E_n\}_{n=0}$ be the sequence of points on the unit circle given by

$$E_n \quad (\ \cos(\tfrac{\phi}{2} + \tfrac{n\pi}{3}),\ \sin(\tfrac{\phi}{2} + \tfrac{n\pi}{3})\ ). \tag{9.25}$$

Show that $E_0, \ldots, E_5$ are the vertices for a star of David figure and that $E_{n+6} = E_n$, for all n. Argue that

$$\cos(p_n) \quad = \quad \cos\left[\sum_{j=0}^{n-1} \tfrac{1}{2}(\ _n - \tfrac{2\pi}{3})\right] + \tfrac{n\pi}{3}$$

$$\cos(\tfrac{\phi}{2} + \tfrac{n\pi}{3}),$$

for large n. Similarly, $\sin(p_n) \quad \sin(\tfrac{\phi}{2} + \tfrac{n\pi}{3})$, for large n. Use these approximations to argue that the sequence $\{P_n\}_{n=0}$ asymptotically approaches the sequence $\{E_n\}_{n=0}$.

(f) Experimentally confirm the results in part (e) by choosing several sets of initial central angles $_0$, $_0$, $_0$ for T_0 and running the animation for each triangle sequence. For each choice of $_0$, $_0$, $_0$: (1) use formula (9.22) to compute the angle , (2) estimate from the animation the angle that E_0 makes with the x-axis, and (3) compare the values found for and .

9.2.3 Project: Sequences of Quadrilaterals

There are natural generalizations of the above method for producing sequences of triangles. One would be to produce a sequence of inscribed quadrilaterals, starting from an initial inscribed quadrilateral. The project in this subsection is for this. An ensuing project in the next subsection does a similar thing, starting with an initial inscribed m-gon (a polygon with m sides). Of course the m-gon case contains the triangle ($m = 3$) and quadrilateral ($m = 4$) cases, but you still might want to do the quadrilateral case here before tackling the general case.

Starting with a given circle, we produce a sequence $\{H_n\}_{n=0}$ of inscribed quadrilaterals by the following method.

The Method: Consider any initial inscribed quadrilateral H_0 with vertices P_0, Q_0, R_0, S_0. See Figure 9.15.

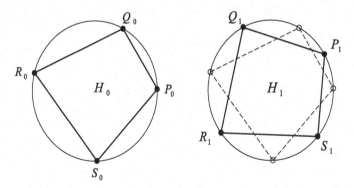

Figure 9.15: *An initial inscribed quadrilateral H_0 and a new one H_1 obtained from it by bisecting arcs.*

Bisect each of the four arcs subtended by the sides of H_0 to obtain four points P_1, Q_1, R_1, S_1 on these arcs. These points form the vertices of the next inscribed quadrilateral H_1, as shown in the figure. Similarly, the second quadrilateral H_2 in the sequence is obtained from H_1 in the same way, i.e., bisect the arcs subtended by the sides of H_1 to obtain the vertices of H_2. Continuing this process indefinitely, we get a sequence $\{H_n\}_{n=0}$ of inscribed quadrilaterals.

Problem: Modify the procedure `trisec` written above for the triangle sequence to get a procedure `quadsec`. Use this procedure to investigate the limiting behavior of the quadrilateral sequences starting from several different initial quadrilaterals. You should observe the following behavior:

(i) If the initial quadrilateral H_0 is equilateral, then it must be a square and the sequence $\{H_n\}_{n=0}$ is an alternating sequence of squares:

$$H_0, H_1, H_0, H_1, \ldots,$$

i.e., $H_{2k} = H_0$ and $H_{2k+1} = H_1$, for all k, and the figure consisting of H_0, H_1 is an eight-sided regular star.

(ii) More generally, for an arbitrary choice of initial quadrilateral H_0, the sequence $\{H_n\}_{n=0}$ converges to an alternating sequence of squares. That is, there is a pair of squares H, H , which form an eight-sided, regular star, and for which

$$H_{2k} \quad H \qquad H_{2k+1} \quad H \ ,$$

for large k. Here the approximations mean that the vertices are approximately the same.

(iii) Each of the sequences of vertices, like $\{P_n\}_{n=0}$, converges to a periodic sequence, of period eight, that cycles counterclockwise through the eight vertices of the limiting eight-sided regular star.

Do a detailed study of quadrilateral sequences, like the one above for triangle sequences. In particular, conduct experimental studies of the above properties and then prove the validity of these properties. For this, do the following:

(1) Let $u_n = (\ _n, \ _n, \ _n, \ _n)$ be the vector whose components are the four central angles subtended by the vertices of H_n. See Figure 9.16. Show that

$$u_{n+1} = Au_n = A^n u_0,$$

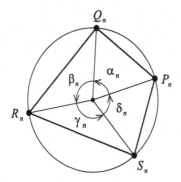

Figure 9.16: *The four central angles subtended by the vertices of quadrilateral* H_n.

where A is the 4×4 matrix

$$A = \begin{matrix} \frac{1}{2} & \frac{1}{2} & 0 & 0 \\ 0 & \frac{1}{2} & \frac{1}{2} & 0 \\ 0 & 0 & \frac{1}{2} & \frac{1}{2} \\ \frac{1}{2} & 0 & 0 & \frac{1}{2} \end{matrix}.$$

(2) Show that $\det(A - \lambda I) = (\lambda - \frac{1}{2})^4 - \frac{1}{16}$ and so the eigenvalues of A are

$$1, \quad \tfrac{1}{2}(1 + \zeta_1), \quad \tfrac{1}{2}(1 + \zeta_2), \quad \tfrac{1}{2}(1 + \zeta_3),$$

where ζ_k, $k = 0, 1, 2, 3$, are the *fourth roots of unity*:

$$\zeta_0 = 1, \quad \zeta_1 = i, \quad \zeta_2 = -1, \quad \zeta_3 = -i.$$

By definition, a root of the equation $x^4 = 1$ is called a fourth root of unity. Show that the above expressions for the fourth roots of unity can also be written as

$$\zeta_k = e^{2\pi k i / 4}, \qquad (k = 0, 1, 2, 3).$$

Show that

$$v_k = (1, \quad \zeta_k, \quad \zeta_k^2, \quad \zeta_k^3,)$$

is an eigenvector corresponding to eigenvalue $\lambda_k = \tfrac{1}{2}(1 + \zeta_k)$, for $k = 0, 1, 2, 3$.

(3) Use the results in part (2) to show that

$$\lim_{n \to} A^n = B \begin{array}{cccc} \frac{1}{4} & \frac{1}{4} & \frac{1}{4} & \frac{1}{4} \\ \frac{1}{4} & \frac{1}{4} & \frac{1}{4} & \frac{1}{4} \\ \frac{1}{4} & \frac{1}{4} & \frac{1}{4} & \frac{1}{4} \\ \frac{1}{4} & \frac{1}{4} & \frac{1}{4} & \frac{1}{4} \end{array}. \tag{9.26}$$

Use this to prove that

$$\lim_{n \to} (\ _n, \ _n, \ _n, \ _n) = (\tfrac{\pi}{4}, \tfrac{\pi}{4}, \tfrac{\pi}{4}, \tfrac{\pi}{4}),$$

and hence, for large n, the quadrilateral H_n is approximately a square.

(4) To analyze the asymptotic behavior of the sequences

$$\{P_n\}_{n=0}, \ \{Q_n\}_{n=0}, \ \{R_n\}_{n=0}, \ \{S_n\}_{n=0}$$

of vertices in the quadrilateral sequence, consider the angles $p_n, \ q_n, \ r_n, \ s_n$ that determine the vertices. See Figure 9.17.

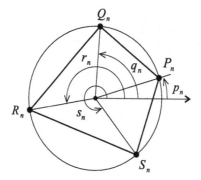

Figure 9.17: *The angles $p_n, \ q_n, \ r_n, \ s_n$ that determine the vertices of quadrilateral H_n.*

These angle sequences will not converge, indeed they get bigger and bigger, but modulo multiples of 2   these sequences can be handled as follows.

(a) Show that for any n 1

$$p_n = p_0 + \tfrac{1}{2}(\ _0 + \ _1 + \cdots + \ _{n-1}).$$

Show that similar formulas hold for q_n, r_n, s_n. Use the divergence test to show that the series $\sum_{n=0}^{\infty} {}_n$ diverges. Similarly, the series $\sum_{n=0}^{\infty} {}_n$, $\sum_{n=0}^{\infty} {}_n$, $\sum_{n=0}^{\infty} {}_n$ diverge.

(b) As in the triangular case, one can show that each of the series

$$= \sum_{n=0}^{\infty} \left({}_n - \tfrac{\pi}{2} \right) \tag{9.27}$$

$$= \sum_{n=0}^{\infty} \left({}_n - \tfrac{\pi}{2} \right) \tag{9.28}$$

$$= \sum_{n=0}^{\infty} \left({}_n - \tfrac{\pi}{2} \right) \tag{9.29}$$

$$= \sum_{n=0}^{\infty} \left({}_n - \tfrac{\pi}{2} \right) \tag{9.30}$$

converges (see Section 9.2.4). Assuming this, find the formulas for , , , . To do this, you may find it easier to proceed by working the following parts:

(c) Let V be the vector $V = (, , ,)$. Show that the four series in equations (9.27) (9.30) can be combined into a single series in vector form:

$$V = (, , ,) = \sum_{n=0}^{\infty} \left({}_n - \tfrac{\pi}{2}, \; {}_n - \tfrac{\pi}{2}, \; {}_n - \tfrac{\pi}{2}, \; {}_n - \tfrac{\pi}{2} \right)$$

$$= \sum_{n=0}^{\infty} A^n (u_0 - v),$$

where $u_0 = (\; {}_0, \; {}_0, \; {}_0, \; {}_0)$ and $v = (\tfrac{\pi}{2}, \tfrac{\pi}{2}, \tfrac{\pi}{2}, \tfrac{\pi}{2})$. To find a formula for the sum of the series, consider the sequence of partial sums

$$V_N \quad \sum_{n=0}^{N} A^n (u_0 - v),$$

for $N = 1, 2, 3, \ldots$. Show that for each N

$$(I - A)V_N = u_0 - A^{N+1}u_0, \tag{9.31}$$

and hence taking N gives the equation

$$(I - A)V = u_0 - Bu_0, \tag{9.32}$$

where B is the matrix in equation (9.26). Note that $Bu_0 = (\tfrac{\pi}{2}, \tfrac{\pi}{2}, \tfrac{\pi}{2}, \tfrac{\pi}{2})$.

(d) Find the solution $V = (\ ,\ ,\ ,\ )$ of the system of equations (9.32), with $\ ,\ ,\ $ each expressed in terms of an arbitrary parameter. Note $(I - A)$ is not invertible, so there is not a unique solution. Show that the $\ ,\ ,\ $ defined by equations (9.27) (9.30) must satisfy $\ +\ +\ +\ = 0$. Use this to eliminate the arbitrary parameter to get the following formulas

$$= \tfrac{3}{2}\ 0 +\ 0 + \tfrac{1}{2}\ 0 - \tfrac{3\pi}{2} \tag{9.33}$$

$$= -\tfrac{1}{2}\ 0 +\ 0 + \tfrac{1}{2}\ 0 - \tfrac{\pi}{2} \tag{9.34}$$

$$= -\tfrac{1}{2}\ 0 -\ 0 + \tfrac{1}{2}\ 0 + \tfrac{\pi}{2} \tag{9.35}$$

$$= -\tfrac{1}{2}\ 0 -\ 0 - \tfrac{3}{2}\ 0 + \tfrac{3\pi}{2}. \tag{9.36}$$

(e) Each sequence of vertices approaches a period eight sequence of points. To show this just consider the sequence $\{P_n\}_{n=0}$ of vertices. Vertex $P_n = (\cos p_n, \sin p_n)$, and we assume for convenience that the initial angle $p_0 = 0$. The nth angle is given by

$$p_n = \sum_{j=0}^{n-1} \tfrac{1}{2}\ j.$$

Let $\ $ be the angle given by equation (9.33), or equivalently equation (9.27), and let $\{E_n\}_{n=0}$ be the sequence of points on the unit circle given by

$$E_n \quad \left(\cos(\tfrac{\phi}{2} + \tfrac{n\pi}{4}),\ \sin(\tfrac{\phi}{2} + \tfrac{n\pi}{4}) \right). \tag{9.37}$$

Show that $E_0, \ldots, E_7$ are the vertices for a regular star of eight sides and that $E_{n+8} = E_n$, for all n. Argue that

$$\cos(p_n) = \cos\left[\sum_{j=0}^{n-1} \tfrac{1}{2}(\ n - \tfrac{\pi}{2}) \right] + \tfrac{n\pi}{4}$$

$$\cos(\tfrac{\phi}{2} + \tfrac{n\pi}{4}),$$

for large n. Similarly, $\sin(p_n)\quad \sin(\tfrac{\phi}{2} + \tfrac{n\pi}{4})$, for large n. Use these approximations to argue that the sequence $\{P_n\}_{n=0}$ asymptotically approaches the sequence $\{E_n\}_{n=0}$.

(f) Experimentally confirm the results in part (e) by choosing several sets of initial central angles $\ 0,\ 0,\ 0,\ 0$ for T_0 and running the animation for each quadrilateral sequence. For each choice of $\ 0,\ 0,\ 0,\ 0$: (1) use formula (9.33) to compute the angle $\ $, (2) estimate from the animation the angle $\ $ that E_0 makes with the x-axis, and (3) compare the values found for $\ $ and $\ $.

9.2.4   Project: Sequences of Polygons

The general project described here involves a sequence $\{G_n\}_{n=0}$ of inscribed polygons with m sides (also called m-gons). The previous cases, $m = 3$ and $m = 4$ for triangles and quadrilaterals, lead naturally to this and suggest how to study the general case. However, now the notation for the vertices and angles will have to be modified since now there are m vertices and m central angles. The sequence of inscribed m-gons is produced by the following method.

The Method: Consider any initial inscribed m-gon G_0 with vertices

$$P_0^1, P_0^2, \ldots, P_0^m.$$

See Figure 9.18.

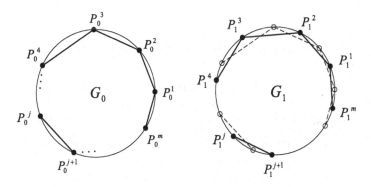

Figure 9.18: *An initial inscribed m-gon G_0 and a new one G_1 obtained from it by bisecting arcs.*

Bisect each of the m arcs subtended by the sides of G_0 to obtain m points $P_1^1, P_1^2, \ldots, P_1^m$ on these arcs. These points form the vertices of the next inscribed m-gon G_1, as shown in the figure. Similarly, the second m-gon G_2 in the sequence is obtained from G_1 in the same way, i.e., bisect the arcs subtended by the sides of G_1 to obtain the vertices $P_2^1, P_2^2, \ldots, P_2^m$ of G_2. Continuing this process indefinitely, we get a sequence $\{G_n\}_{n=0}$ of inscribed m-gons.

Problem: Based on the code for procedure `trisec`, or the procedure `quadsec`, written above, write a similar procedure `polygonsec` to produce graphics (static and dynamic) for the sequence of inscribed m-gons that arises from an initial m-gon. Use this procedure to investigate the limiting behavior of the m-gon sequences starting from several different initial m-gons. You should observe the following behavior:

(i) If the initial m-gon G_0 is equilateral, then it must be *regular* (i.e., all its central angles are the same, namely 2 $/m$), and the sequence $\{G_n\}_{n=0}$ is an alternating sequence of regular m-gons:

$$G_0, G_1, G_0, G_1, \ldots,$$

i.e., $G_{2k} = G_0$ and $G_{2k+1} = G_1$, for all k, and the figure consisting of G_0, G_1 is a $2m$-sided, regular star.

(ii) More generally, for an arbitrary choice of initial m-gon G_0, the sequence $\{G_n\}_{n=0}$ converges to an alternating sequence of regular m-gons. That is, there is a pair of regular m-gons G, G, which form a $2m$-sided regular star, and for which

$$G_{2k} \quad G \qquad G_{2k+1} \quad G,$$

for large k. Here the approximations mean that the vertices are approximately the same.

(iii) Each of the sequences of vertices, like $\{P_n^1\}_{n=0}$, converges to a periodic sequence, of period $2m$, that cycles counterclockwise through the $2m$ vertices of the limiting $2m$-sided regular star.

Do a detailed study of m-gon sequences, like the one above for triangles and quadrilateral sequences. In particular, conduct experimental studies of the above properties and then prove the validity of these properties. For this, do the following:

(1) Let $u_n = (\ \frac{1}{n},\ \frac{2}{n}, \ldots,\ \frac{m}{n})$ be the vector whose components are the central angles subtended by the vertices of G_n. See Figure 9.19. Show that

$$u_{n+1} = Au_n = A^n u_0,$$

where A is the $m \times m$ matrix

$$A = \begin{pmatrix} \frac{1}{2} & \frac{1}{2} & 0 & 0 & \cdots & 0 \\ 0 & \frac{1}{2} & \frac{1}{2} & 0 & \cdots & 0 \\ & & & \ddots & & \\ 0 & 0 & \cdots & 0 & \frac{1}{2} & \frac{1}{2} \\ \frac{1}{2} & 0 & \cdots & 0 & 0 & \frac{1}{2} \end{pmatrix}$$

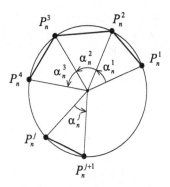

Figure 9.19: *The central angles,* $\frac{1}{n}, \frac{2}{n}, \ldots, \frac{m}{n}$, *subtended by the vertices of m-gon* G_n.

(2) Show that

$$\det(A - \lambda I) = (-1)^m \left[(\lambda - \tfrac{1}{2})^m - (\tfrac{1}{2})^m \right],$$

and so the eigenvalues of A are

$$\lambda_k = \tfrac{1}{2}(1 + \omega_k),$$

$k = 0, 1, \ldots, m - 1$, where ω_k are the *mth roots of unity:*

$$\omega_k = e^{2\pi k i/m}, \qquad (k = 0, 1, \ldots, m - 1).$$

By definition, a root of the equation $x^m = 1$ is called an mth root of unity. Show that besides $\omega_0 = 1$, all the other mth roots of unity ω_k are complex and are the roots of the equation

$$x^{m-1} + \cdots + x^2 + x + 1 = 0.$$

Show that
$$v_k = (1, \omega_k, \omega_k^2, \ldots, \omega_k^{m-1}),$$
is an eigenvector corresponding to eigenvalue $\lambda_k = \tfrac{1}{2}(1 + \omega_k)$, for $k = 0, 1, \ldots, m - 1$.

Let $M = [v_0, v_1, \ldots, v_{m-1}]$ be the $m \times m$ matrix with the eigenvectors v_k as its columns. Show that $AM = MD$, where $D = \text{diag}(\lambda_1, \lambda_1, \ldots, \lambda_{m-1})$ is the $m \times m$ diagonal matrix with the eigenvalues of A along the diagonal.

Show that M is an invertible matrix and find its inverse M^{-1}. For this, let the entries of of M be denoted by $M = \{M_{jk}\}_{j,k=0,1,\ldots,m-1}$, so that M_{jk} is the entry in the jth row and kth column.[1] Show that

$$M_{jk} = \begin{smallmatrix} j \\ k \end{smallmatrix} = e^{2\pi ijk/m},$$

for all $j, k = 0, 1, \ldots, m - 1$. Then let

$$M_{jk}^{-1} = \tfrac{1}{m} \begin{smallmatrix} j \\ m-k \end{smallmatrix} = \tfrac{1}{m} e^{-2\pi ijk/m},$$

and show that the inverse matrix is $M^{-1} = \{M_{jk}^{-1}\}_{j,k=0,1,\ldots,m-1}$.

(3) Use the results in part (2) to show that

$$\lim_{n \to} A^n = B \begin{bmatrix} \tfrac{1}{m} & \tfrac{1}{m} & \cdots & \tfrac{1}{m} \\ \tfrac{1}{m} & \tfrac{1}{m} & \cdots & \tfrac{1}{m} \\ \vdots & \vdots & & \vdots \\ \tfrac{1}{m} & \tfrac{1}{m} & \cdots & \tfrac{1}{m} \end{bmatrix}. \tag{9.38}$$

Use this to prove that

$$\lim_{n \to} \left(\begin{smallmatrix} 1 \\ n \end{smallmatrix}, \begin{smallmatrix} 2 \\ n \end{smallmatrix}, \ldots, \begin{smallmatrix} m \\ n \end{smallmatrix} \right) = \left(\tfrac{2\pi}{m}, \tfrac{2\pi}{m}, \ldots, \tfrac{2\pi}{m} \right),$$

and hence, for large n, the m-gon G_n is approximately a regular m-gon.

(4) To analyze the asymptotic behavior of the sequences

$$\{P_n^1\}_{n=0}, \ \{P_n^2\}_{n=0}, \cdots \{P_n^m\}_{n=0}$$

of vertices in the m-gon sequence, consider the angles

$$p_n^1, \ p_n^2, \ldots, \ p_n^m,$$

that determine the vertices. See Figure 9.20.

Analyze the behavior of these angles as follows.

(a) Show that for all $j = 1, 2, \ldots, m$ and all $n \geq 0$,

$$p_{n+1}^j = p_n^j + \tfrac{1}{2} \begin{smallmatrix} j \\ n \end{smallmatrix}, \qquad (j = 1, 2, \ldots, m).$$

---

[1] For convenience the indexing begins with 0 instead of 1.

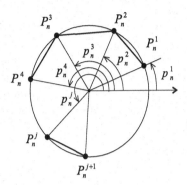

Figure 9.20: *The angles that determine the vertices in the m-gon G_m.*

(b) Use the equation in part (a) to show that for all $j = 1, 2, \ldots, m$, and for any $n \geq 1$,

$$p_n^j = p_0^j + \tfrac{1}{2}\left(\theta_0^j + \theta_1^j + \cdots + \theta_{n-1}^j \right).$$

Use the divergence test to show that the series $\sum_{n=0}^{\infty} \theta_n^j$ diverges.

(c) Show that each of the series

$$\sum_{n=0}^{\infty} \left(\theta_n^j - \tfrac{2\pi}{m} \right) \qquad (j = 1, 2, \ldots, m) \qquad (9.39)$$

converges and find the formulas for σ^j, $j = 1, \ldots, m$, i.e., the value of the sum of the series. You can find the formulas exactly like you did in in the cases $m = 3$ (triangles) and $m = 4$ (quadrilaterals). To show convergence of each series proceed as follows.

Let V be the vector $V = (\sigma^1, \sigma^2, \ldots, \sigma^m)$. Show that the series in equations (9.39) can be combined into a single series in vector form:

$$V = \sum_{n=0}^{\infty} \left(\theta_n^1 - \tfrac{2\pi}{m}, \, \theta_n^2 - \tfrac{2\pi}{m}, \ldots, \, \theta_n^m - \tfrac{2\pi}{m} \right) \qquad (9.40)$$

$$= \sum_{n=0}^{\infty} A^n (u_0 - v), \qquad (9.41)$$

where $u_0 = (\theta_0^1, \theta_0^2, \ldots, \theta_0^m)$ and $v = (\tfrac{2\pi}{m}, \tfrac{2\pi}{m}, \ldots, \tfrac{2\pi}{m})$. To show convergence of this series of vectors, consider the sequence of partial sums

$$V_N \quad \sum_{n=0}^{N} A^n(u_0 - v).$$

for $N = 1, 2, 3, \ldots$. Write A as $A = MDM^{-1}$ and show that $A^n = MD^n M^{-1}$, for all n. Let $f = M^{-1}(u_0 - v)$ and denote the components of f by $f = (f_0, f_1, \ldots, f_{m-1})$. Then show that $f_0 = 0$, and consequently

$$
\begin{aligned}
V_N &= \sum_{n=0}^{N} A^n(u_0 - v) \\
&= \sum_{n=0}^{N} MD^n M^{-1}(u_0 - v) \\
&= \sum_{n=0}^{N} MD^n f \\
&= M \begin{bmatrix} 0 \\ \sum_{n=0}^{N} \quad_1^n \; f_1 \\ \vdots \\ \sum_{n=0}^{N} \quad_{m-1}^n \; f_{m-1} \end{bmatrix}.
\end{aligned}
$$

Use this and apply the geometric series $\sum_{n=0}^{} r^n$ to $r = \quad_k$, $k = 1, \ldots, m-1$, to conclude that the sequence of $\{V_N\}_{N=1}$ partial sums converges.

(d) Let ^{1} be the angle given by equation (9.39) and let $\{E_n\}_{n=0}$ be the sequence of points on the unit circle given by

$$E_n \quad \left(\cos\left(\tfrac{\phi^j}{2} + \tfrac{n\pi}{m}\right), \; \sin\left(\tfrac{\phi^j}{2} + \tfrac{n\pi}{m}\right) \right). \tag{9.42}$$

Show that $E_0, \ldots, E_{m-1}$ are the vertices for a regular star with $2m$ points and that $E_{n+2m} = E_n$, for all n. Assume that $p_0^1 = 0$ an argue that

$$\cos(p_n^1) = \cos\left[\sum_{j=0}^{n-1} \tfrac{1}{2}\left(\quad_n^1 - \tfrac{2\pi}{m}\right) \right] + \tfrac{n\pi}{m}$$

$$\cos\left(\tfrac{\phi^j}{2} + \tfrac{n\pi}{m}\right),$$

for large n. Similarly, $\sin(p_n^1) \quad \sin\left(\tfrac{\phi^j}{2} + \tfrac{n\pi}{m}\right)$, for large n. Use these approximations to argue that the sequence $\{P_n^1\}_{n=0}$ asymptotically approaches the sequence $\{E_n\}_{n=0}$.

9.3 Projects on Random Walks

Consider the following situation. You are standing at the origin in the Cartesian plane with a *four-sided* die. You roll the die, getting a 1, 2, 3, or 4, with each outcome equally likely.[2] If you roll a one you take one step in the positive x direction. If you roll a two take a step in the positive y direction, a three take a step in the negative x direction, and if you roll a four take a step in the negative y direction. Once you have taken a step in one of the four coordinate directions, roll the die again and take another step according to the same rules. Continue to do this for a sequence of N steps. Your resulting path is called a *random walk* of length N in the plane or a 2-*dimensional random walk*. The walk is random because the direction of each step is determined by the die, and cannot be known before the die is rolled.

A random walk could be simulated on the computer if we could find a way of producing the numbers $1, 2, 3, 4$ randomly. This would require having a computer algorithm for producing a random number. Of course it is impossible to write a procedure to produce a truly random number since any procedure is a sequence of instructions and the resulting output is exactly reproducible. However, procedures to produce numbers which are called *pseudorandom numbers* have been written. Pseudo means that if the procedure is used to produce a sequence of numbers, then this sequence appears to have been randomly generated. Such a procedure is called a random number generator (even though generation of truly random numbers is not possible on a computer).[3]

Most random number generators are *linear congruence generators* (LCGs) and are based on the judicious selection of positive integers a, b, and m, which are fixed and used to generate a sequence of pseudorandom integers. The generation of these integers starts with an initial integer x_0, called the *seed*, and then recursively computes integers $x_1, x_2, x_3, \ldots$, where x_{n+1} is defined by

$$x_{n+1} = ax_n + b \quad (\mathrm{mod}\ m)$$

This means that x_{n+1} is the remainder when $ax_n + b$ is divided by m. So each x_n is one of the integers $0, 1, 2, \ldots, m - 1$.

---

[2] One says that the *probability* of each outcome is $1/4$.

[3] It is often convenient, in our discussion related to the computer, not to distinguish between random numbers (which are mathematical idealizations) and pseudorandom numbers. Thus, we will often just say random when we mean pseudorandom.

9.3.1 Sample Project: Simulating a Basic Random Walk

Now we will construct a Maple procedure to produce a basic random walk. (See [Ber], [Ros1], [Ros2] for a background on random walks.)

Problem: Write a procedure to simulate a 2-dimensional random walk of length N, where at each step the walker moves one unit with equal probability in one of the four coordinate directions. Produce an animation of the walk.

Analysis of the Problem: Maple has a procedure rand(p..q) which, when $p < q$ are integers, produces a pseudorandom integer between p and q. Thus, we can use rand(1..4) to play the role of the four-sided die in the random walk above.

When we make the command rand(1..4), the output is a procedure as shown below.

```
proc()
local t; global _seed;
_seed := irem(427419669081*_seed,999999999989);
t := _seed; irem(t,4)+1
end
```

As you can see, rand(1..4) defines a procedure with no parameters and a single output, the result of the last statement executed, i.e., irem(t,4)+1. Here irem is Maple's modulo function.[4] Thus, the output will be $1, 2, 3$, or 4. Also note that _seed x_0 is a global variable which has some integer value when the procedure is invoked. The procedure then computes a new value x_1 for the seed from the formula $x_1 = ax_0 \pmod{m}$, where $a = 427419669081$ and $m = 999999999989$. Thus, the seed changes in this way at each invocation of the procedure, and the theory is that this type of formula will produce a seemingly random sequence from a sequence of successive invocations.

For example, suppose we invoke the procedure 20 times in succession. To put all the output on a single line, we use the seq command. Also note that the command rand(1..4) only defines the generator. We invoke this the random number generator with the command rand(1..4)(). Thus,

```
seq(rand(1..4)(),i=1..20);
```

---

[4]In this case irem(t,4) gives the remainder when the integer t is divided by 4. Mathematically, we would write this as $t \pmod 4$ and the result is 0, 1, or 2.

produces the output

3, 2, 1, 3, 2, 3, 4, 4, 3, 2, 4, 3, 2, 1, 1, 2, 2, 3, 3, 1

Note that we would expect that about one fourth of the numbers generated would be ones, one fourth of them twos, etc. This is approximately the case (There are four ones, six twos, seven threes, and three fours). Also note that if you execute this command, the output may be different. This is because the initial seed may be different. With this understanding of how (pseudo) random numbers are generated by Maple, we can now design the code as follows.

Design of the Code: Here there are two things to do: (1) generate the steps in the random walk, and (2) produce the animation. The procedure is called `randomwalk` and has the number of steps N as input. The coordinates of the steps are stored in an array called `position` and the variable `step` is the step counter (it is incremented by 1 before each step). The first part of the procedure will use `rand(1..4)()` to determine the direction of each step. The second part will generate the plots of the path the walker has taken up through step k, and then will store these in an array `path`.

The output from the procedure is a graphic illustrating the walk. There are three types of graphics that can be produced, based on the value of the input parameter `typ`. For `typ=0`, the output is a static picture showing all the points in the plane that the walker visits. These points are indicated with a dot (a small disk) with radius specified by the input parameter `dotsize`. The graphic also includes the line segments joining each successive step in the walk,

Specifying `typ=1` for the type of graphic produces an animation showing the line segments between successive steps. The segments persist as the walk progresses, leaving a track or trace of the walk. The present position is marked by a dot.

The `typ=2` produces an animation also, but takes far less memory since each frame of the movie consists of a step, marked with a dot and connected with a line segment to the prior step.

Code for a 2-Dimensional Random Walk

```
randomwalk:=proc(N,typ,dotsize)
   local position,step,randnum,prob,k,path;
   position:=array(0..N);
   step:=0;position[0]:=[0,0];
```

```
while step<N do
step:=step+1;
randnum:=rand(1..4)();
if randnum=1 then
        position[step]:=position[step-1]+[1,0];
    elif randnum=2 then
        position[step]:=position[step-1]+[0,1];
    elif randnum=3 then
        position[step]:=position[step-1]+[-1,0];
    else
        position[step]:=position[step-1]+[0,-1];
end if;
end do;
with(plots,display):with(plottools,disk):
for k from 1 to N do
  if typ=1 then
    path[k]:=display({disk(position[k],dotsize,color=black),
                plot([seq(position[i],i=0..k)],thickness=2)});
    elif typ=0 or typ=2 then
    path[k]:=display({disk(position[k],dotsize,color=black),
                plot([seq(position[i],i=k-1..k)],thickness=2)});
  end if;
end do;
if typ=1 or typ=2 then
  display(seq(path[k],k=1..N),scaling=constrained,
          insequence=true);
elif typ=0  then
  display(seq(path[k],k=1..N),scaling=constrained);
end if;
end proc:
```

Figure 9.21 shows a random walk consisting of 200 steps.

The animation on the CD-ROM will give you a better understanding of the figure and how the walker visits some positions more than once.

9.3.2 Sample Project: Random Walk in a Square

An interesting variation of the random walk is the inclusion of a boundary. For instance, forcing the random walker to stay within a C by C square with corners $[C, C]$, $[-C, C]$, $[-C, -C]$, and $[C, -C]$. So the random walk takes place as described above, but if the walker hits the boundary at any

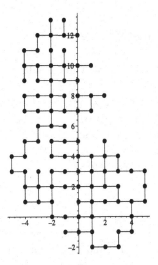

Figure 9.21: *A random walk produced by* `randomwalk(200,0,.13)`.

time, then she is re"ected away from the boundary back into the square. To do this we must decide on the rule defining the re"ection. A simple method would be for the walker to do the following: If a step would take her to or beyond the boundary, then the next step is to the point on the boundary where she would either hit the boundary or cross it. The ensuing step after that is one unit in the reverse direction.

We write a separate procedure for handling the re"ection. This procedure will be called from the main procedure whenever the walker hits or crosses the boundary. The input to the re"ection procedure is: (1) the next position (called **point**) resulting from a step, which if the walker took it would cause her to either hit the boundary or go beyond it, and (2) the location of the square's boundary, specified by **bdry** (which is assumed to be a positive number the C in the above discussion). The output is: (1) the boundary point, called **nextpoint1**, where the step crosses or meets the boundary, and (2) the location of the walker after the re"ection, called **nextpoint2** (one unit in the reverse direction).

Note: When re"ection occurs, the walker's step is split into two steps a step to a boundary point followed by a reverse step of 1 unit back into the square. Since we assume that $C > 1$, we are assured that the point **nextpoint2** lies strictly within the square.

```
reflect:=proc(point,bdry,nextpoint1::evaln,nextpoint2::evaln)
```

```
local a,b,C;
a:=eval(point[1]);b:=eval(point[2]);C:=eval(bdry);
#--------------------------------------------------------
#First case: Check if the walker is on the boundary
#--------------------------------------------------------
if a<=C and a>=-C and b<=C and b>=-C then
   if a=C then nextpoint1:=[a,b];nextpoint2:=[a-1,b]
     elif a=-C then nextpoint1:=[a,b];nextpoint2:=[a+1,b]
     elif b=C then nextpoint1:=[a,b];nextpoint2:=[a,b-1]
     else nextpoint1:=[a,b];nextpoint2:=[a,b+1] ;
   end if;
   RETURN();
end if;
#--------------------------------------------------------
#Second case: Check if walker is beyond the boundary
#--------------------------------------------------------
if b>=-C and b<=C then
  if a>C then nextpoint1:=[C,b];nextpoint2:=[C-1,b] end if;
  if a<-C then nextpoint1:=[-C,b];nextpoint2:=[-C+1,b] end if;
end if;
if a<=C and a>=-C then
  if b>C then nextpoint1:=[a,C];nextpoint2:=[a,C-1] end if;
  if b<-C then nextpoint1:=[a,-C];nextpoint2:=[a,-C+1] end if;
end if;
RETURN();
end proc:
```

Now we adjust the random walk procedure from the previous section to check at each step to see if the walker hits or crosses the boundary. Whenever the walker hits/crosses the boundary, the `reflect` procedure is called.

The new procedure is called `Rwalkwbdry1` and, as before, has as input the number of steps in the walk, the boundary location `bdry` (specified by C above), the type of graphic output (specified by `typ`), and the dot size for the dots marking each point the walker visits during the walk.

Code for a Random Walk in a Square

```
Rwalkwbdry1:=proc(N,bdry,typ,dotsize)
   local position,C,step,randnum,k,boundary,path;
   position:=array(0..N+1);
```

```
C:=eval(bdry);
step:=0;position[0]:=[0,0];
while step<N do
  step:=step+1;
  randnum:=rand(1..4)();
  if randnum=1 then
     position[step]:=position[step-1]+[1,0];
    elif randnum=2 then
       position[step]:=position[step-1]+[0,1];
    elif randnum=3 then
       position[step]:=position[step-1]+[-1,0];
    else
       position[step]:=position[step-1]+[0,-1];
  end if;
  if (position[step][1]>=C or position[step][1]<=-C or
             position[step][2]>=C or position[step][2]<=-C)
    then
       reflect(position[step],C,nextpoint1,nextpoint2);
       position[step]:=nextpoint1;
       position[step+1]:=nextpoint2;
       step:=step+1
  end if;
  end do;
with(plots,display):with(plottools,disk):
boundary:=plot([[-C,-C],[-C,C],[C,C],[C,-C],[-C,-C]],
                color=black):
for k from 1 to N do
 if typ=1 then
   path[k]:=display({boundary,
               disk(position[k],dotsize,color=black),
               plot([seq(position[i],i=0..k)],thickness=2)});
  elif typ=0 or typ=2 then
   path[k]:=display({boundary,
               disk(position[k],dotsize,color=black),
               plot([seq(position[i],i=k-1..k)],thickness=2)});
  end if;
end do;
if typ=1 or typ=2 then
  display(seq(path[k],k=1..N),scaling=constrained,
          insequence=true);
```

```
elif typ=0  then
   display(seq(path[k],k=1..N),scaling=constrained);
end if;
end proc:
```

Figure 9.22 shows the result of the call `Rwalkwbdry1(200,3.2,0,0.1)`.

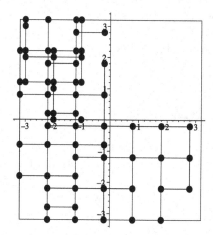

Figure 9.22: *Random walk of* 200 *steps in a square with* $C = 3.2$.

Observe that when there is no boundary, as in the case of the basic random walk discussed first, the walker remains on the lattice, or grid, $\mathbb{Z} \times \mathbb{Z}$ of unit spacing in $\mathbb{R}^2$. However, in the present case, the imposition of the boundary can cause the walker to get off this lattice, as Figure 9.22 clearly shows.

9.3.3 Sample Project: Basic Random Walk in a Disk

We can confine our random walk within other boundary shapes provided we can detect when the walker would hit or pass beyond the boundary and decide on the re"ection rules.

We consider a circular boundary of the form $x^2 + y^2 = r^2$ and assume the walker starts at the origin. Assume that $r > 1$, since our step length is 1. It is an easy matter to decide if the walker's next step would take him to or beyond the boundary since we have the walker's coordinates.

We will use a similar re"ection rule to that for the square   step to the boundary and then step one unit in the reverse direction   even though now the rule may seem artificial. For the case of a square, any step that would

hit or cross the boundary makes a path that strikes the boundary perpendicularly. So reversal of the direction in this case makes sense physically.
For the circular boundary, the steps hardly ever strike the boundary perpendicularly, and so reversal of direction is not a physical re"ection. Some
of the projects below look at the truly re"ected case.

Re ection Rule: Suppose the walker is at location $P = (P_1, P_2)$ inside
the circle at the $(k-1)$st step, but the next step would be to a position
$Q = (Q_1, Q_2)$ that is either on the boundary or beyond the boundary. Thus,
$P_1^2 + P_2^2 < r^2$ and $Q_1^2 + Q_2^2 \quad r^2$. We divide the contemplated step from P
to Q into two steps: (1) an actual step (the kth step) to the boundary point
P', where the segment $\overline{PQ}$ intersects the circle, and then (2) a contemplated
step of one unit in the reverse direction to a point Q'. If Q' is inside the
circle, then the $(k+1)$st step will be to Q', and the re"ection process is
over. However, if Q' is outside the circle, we apply the re"ection process
again, i.e., divide the contemplated step from P' to Q' into two steps: (1')
an actual step (the $(k+1)$st step) to the boundary point P'', where the
segment $\overline{P'Q'}$ intersects the circle, and then (2') a contemplated step of one
unit in the reverse direction to a point Q''. See Figure 9.23.

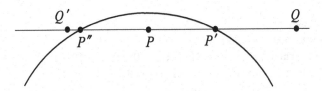

Figure 9.23: *If a two successive re ections occur, then the walker enters a
continual loop of re ections between the points P' and P''.*

It is easy to see that in this latter case, the re"ection process continues
cyclically with steps back and forth between P' and P''.

We write a procedure, called `circreflect`, to do one basic re"ection
(labeled (1) and (2) in the above re"ection rule). Then this procedure can
be called repeatedly if multiple re"ections occur. The procedure works like
this. At the $(k-1)$st step the walker is at a point $P = (P_1, P_2)$ inside the
circle, and the contemplated next step takes him to a point $Q = (Q_1, Q_2)$
outside the circle. Since the steps are always vertical or horizontal, we have
one of the following cases:

(a) (Vertical Step to Q) In this case $P_1 = Q_1$ and Q is either above or
below P. That is, either $P_2 < Q_2$ or $Q_2 < P_2$. If we let $y = \quad \overline{r^2 - P_1^2}$,

then the boundary point is either $P' = (P_1, y)$ or $P' = (P_1, -y)$, in the above or below case, respectively.

(b) (Horizontal Step to Q) In this case $P_2 = Q_2$ and Q is either to the right or to the left of P. That is, either $P_1 < Q_1$ or $Q_1 < P_1$. If we let $x = \sqrt{r^2 - P_2^2}$, then the boundary point is either $P' = (x, P_2)$ or $P' = (-x, P_2)$, in the right or left case, respectively.

The following procedure `circreflect` implements the above cases. It takes the points P, Q and radius r of the circular boundary as inputs (with formal parameter names `PP`,`QQ`, `rr` respectively). The output is the boundary crosspoint P' stored in `nextP` and the re"ected point Q' stored in `nextQ`.

```
circreflect:=proc(PP,QQ,rr,nextP::evaln,nextQ::evaln)
  local r,P,Q,x,y;
  P:=PP;Q:=QQ; r:=rr;
  if P[1]=Q[1] then
     y:=evalf(sqrt(r^2-P[1]^2));
     if   P[2]<Q[2] then nextP:=[P[1],y]; nextQ:=[P[1],y-1];
     elif Q[2]<P[2] then nextP:=[P[1],-y];nextQ:=[P[1],-y+1];
     end if;
  elif P[2]=Q[2] then
     x:=evalf(sqrt(r^2-P[2]^2));
     if   P[1]<Q[1] then nextP:=[x,P[2]];nextQ:=[x-1,P[2]];
     elif Q[1]<P[1] then nextP:=[-x,P[2]];nextQ:=[-x+1,P[2]];
     end if;
  end if;
end proc:
```

Now we get the main procedure which we call `Rwalkwbdry2` by adjusting the `Rwalkwbdry1` procedure. This procedure has as input the number of steps, `N`, in the walk, the radius, `bdry`, of the boundary circle, and the type of graphical output, `typ`. We dispense with specifying the dot size and take it always to be $r/50$.

$\boxed{\text{Code for a Random Walk in a Disk}}$

```
Rwalkwbdry2:=proc(N,bdry,typ)
  local position,r,step,randnum,boundary,k,path;
  position:=array(0..N+1);
  r:=eval(bdry);
```

```
step:=0;position[0]:=[0,0];
while step<N do
  step:=step+1;
  randnum:=rand(1..4)();
     if randnum=1 then
       position[step]:=position[step-1]+[1,0];
     elif randnum=2 then
       position[step]:=position[step-1]+[0,1];
     elif randnum=3 then
       position[step]:=position[step-1]+[-1,0];
     else
       position[step]:=position[step-1]+[0,-1];
     end if;
#----------------------------------------------------------------
#Check to see if walker is on or beyond the boundary circle.
#If so, call the reflection procedure (maybe many times)
#----------------------------------------------------------------
    while position[step][1]^2+position[step][2]^2>=r^2
         and step<N do
       circreflect(position[step-1],position[step],r,
                   nextpoint1,nextpoint2);
       position[step]:=nextpoint1;
       position[step+1]:=nextpoint2;
       step:=step+1
    end do;
#----------------------------------------------------------------
#Loop back and generate another random number
#----------------------------------------------------------------
  end do;
#----------Now do the Graphics---------------------------------
  with(plots,display,implicitplot):with(plottools,disk):
  boundary:=implicitplot(x^2+y^2=r^2,x=-r..r,y=-r..r,
                         color=black):
  for k from 1 to N do
    if typ=1 then
      path[k]:=display({boundary,
                disk(position[k],r/50,color=black),
                plot([seq(position[i],i=0..k)],thickness=2)});
    elif typ=0 or typ=2 then
      path[k]:=display({boundary,
```

```
            disk(position[k],r/50,color=black),
            plot([seq(position[i],i=k-1..k)],thickness=2)});
   end if;
 end do;
 if typ=1 or typ=2 then
   display(seq(path[k],k=1..N),scaling=constrained,
           insequence=true);
 elif typ=0  then
   display(seq(path[k],k=1..N),scaling=constrained);
 end if;
end proc:
```

Figure 9.24 shows the result of a call Rwalkwbdry2(200,3.2,0).

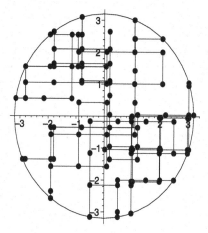

Figure 9.24: *Random walk of* 200 *steps in a disk of radius* $r = 3.2$.

9.3.4   Project: Random Walks -- Other Boundaries and Variations

There are many variations and extensions to the random walks described above. Here, we still consider 2-dimensional walks with horizontal/vertical steps, but we vary either the re"ection rule or the geometry of the boundary. After studying the previous sample projects do the following parts (whichever your instructor assigns).

  1. Write a procedure circreflect2 to change the re"ection rule for the random walk inside a circle to the following. Suppose the walker's

position is P (inside the circle) and the next step would be Q, where Q is either on or beyond the boundary. Let P' be the point where $\overline{PQ}$ intersects the boundary. If $\overline{PQ}$ is perpendicular to the tangent line to the boundary circle,[5] then the previous re"ection rule is applied: take one step to the point P', and then take a unit step in the reverse direction. If, however, $\overline{PQ}$ is not perpendicular to the tangent line, then take a step to the point P' on the boundary, and then take a unit step to the point Q' which is in a direction that is (a) perpendicular to the direction of the previous step and (b) toward the interior of the circle. In some cases, this latter step gives a point Q' that is outside the circle. In this case, apply the re"ection process again (and continually if necessary). See Figure 9.25.

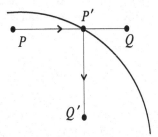

Figure 9.25: *New re ection rule for a walk in a disk.*

Is oscillation possible with this re"ection rule? That is, will the steps begin to cycle between various points on the boundary? Now write a procedure `Rwalkwbdry3` to simulate a random walk inside a circle using this re"ection rule.

2. **(Random Walk in a Diamond)** Instead of restricting the walk to the interior of a disk, consider restricting it to the interior D of a diamond. Specifically, D is the region bounded by the four lines

$$x + y = r, \quad x + y = -r, \quad x - y = r, \quad x - y = -r,$$

where $r > 1$ is a given number. As in Sample Project 9.3.3 for the disk, write a procedure to produce graphics like the ones there. The re"ection rule is the same as in that project   step to the boundary

---

[5]There are only four places where this can happen.

and then a unit step in the reverse direction — except now the boundary has a different geometry. Is oscillation between boundary points possible here?

3. **(Random Walk in a More General Region)** Suppose D is a region in $\mathbb{R}^2$ given by

$$D = \{\, (x,y) \quad \mathbb{R}^2 \mid H(x,y) < 0 \,\},$$

where $H : \mathbb{R}^2 \quad \mathbb{R}$ is a function of two variables. The boundary of D is the curve C defined implicitly by the equation $H(x,y) = 0$. We assume that the region D has the property: If P is in D and Q is outside of D, then the line segment $\overline{PQ}$ intersects the boundary C in one and only one point. It will be helpful to phrase this property analytically as follows. If we let $f : \mathbb{R} \quad \mathbb{R}^2$ be the parametrization:

$$f(t) \quad P + t\,\overrightarrow{PQ}$$

of the line through P and Q, then the equation $H(f(t)) = 0$ has one and only one solution t in the interval $(0, 1)$.

As in Sample Project 9.3.3 for the disk, write a procedure to produce graphics like the ones there. The re"ection rule is the same as in that project step to the boundary and then a unit step in the reverse direction except now the boundary has a different geometry. Is oscillation between boundary points possible here?

9.3.5   Random Walks in 3-Dimensions

A random walk in $\mathbb{R}^3$ is the natural analog of a random walk in $\mathbb{R}^2$. The walker starts at the origin and takes a unit step in one of the six coordinate directions, based on the outcome of the roll of a six-sided die. As in the sample projects, develop the code to animate and study such walks without and with boundaries. Specifically do the following parts that you are assigned:

1. **(The Basic Random Walk in $\mathbb{R}^3$)** Here there is no boundary and the walk is unrestricted.

2. **(A Random Walk in a Cube)** Here the walk is restricted to the interior of the cube $[-C, C] \times [-C, C] \times [-C, C]$ and the re"ection rule is the exact analog of the corresponding walk in the square $[-C, C] \times [-C, C]$.

3. **(A Random Walk in a Ball)** Here the walk is restricted to the interior of the ball $B(0, r) \quad \{(x, y, z) \mid x^2 + y^2 + z^2 \quad r^2\}$ of radius r and center at the origin. The re"ection rule is the exact analog of the corresponding walk in the disk of radius r and center at the origin.

9.3.6 Project: Walking in a Random Direction

So far we have looked at random walks where the walker moves only in directions parallel to the coordinate directions. That is, at each step one of the four coordinate directions was chosen at random as the direction of the next step. In this project, the walker again starts at the origin but an angle in the interval $(0, 2\,]$ is chosen at random as the direction of the next step. This angle is measured from the positive x-axis. The walker moves one unit in this direction, stops, chooses another angle at random, and then moves one unit in that direction. See Figure 9.26.

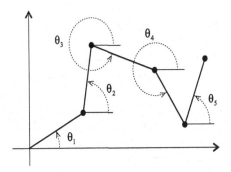

Figure 9.26: *A five-step random walk where the direction is random.*

Now we can no longer use the basic random number generator `rand(1..n)`, because this only generates random integers in the range $1, 2, \ldots, n$, with each such integer being equally likely of being generated, i.e., each has probability $1/n$ of occurring. Now we need to generate random numbers in the interval $[0, 2\,]$ (there are infinitely many such numbers), and each needs to be equally likely of being generated. Maple has a procedure for generating numbers uniformly distributed in a given interval. It is (see [Mon2, p. 61]):

```
uniform:=proc(r::constant..constant)
  proc()
    local intrange,f;
    intrange:=map(x->round(x*10^Digits),evalf(r));
```

```
    f:=rand(intrange);
    evalf(f())/10^Digits;
  end proc:
end proc:
```

Then to generate a random number between 0 and 2 , we simply make the command

```
uniform(0..2*Pi)():
```

Using this generator to produce random angles   [0, 2], do the following parts of this project. In each case produce graphical output like that in the sample projects.

1. **(A Basic Walk with Random Directions)** Write a procedure to simulate a random walk of length N where the direction is random. Each step, however, has unit length.

2. **(A Random Direction Walk in a Vertical Strip)** Adjust the procedure in part 1, so that the walk is confined to the region between the two vertical lines $x = -C$ and $x = C$. We assume that $C > 1$. Use the following re"ection rule when the walker hits the boundary.

   Re ection Rule: Suppose the walker is at location $P = (P_1, P_2)$ inside the strip at the $(k - 1)$st step, but the next step would be to a position $Q = (Q_1, Q_2)$ that is either on the boundary or beyond the boundary. Thus, $-C < P_1 < C$ and $|Q_1|$ C. We divide the contemplated step from P to Q into two steps: (1) an actual step (the kth step) to the boundary point P', where the segment $\overline{PQ}$ intersects the boundary (i.e., one of the lines $x = \pm C$), and then (2) a step (the $(k + 1)$st) of one unit to a point Q' inside the strip, such that $\_PP'Q' = \quad - 2$. Here,   is the angle that $\overline{PP'}$ makes with the vertical boundary line. See Figure 9.27.

3. **(Random Direction Walk in a Disk)** Write a procedure to simulate a random walk of length N where the direction is random and the walker is confined to a disk of radius $r > 1$ centered at the origin. Use the re"ection rule in part 2 above, except now the vertical line forming the boundary there is replaced by a tangent line to the circle here. Specifically, let P' be the point where the walker hits or would cross the boundary and   the angle of incidence of the walker's path with the tangent line to the circle at P'. The walker takes a step (the

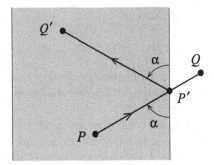

Figure 9.27: *Re ection rule for Problem 2.*

kth step) to P' and then takes a step (the $(k + 1)$st) back into the circle on a path which forms an angle   with the tangent line at P'. See Figure 9.28.

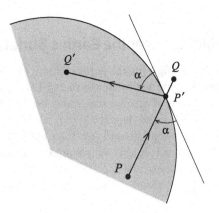

Figure 9.28: *Re ection rule for Problem 3.*

9.4 Projects on Newton s Second Law

The projects in this section deal with the motion of one or more bodies governed by Newton's 2nd Law from physics. This law says the change in momentum for each body is due to the force acting on the body. Thus, if a body has mass m (which we assume does not vary in time) and velocity v, then Newton's 2nd Law has the simple form

$$m\frac{dv}{dt} = F.$$

Since the acceleration of the body is the rate of change of velocity, this law is also interpreted as saying that the mass times the acceleration of the body is equal to the force acting on the body.

Despite the simple form of this law, the complexity and predictive power of it become apparent only when applied to study specific motions of one or more bodies. Typically the forces are known and the motions are to be determined from the 2nd Law. Thus, the 2nd Law is really a differential equation, or system of differential equations, when there are several bodies. To solve these differential equations, and thus determine the motion of the bodies, is usually impossible to do exactly. However, one can always resort to approximate numerical methods to determine the motion.

The projects here illustrate particular cases of motions governed by Newton's 2nd Law and how to write some simple code to approximate numerically the actual motions.

9.4.1   Sample Project: Motion Near the Earth s Surface

Motion of any body near the earth's surface (or the surface of any planet) is a particularly simple case of motion governed by Newton's 2nd Law. The simplifying assumption is that the force of gravity accelerates each body uniformly in a vertical direction with force of magnitude mg, where m is the mass of the body and $g = 32$ feet/second/second (for the earth). You probably studied such motions in Calculus I. In this sample project, we also incorporate an additional force, air resistance, acting on the body during its motion. This force is assumed to be proportional to the velocity of the body and directed in the opposite direction.

Problem: Write a program to numerically compute the position of a body moving near the earth's surface and subject only to the forces of gravity and air resistance. In addition to static plots of the successive positions, also write the code to animate the motion. Analyze the effect of air resistance on the motion and, for a given air resistance, experimentally determine how to project an object with given initial speed so that it travels the greatest horizontal distance.

Analysis of the Problem: The body moves in a vertical plane, which we take to be the x-y plane with origin located at the surface of the earth and y-axis vertical. See Figure 9.29. Relative to this coordinate system,

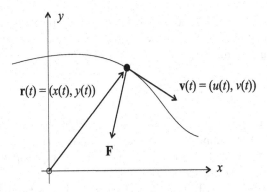

Figure 9.29: *Position and velocity of a body moving in a vertical plane near the earth s surface.*

the position vector $\mathbf{r}$ and velocity vector $\mathbf{v}$ for the body have coordinate expressions

$$\begin{aligned} \mathbf{r}(t) &= (x(t), y(t)) \\ \mathbf{v}(t) &= (u(t), v(t)) = (x'(t), y'(t)), \end{aligned}$$

and the force $\mathbf{F}$ acting on the body has expression

$$\mathbf{F}(t) = -mg\mathbf{j} - c\mathbf{v}(t) = (-cu(t), -mg - cv(t)).$$

With this notation, we can express the 2nd Law, $m\mathbf{v}' = \mathbf{F}$, in component form by the following pair of differential equations

$$mu' = -cu \tag{9.43}$$
$$mv' = -mg - cv. \tag{9.44}$$

When $c = 0$ (i.e., there is no air resistance) you can determine u and v by direct integration and then x and y can be determined from

$$x' = u \tag{9.45}$$
$$y' = v, \tag{9.46}$$

by direct integration also. You probably did this in your calculus course. When $c \neq 0$, you can use techniques from a differential equations course to determine u, v, x, y as well. We will not do this here, but rather discuss a

numerical scheme, or method, to approximate the velocities and positions
at a specified sequence of times.

This numerical method is known as *Euler s method* and applies to all
types of differential equations. The method is based on the fact that the
time derivatives are limits of difference quotients; for example,

$$u'(t) = \lim_{t \to 0} \frac{u(t + \Delta t) - u(t)}{\Delta t}.$$

Thus, for Δt small, the time derivative $u'(t)$ is approximated by the quotient
$(u(t + \Delta t) - u(t))/\Delta t$. If we suppose that the initial positions and velocities are known, say at time $t = 0$, then we can calculate the approximate
positions and velocities at other times as follows.

We divide a given time interval $[0, T]$, into K equal subintervals, each
of duration $\Delta t = T/K$, giving a sequence of times $t_n = n\Delta t$, $n = 0, \ldots, K$.
The initial velocities and positions are given

$$u(0) = u^0, \; v(0) = v^0, \; x(0) = x^0, \; y(0) = y^0,$$

and we denote the velocities and positions at the nth time t_n by

$$u^n \quad u(t_n), \; v^n \quad v(t_n), \; x^n \quad x(t_n), \; y^n \quad y(t_n),$$

for $n = 1, \ldots, K$. Next, in equations (9.43) (9.44), we replace the derivatives by the respective difference quotients and the right-hand sides by the
velocities at time t_n and get

$$\frac{u^{n+1} - u^n}{\Delta t} = -\frac{c}{m} u^n \tag{9.47}$$

$$\frac{v^{n+1} - v^n}{\Delta t} = -g - \frac{c}{m} v^n \tag{9.48}$$

We do a similar thing in equations (9.45) (9.46), except now for technical
reasons we replace the right-hand sides by the velocities at time t_{n+1}, and
get

$$\frac{x^{n+1} - x^n}{\Delta t} = u^{n+1} \tag{9.49}$$

$$\frac{y^{n+1} - y^n}{\Delta t} = v^{n+1}. \tag{9.50}$$

Finally by rearranging equations (9.47) (9.48) and equations (9.49) (9.50),
we arrive at the desired equations:

Euler s Scheme for the Problem:

$$u^{n+1} = (1 - \frac{c}{m}\Delta t)u^n \tag{9.51}$$

$$v^{n+1} = -g\Delta t + (1 - \frac{c}{m}\Delta t)v^n \tag{9.52}$$

$$x^{n+1} = x^n + u^{n+1}\Delta t \tag{9.53}$$

$$y^{n+1} = y^n + v^{n+1}\Delta t \tag{9.54}$$

The scheme allows us to determine the approximate velocities and positions at time t_{n+1} from those at time t_n. Starting with $n = 0$, when the velocities and positions are given, and iteratively using the scheme, we can get the approximations to the velocities and positions at all the times $t_1, t_2, \ldots, t_K = T$. Note that we must first compute u^{n+1}, v^{n+1} from equations (9.51) (9.52) before x^{n+1}, y^{n+1} can be computed from equations (9.53) (9.54).

Design of the Code: We write a procedure that takes as input the mass m of the body, the coe cient c of air resistance, the initial positions and velocities x^0, y^0, u^0, v^0, the time step size Δt, and the number of time steps K. The resulting output will be a list

$$[[x^0, y^0], [x^1, y^1], \ldots, [x^K, y^K]],$$

of the approximate positions at each of the discrete times $t_n = n\Delta t$. The procedure is a simple translation of the Euler Scheme (9.51) (9.54) into a Maple do loop:

Code for Gravitational Motion with Air Resistance

```
gravmot:=proc(m,c,x0,y0,u0,v0,dt,K,traj::evaln)
  local X,Y,U,V,g,n;
  X:=array(0..K);Y:=array(0..K);U:=array(0..K);V:=array(0..K);
  X[0]:=x0;Y[0]:=y0;U[0]:=u0;V[0]:=v0;
  g:=32.0;
  for n from 0 to K-1 do
    U[n+1]:=U[n]*(1-c*dt/m);
    V[n+1]:=V[n]*(1-c*dt/m)-g*dt;
    X[n+1]:= X[n]+U[n+1]*dt;
    Y[n+1]:= Y[n]+V[n+1]*dt;
  end do:
  traj:=[seq([X[n],Y[n]],n=0..K)];
  RETURN();
end proc:
```

We can now apply the code to study an old subject: *ballistics*, which is the study of the dynamics of projectiles. Galileo gave this subject a definitive treatment in the 1500s with the primary application being the motion of cannon balls.

Thus, suppose we fire a 16-lb cannon ball from ground level with initial velocity $\mathbf{v} = (50, 50)$. The mass is $m = 16/32 = 0.5$, and we consider the trajectories for motion without air resistance, $c = 0$, and with air resistance, $c = 0.2$. The following code computes the trajectories and displays them in a common figure.

```
gravmot(0.5,0,0,0,50,50,.01,320,path1);
gravmot(0.5,0.2,0,0,50,50,.01,320,path2);
plot({path1,path2});
```

See Figure 9.30. While both trajectories of the cannon ball appear to be

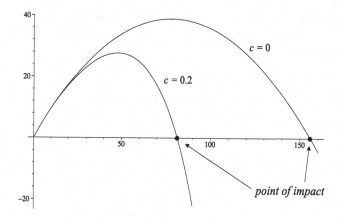

Figure 9.30: *Trajectories of a 16-pound cannon ball fired from ground level* $\mathbf{r} = (0, 0)$ *with velocity* $\mathbf{v} = (50, 50)$. *The top curve is the path taken when there is no air resistance,* $c = 0$, *while the bottom curve is the path taken when there is air resistance,* $c = 0.2$.

parabolas, one can show that only for the case of no air resistance is the trajectory a parabola (exercise).

We can produce an animation of the "ight of the cannon ball with no air resistance by using the following code. There are 320 frames FR[j] in the animation, with the jth frame consisting of the plot of the points $\{(x^n, y^n)\}_{n=1}^j$, which are the positions at all the times up to time t_j. This would be a lot of information to animate, so when we do the animation we only include every tenth frame in the movie. For convenience we use a 0th frame.

```
with(plots,animate):
K:=320:
fq:=10:
FR:=array(0..K):
FR[0]:=CURVES([path1[1]]):
for j to K do
  FR[j]:=CURVES([seq(path1[n],n=1..j)])
end do:
PLOT(ANIMATE(seq([FR[fq*i]],i=0..K/fq)),STYLE(POINT));
```

A standard topic in ballistics is to determine, for a given air resistance, how the cannon should be aimed so that the cannon ball goes the farthest, i.e. greatest horizontal distance. Aiming amounts to adjusting the angle that the barrel makes with the ground. In the example above the initial velocity was arbitrarily taken as $\mathbf{v} = (50, 50)$, which ammonts to aiming at an angle of $/4$. The speed, or muzzle velocity, with which the cannon ball leaves the barrel is s $|\mathbf{v}| = \overline{50^2 + 50^2} = 50 \overline{2}$. If we wish to fire the cannon at any other given angle , with the same muzzle velocity $s = 50 \overline{2}$, then the initial velocity should be taken to be $\mathbf{v} = (50 \overline{2} \cos , 50 \overline{2} \sin)$. Thus, mathematically, the question is how to choose so as to maximize the distance, for a given value of c.

For no air resistance it is not too hard to show that $= /4$ will maximize the distance (exercise). However, in the case of air resistance $(c > 0)$, the determination of the maximizing angle takes more work (exercise). Figure 9.31 shows the trajectories of a cannon ball fired with muzzle velocity $s = 50 \overline{2}$, in air with resistance $c = 0.2$, and angles $= (\frac{1}{2} - \frac{n}{20})$, for $n = 1, \ldots, 9$.

Thus, empircally it appears that $= 36$ approximates that angle that gives the greatest distance for this value of c. The material on the CD-ROM discusses this and other aspects of ballistics further.

9.4.2 Project: 1-Body Planar Motion

The sample project above is a special case of the general problem of motion of a body in the x-y plane subject to a given force $\mathbf{F} = (f, g)$, where the components f, g can depend on the time, the positions, and the velocities.

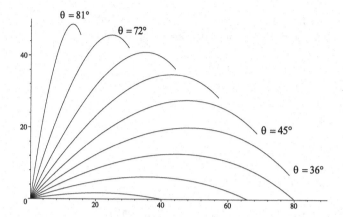

Figure 9.31: *The trajectories of a cannon ball fired at various angles with air resistance coefficient $c = 0.2$.*

Newton's 2nd Law for this general case is

$$mu' = f(t, x, y, u, v) \tag{9.55}$$
$$mv' = g(t, x, y, u, v) \tag{9.56}$$

An interesting project is to extend the code for the Euler method in Sample Project 9.4.1 to this more general situation. Specifically:

(1) Alter the Euler code to incorporate the functions f and g as input to the procedure. *Note*: Here g no longer stands for the constant acceleration of gravity near the earth's surface, but rather the second component of the given force **F**.

(2) **Central Force Problem:** Consider the special case where the force **F** is an attraction toward the origin with magnitude that only depends on the distance $r = \sqrt{x^2 + y^2}$ from the origin.

  (a) Show that the components of the central force have the form

$$f(x, y) = -H(r)\frac{x}{r}$$
$$g(x, y) = H(r)\frac{y}{r},$$

  where $H(r)$ is a function of $r = \sqrt{x^2 + y^2}$.

(b) **(Inverse Square Law)** For $H(r) = r^{-2}$ the magnitude of attraction is called an inverse square law. Study the motions of a body of mass $m = 1$, initial position $x(0) = 1, y(0) = 0$ and initial velocities (i) $u(0) = -0.5, v(0) = 0.5$, (ii) $u(0) = -1, v(0) = 1$, (iii) $u(0) = -1, v(0) = 1.25$.

(c) **(Inverse Cube Law)** For $H(r) = r^{-3}$ the magnitude of attraction is called an inverse cube law. Study the motions of a body of mass $m = 1$, initial position $x(0) = 1, y(0) = 0$ and initial velocities (i) $u(0) = 0, v(0) = 1$, (ii) $u(0) = -0.1, v(0) = 1$.

(d) For the studies in parts (b) and (c) produce static pictures of the trajectory for each motion and two types of moving pictures (animations). In the first animation each frame should contain only the point representing the position of the body at that time (and axes, labels, etc.). This movie will exhibit how the motion would actually appear. In the second animation, each frame should contain all the positions of the body from time 0 up to the time corresponding to the frame. This movie will show the motion traced out leaving a track.

9.4.3 Project: N-Body Planar Motion

Write a program to calculate the approximate motion of a system of N bodies using the Euler numerical method. To keep things simple, you may assume that the motion is planar and takes place in the x-y plane. Then the forces acting on the bodies must be planar as well and, in general, the force on each body can depend on the positions and velocities of all the other bodies (but, for simplicity, assume it does not depend on the time t).

Physics and Math Background: The motion of the system is governed by Newton's 2nd Law, which in mathematical form is a system of differential equations. Denote the positions of the bodies at time t by $\mathbf{r}_1(t), \mathbf{r}_2(t), \ldots, \mathbf{r}_N(t)$. These are vector-valued functions, which in terms of coordinates (for the planar case) are

$$\mathbf{r}_i(t) = (x_i(t), y_i(t)) \qquad \text{(for } i = 1, \ldots, N).$$

The first derivatives of these vector-valued functions are the velocity functions and, with the notation $\mathbf{v}_i = \mathbf{r}_i$, the velocity of the ith body at time t is

$$\mathbf{v}_i(t) = (u_i(t), v_i(t)) = (x_i'(t), y_i'(t)).$$

For convenience of notation let $\mathbf{r}(t) = (\mathbf{r}_1(t), \ldots, \mathbf{r}_N(t))$ denote the vector consisting of all the positions and similarly $\mathbf{v}(t) = (\mathbf{v}_1(t), \ldots, \mathbf{v}_N(t))$ will denote the vector of all the velocities. You can think of $\mathbf{r}(t), \mathbf{v}(t)$ as giving the *state* of the system at time t (all the positions and all the velocities). The positions $\mathbf{r}_i$ and velocities $\mathbf{v}_i$ of the system are required to satisfy the following system of differential equations:

$$m_i \mathbf{v}_i' = F_i(\mathbf{r}, \mathbf{v}) \qquad (9.57)$$
$$\mathbf{r}_i' = \mathbf{v}_i, \qquad (9.58)$$

for $i = 1, \ldots, N$. The first equation says that the mass m_i times the change in velocity $\mathbf{v}_i$ of the ith body must be equal to the total force F_i acting on it. This is Newton's 2nd Law. From the theory of differential equations, we know that if initial positions and velocities $(\mathbf{r}(0), \mathbf{v}(0))$ are given, then the system (9.57) (9.58) of differential equations has a unique solution $(\mathbf{r}(t), \mathbf{v}(t))$ for t in an interval $[0, T]$. This is a theoretical result and relies on reasonable assumptions about the system of given forces F_i, $i = 1, \ldots, N$. Only in rare cases can we actually solve the system of DEs to obtain explicit solutions. However, as this project shows, it is fairly easy to obtain *approximate* solutions using numerical methods.

Euler s Numerical Scheme: To replace the system (9.57) (9.58) by a discrete, finite difference system, we proceed as in Sample Project 9.4.1. Divide the time interval $[0, T]$ into K equal subintervals, each of length $\Delta t = T/K$.

The Euler scheme generates a discrete sequence

$$\mathbf{v}_i^n = (u_i^n, v_i^n)$$
$$\mathbf{r}_i^n = (x_i^n, y_i^n),$$

$n = 1, \ldots, K$, for the ith body at the times $t_n = 0, \Delta t, 2\Delta t, \ldots, K\Delta t = T$. Thus, the approximate positions of the ith body are

$$\mathbf{r}_i^1, \mathbf{r}_i^2, \ldots, \mathbf{r}_i^K,$$

and the approximate velocities of the ith body at these positions are

$$\mathbf{v}_i^1, \mathbf{v}_i^2, \ldots, \mathbf{v}_i^K.$$

As above, the equations for the Euler scheme come from replacing the derivatives $\mathbf{v}_i'$ and $\mathbf{r}_i'$ at time $t_n = n\Delta t$ by the approximating finite differences:

$$\mathbf{v}_i' \quad \frac{\mathbf{v}_i^{n+1} - \mathbf{v}_i^n}{\Delta t}, \qquad \mathbf{r}_i' \quad \frac{\mathbf{r}_i^{n+1} - \mathbf{r}_i^n}{\Delta t}.$$

Then the discrete analog of the system of DEs (9.57) (9.58) is the following system of *di erence* equations

$$m_i \, \frac{\mathbf{v}_i^{n+1} - \mathbf{v}_i^n}{\Delta t} \;=\; F_i(\mathbf{r}^n, \mathbf{v}^n) \tag{9.59}$$

$$\frac{\mathbf{r}_i^{n+1} - \mathbf{r}_i^n}{\Delta t} \;=\; \mathbf{v}_i^{n+1} \tag{9.60}$$

Here $\mathbf{r}^n = (\mathbf{r}_1^n, \dots, \mathbf{r}_N^n)$ is the vector of the approximate positions of the N bodies at time $n\Delta t$ and $\mathbf{v}^n = (\mathbf{v}_1^n, \dots, \mathbf{v}_N^n)$ is the vector of approximate velocities. A *solution* of the finite difference system is a sequence $\{(\mathbf{r}^n, \mathbf{v}^n)\}_{n=0}^K$ of position and velocity vectors. Given the initial position and velocity vectors $(\mathbf{r}^0, \mathbf{v}^0)$ for all the bodies, it is easy to solve the system (9.59) (9.60) of difference equations. This is so because it can be rewritten in the following more convenient form:

$$\mathbf{v}_i^{n+1} \;=\; \mathbf{v}_i^n + \frac{\Delta t}{m_i} F_i(\mathbf{r}^n, \mathbf{v}^n) \tag{9.61}$$

$$\mathbf{r}_i^{n+1} \;=\; \mathbf{r}_i^n + \Delta t \mathbf{v}_i^{n+1}, \tag{9.62}$$

for $i = 1, \dots, N$. This explicitly gives the algorithm for determining the approximate positions and velocities at each of the discrete time steps. Knowing $\mathbf{v}^n, \mathbf{r}^n$ at the nth time step, equations (9.61) (9.62) allow us to compute $\mathbf{v}^{n+1}, \mathbf{r}^{n+1}$ at the next time step. Note that $\mathbf{v}^{n+1}$ must be computed first in (9.61), so that its value can be used in computing $\mathbf{r}^{n+1}$ in (9.62). Thus, from the initial positions and velocities, we can generate a solution $\{(\mathbf{r}^n, \mathbf{v}^n)\}_{n=0}^K$ of the finite difference system.

Programming Speci cs: Write programs in Maple to implement the Euler method described above. Specifically:

(1) Write a procedure for the important special case where the forces on the ith body arise from it being attracted to each of the other bodies by forces whose magnitudes are proportional to the reciprocal of the square of the distances between the ith body and all the others. That is,

$$F_i(\mathbf{r}) = \sum_{j=i} \frac{G}{r_{ij}^3} (\mathbf{r}_j - \mathbf{r}_i),$$

where $r_{ij} = |\mathbf{r}_i - \mathbf{r}_j| = \sqrt{(x_i - x_j)^2 + (y_i - y_j)^2}$, is the distance between the ith and jth bodies. *Note:* The notation $\sum_{j=i}$ denotes summation on $j = 1, \dots, N$, with $j = i$ omitted (the i body is not attracted to itself). The constant G is the gravitational constant. Have it be a global variable in your code.

(2) Write Maple code to produce the following:

(a) A static picture of the trajectories for each body. Each trajectory should be plotted just as the sequence of computed points representing the positions at the discrete times. Plot all trajectories in the same picture.

(b) Three moving pictures. In the first of these, each frame should display the points representing the positions of the bodies at the corresponding discrete instant in time. This moving picture will be how the actual motion would appear. In the second, each frame should display, for each body, *all* the points representing the positions of the body from time 0 up to the time corresponding to that frame. If the points on each trajectory are close enough together, this movie would show how the motion would appear if each body left a track in the plane as it moved. In the third movie, show the motion of the line joining the bodies and the track made by the center of mass of the system of bodies.

Do parts (a) and (b) just for $N = 2$ and $N = 3$ bodies and some interesting choice for the masses and the initial positions and velocities.

Appendix A

Maple Reference

There are numerous manuals and tutorials on Maple (cf. [BM], [Hec], [HHR], [Mon1], [Mon2], [Mon3], [Red]), but as an aid to mastering Maple, or at least what is needed to work the exercises in this text, this appendix provides an abbreviated reference manual. In addition there are many extra features and tips about using Maple presented on the CD-ROM accompanying this text. Anyone with extra time to devote to learning more about Maple, will well benefit by working through the material there.

The interface with Maple, i.e., the way in which one enters commands to and receives output from Maple has been in constant "ux (and improvement) over the various releases of the software. This will be the case in the future as well, and so no attempt will be made to describe the interface here. The various features of Maple presented below are common to Maple V, Release 5, Maple 6, and Maple 7, with some minor differences between releases. The main major difference is that Maple 6 and Maple 7 use **end if**, **end do**, and **end proc** to terminate conditional **if** statements, **do** loops, and procedure statements, respectively. Maple V uses **fi**, **od**, and **end** for this.

A.1 Expressions and Functions

The symbols for the basic arithmetical operation are shown in Table A.1 and algebraic expressions are built from these in a way which is common to many programming languages. Thus, for example, the expression:

$$\frac{a[(x^2 + y^2)^3 - 5y]}{x^{-1} + y^{1/3}},$$

can be written in Maple code as:

```
a*((x^2+y^2)^3-5*y)/(x^(-1)+y^(1/3))
```

symbol	operation	example
+	addition	a+b
-	subtraction	a-b
*	multiplication	a*b
/	division	a/b
^	exponentiation	a^b

Table A.1: *Symbols for the arithmetical operators.*

This is known as a Maple *expression* and the symbols a,x,y have no values or data types associated with them. The whole expression is built from simpler expressions using the arithmetical operators.

Note: grouping is accomplished in Maple by using parentheses: ( ). Brackets: [ ] and braces: { } are used for other purposes in Maple and are *not* permitted for grouping in arithmetical expressions. See Table A.2.

symbol	function	example
( )	grouping	m:=(a+b)*(c+d)
[ ]	selection	for n to 20 do A[n]:=n^2 end do
[ ]	list delimiter	L:=[a,a,b,b,b]
{ }	set delimiter	S:={a,b,c}

Table A.2: *Parentheses, brackets, and braces and their functions.*

The operators in Table A.3 allow you to manipulate expressions and build more complicated expressions. The operators =,<,>,+<,>=,<> are called comparison operators and are used primarily in logic conditions as shown in the examples in the table. The equal symbol = is also often used in building equation expressions such as eq1:=5x^3=x*(x+y)^2, to be used in solve commands or other commands, for example: solve(eq1,x).

The assignment operator := allows you to assign a name to an expression, as, for example,

    r := a*((x^2+y^2)^3-5*y)/(x^(-1)+y^(1/3));

You can now use r as a shorthand way of referring to the expression on the right side of the above assignment statement.

The arrow operator -> has the same meaning and use as   (or sometimes) does in mathematics. Namely, in mathematics we usually indicate (or

symbol	meaning	example
=	equal	`if x=y then`
<	less than	`if x<y then`
<=	less than or equal	`while n<=100 do`
>	greater than	`while x>y do`
>=	greater than or equal	`if x >= y then`
<>	not equal	`while n <> 0 do`
:=	assignment operator	`a:=1`
->	arrow operator	`f:=x->5*x+x^2`

Table A.3: *The assignment, arrow, and comparison operators.*

define) a function of x by something like: $x \quad 1/(1+x^2)$, and we often name this function something like f. The definition of such a function in Maple is accomplished by the command:

```
f := x -> 1/(1+x^2);
```

Maple also uses the commonly accepted functional notation, i.e., `f(1)` denotes the value of the function `f` at $x = 1$. Note that, by contrast, the above assignment for `r` does *not* make `r` a function of `x` and `y`. Thus the evaluation `r(x,y)` makes no sense. To properly define a real-valued function of two variables, you should use the arrow operator (or a procedure statement as discussed later). Thus, for example,

```
f := (x,y) ->  a*((x^2+y^2)^3-5*y)/(x^(-1)+y^(1/3));
```

defines such a function, which in this case also contains an unspecified parameter `a`. Various manipulations with `f` (such as plotting its graph) will require that the parameter `a` be assigned a value first.

Vector-valued functions of one or more variables require a slightly different syntax in their definition. For example the statement

```
g := (x,y) -> [x-x^3*y, x^2*y^4, 1-x^2-y^2];
```

defines a vector-valued function $g : \mathbb{R}^2 \quad \mathbb{R}^3$. Maple's notation for a vector (in 3-D) is `[a,b,c]`.

Maple has all the standard mathematical functions built into its library and in most cases uses the usual notation for these functions. Table A.4 shows some of the functions in Maple's library. Not withstanding the standard functional notation, it has become customary to omit the parentheses

symbol	meaning
`exp`	the natural exponential function
`ln`	the natural log function
`sin, cos, tan, cot, sec, csc`	the trig functions
`arcsin, arccos, arctan`	the inverse trig functions
`sinh, cosh, tanh, coth, sech, csch`	the hyperbolic functions
`sqrt`	the square root function
`abs`	the absolute value function

Table A.4: *Some of Maple s library functions.*

in certain cases when using trig, hyperbolic, and log functions (as we have done throughout the text). Thus, for example, one writes $\sin x$ and $\ln x$ for the more correct expressions $\sin(x)$ and $\ln(x)$. Beware: In Maple this is *not* permitted. For example, you must use `sin(x)` for the value of the sine function at x. If you happen to use `sin x` in a Maple expression, it will not be correctly interpreted (Maple thinks it's a new name for something else).

The comparison operators in Table A.3 are used to build logical expressions which can be either true or false (depending on the values of x and y, in the examples shown).

A.2   Plotting and Visualization

Maple has many powerful features for plotting and constructing complicated graphical images in two and three dimensions. Table A.5 shows just a few of the basic commands which will su ce for doing the scientific visualization in this text. All of these commands, except `plot` and `plot3d`, are part of the `plots` package.

`plot(`*expression, ranges, options*`)`

This command can be used to plot the graph of a real-valued function: $f : [a, b]$ $\mathbb{R}$ or of a vector-valued function: $g : [a, b]$ $\mathbb{R}^2$ of a single variable x $[a, b]$. The latter is often called a curve or a parametrization for a curve in the plane and, mathematically, the plot will be of the set of points $g[a, b]$, often called the trace of the curve. The former is a special case of the latter, since f determines a vector-valued function g, via: $g(x) = (x, f(x))$. *Note*: For the sake of explanation here, we use x for the independent variable name. You can use any name you wish (like s or `theta`)

command	function
plot	plots the graph of a function of a single variable
plot3d	plots the graph of a function of two variables
implicitplot	plots the graph of a curve given implicitly by an equation in two variables
implicitplot3d	plots the graph of a surface given implicitly by an equation in three variables
spacecurve	plots a curve in $\mathbb{R}^3$ which is given parametrically
contourplot	plots level curves for a function of two variables
display	displays several 2-D plots simultaneously
display3d	displays several 3-D plots simultaneously
animate	simulates the motion of the graph of $u(x,t)$
animate3d	simulates the motion of the graph of $u(x,y,t)$

Table A.5: *A few commands for graphing and visualizing various structures.*

in applications. In the syntax for this command, expr can be either (i) an expression involving x, or (ii) something like f(x), where f is the name for a real-valued function of a real variable. In either case there must be no other variables or parameters in expr. The ranges consists of two ranges, one like x=a..b for the independent variable and one like c..d or y=c..d for the dependent variable. The latter range is optional but convenient when f has vertical asymptotes in its graph. The options enable you to control the way the graph looks and consists of a list of items such as color=black and numpoints=201 with commas separating the items in the list. Each option has a default value which is used if you do not specify that option. For example, numpoints has default value 49, which means that the plot is executed by evaluating expr at a *minimum* of 49 equally spaced points $\{x_n\}_{n=1}^{49}$ in the interval $[a,b]$, with $x_1 = a$ and $x_{49} = b$. *Note*: Maple uses an adaptive method for this type of plot, so it may actually use more points than specified in numpoints.

You can consult the help menu to see what the other options are or you can select most of them from the plot window after executing a basic plot with no options.

The plot command works differently (and a little unexpectedly) for parametrized curves in the plane: $g(x) = (h(x), k(x))$, x $[a,b]$, where h and k are the component functions. The proper syntax for the plot is:

```
plot([h(x),k(x),x=a..b])
```

Multiple plots in the same picture can be achieved by using a list of valid expressions: {expr1,expr2,...,exprn} in place of the single expression expr. The ranges must be the same for each of these expressions.

---
plot3d(*expression, ranges, options*)

This command can be used to plot the graph of a real-valued function f : $[a,b] \times [c,d]$ $\mathbb{R}$, or a vector-valued function $g : [a,b] \times [c,d]$ $\mathbb{R}^3$ of two variables (x,y) $[a,b] \times [c,d]$. The former is a special case of the latter, since f determines a vector-valued function g, via: $g(x,y) = (x,y,f(x,y))$. In the plot command syntax, expr can be either (i) a single expression involving two independent variables, say, x and y, or (ii) something like f(x,y), where f is a name for a real-valued function of two variables, or (iii) an expression list of the form [h(x,y),k(x,y),l(x,y)], where h,k,l are names for real-valued functions of two variables, or (iv) something like g(x,y), where g is a name for a function of two variables with values in $\mathbb{R}^3$. The ranges part of the plot3d command has the form: x=a..b,y=c..d (there is no optional z range). An example of one of the options in the plot3d command is grid=[51,61], which causes the rectangle $[a,b] \times [c,d]$ to be divided into $50 \times 60 = 3000$ equal subrectangles by choosing 51 equally spaced points x_n in $[a,b]$ (starting at a and ending at b) and 61 equally spaced points y_m in $[c,d]$. The plot is executed by evaluating expr at all the points $(x_n,y_m), n = 1,\ldots,51, m = 1,\ldots,61$. The default value is grid=[25,25]. Other possible options can be explored by consulting the help menu, or can be selected from the plot window once a basic plot is executed. We find it more convenient to do the latter, changing the style, selecting color and lighting, adding the axes, and selecting the view and perspective, after doing a basic plot with no options.

Multiple plots in the same picture can be achieved by using a list of valid expressions: {expr1,expr2,...,exprn} in place of the single expression expr. The ranges must be the same for each of these expressions.

---
implicitplot(*equations, ranges, options*)

This command is used to plot one or more equations of the form $f(x,y) = c$, where c is a constant and f is a real-valued function of two variables. For each equation, the plot can be thought of as: (a) the graph of a curve in the plane, or (b) the graph of a level curve for the function f, or (c) the graphs of the various functions y of x defined implicitly by $f(x,y) = c$. In the syntax for this command, **equations** is either a single equation or a set of equations, e.g.: {x^2+y^{2}=1,x-y=1}. The designation **ranges** has the form x=a..b,y=c..d, which represents the viewing rectangle $[a, b] \times [c, d]$ for the plot. One commonly used option is **grid**, which allows you to obtain a finer resolution in the plot. The default is **grid=[25,25]**.

---
implicitplot3d(*equations, ranges, options*)

This command is used to plot one or more equations of the form $f(x, y, z) = c$, where c is a constant and f is a real-valued function of three variables. The plot of a single such equation can be thought of as (a) the graph of a surface in 3-space, or (b) the graph of a level surface for the function f, or (c) the graphs of the various functions z of (x, y) defined implicitly by $f(x, y, z) = c$. In the syntax for this command, **equations** is either a single equation or a set of equations, e.g.: {x^2+y^{2}+z^{2}=1,x-y+z=1}. The designation **ranges** has the form x=a..b,y=c..d,z=m..n, which represents the viewing box $[a, b] \times [c, d] \times [m, n]$ for the plot. One commonly used option is **grid**, which allows you to obtain a finer resolution in the plot. The default is **grid=[10,10,10]**.

---
spacecurve(*expression, range, options*)

This command is used to plot the trace of a vector-valued function g : $[a, b]$ $\mathbb{R}^3$, which is often called a curve or a parametrized curve in 3-space. In the syntax, **expr** can be either (i) an expression list of the form: [h(x),k(x),l(x)], where h,k,l are real-valued functions of a single variable x, or (ii) g(x), where g is a function of a single variable with values in $\mathbb{R}^3$ (a vector-valued function). Some useful options are **color=black** and **numpoints=200**. The default value for **numpoints** is 50, and, unlike the use of this option in the **plot** command, the number of points specified is the actual number of points used. Thus **numpoints=3** produces a plot of three points connected by two straight line segments.

---
contourplot(*expression, ranges, options*)

This command is *not* very useful since you can accomplish the same thing with **plot3d**. It is mentioned here because contour plots are important. The command is used to plot a sequence of level curves: $f(x, y) = c_n, n = 1, \ldots, N$ for a real-valued function of two variables. Mathematically this is the same as slicing the graph of f (a surface) by a sequence of horizontal planes $z = c_n$ and projecting the resulting curves on the x-y plane. In the syntax, expr can be either (i) an expression involving two variables (say x,y) and numbers but no other variables or parameters, or (ii) an expression like f(x,y) where f is a real-valued function of two variables. ranges has the form x=a..b,y=c..d. As mentioned, you can get the same graphic as contourplot(expr,ranges) by using plot3d(expr,ranges), choosing: style=contour from the plot window, and then rotating the view to one looking down on the x-y plane.

display3d(*list*)

This command is used to display the results of several 3-d plot structures in the same picture. In the syntax, list has the form: {p1,p2,...,pn}, where each of the p's in the list is a (3-D) plot structure. A plot structure is a name to which has been assigned the output of one of Maple's plotting commands. For example, the commands

```
p1:=plot3d(x^2+y^2,x=-1..1,y=-1..1)
line:=spacecurve([t,t+1,t+2],t=-1..1,color=black)
```

create two plot structures named p1 and line. Then the command

```
display3d({p1,line})
```

will display the surface and the line in the same picture. *Note*: When creating a plot structure, you should end the assigment statement p1:=... with a colon before entering the command. This prevents the result from being written back to the monitor screen.

implicitplot3d(*list*)

This command works the same way as display3d, except the list contains 2-d plot structures.

animate($u(x,t)$,*ranges,options*)

In this command, u is a real-valued function of two variables and ranges has the form: x=a..b,t=c..d. What is displayed in the animation window is a succession of plots of: $u(\cdot,t_1),u(\cdot,t_2)\ldots,u(\cdot,t_n)$, as functions of x. Here $t_i,i=1,2,\ldots,n$ are equally spaced times in which $t_1=a$ and $t_n=b$. Also n is the number of frames (an option). The default is frames=8.

animate3d($u(x,y,t)$,*ranges,options*)

This command is entirely similar to animate except that now u depends on 3 variables.

A.3 Programming

Here is a list of the main programming constructs in Maple and a brief description of them.

```
              if C then S end if
```

This works pretty much as expected: If the condition C is true, then the statements S are executed. Otherwise the statements S are skipped and control passes to the next statement after **end if**.

```
           if C then S_1 else S_2 end if
```

Similar to the above: if the condition C is true, then the statements S_1 are executed. Otherwise the statements S_2 are executed. In either event control passes afterward to the next statement after **end if**.

```
        if C_1 then S_1 elif C_2 then S_2 end if
```

This statement is a combination of an **if-then-else** statement and an **if-then** statement. Specifically, it is equivalent to

$$\texttt{if } C_1 \texttt{ then } S_1 \texttt{ else if } C_2 \texttt{ then } S_2 \texttt{ end if end if}$$

In essence the **else** and **if** are contracted to **elif** and the extra **end if** at the end is omitted. This form is particularly useful when many different conditions are to be tested and the multiple nesting of **if-then-else** statements is confusing.

```
   for I from I_1 to I_2 by K do statements end do
```

This is the basic do loop in Maple and works as follows. I is the index that controls the loop and is successively assigned integer values in the range

$$I_1, I_1 + K, I_1 + 2K, \ldots, I_1 + nK, \ldots$$

I_1, I_2 are any integers. If $I_1 \quad I_2$, then the increment K must be positive and the looping continues until I is assigned a value that exceeds I_2. If $I_1 \quad I_2$, then K must be negative (so it's a decrement) and the looping continues until I is assigned a value that is less than I_2. Each time I is assigned a value and this value is in the correct range, all the Maple statements denoted by *statements* are executed and then the process is repeated with I being assigned the next value.

The optional items are `from` and `by`. When they are omitted, Maple assumes that $I_1 = 1$ and $K = 1$.

---

for I from I_1 to I_2 by K while C do *statements* end do

This works exactly like the above `for` statement except that the `while` clause causes the loop to terminate whenever the condition C is false.

---

while C do *statements* end do

This is an optional form of the last do loop and has the looping controlled entirely by the condition C.

---

for x in S while C do *statements* end do

This type of do loop, called a `for-in` loop in order to distinguish it from the `for-from` loops above, is useful when a sequence of actions must be performed on every element x in a set S. It is equivalent to the statement

for i from 1 to nops(S) while C do x:= S[i]; *statements* end do

Despite this equivalence, the `for-in` loop often conveys more clearly the nature of the algorithm.

F:=proc($I_1, \ldots, I_k, O_1, \ldots, O_m$)
   local $L_1, \ldots, L_r$;
   global $G_1, \ldots, G_s$;
   option O;
   description D;
   statement1; *statement2*; ... ; *statementN*
   end proc:

This is the general form for the procedure statement in Maple. It is the analog of a subroutine in FORTRAN and of a function in C. The procedure is *defined* by this statement, which uses formal input parameters $I_1, \ldots, I_k$ and formal output parameters $O_1, \ldots, O_m$. When the procedure is *called*, or invoked, by using $F(i_1, \ldots, i_k, o_1, \ldots, o_m)$, the statements: *statement1; statement2;...;statementN* are executed using the values $i_1, \ldots, i_k$ of the actual input parameters. The execution of the statements in the procedure results in values being assigned to the actual output parameters $o_1, \ldots, o_m$.

The precise functioning of procedures cannot be described brie"y, so you should see Chapter 5 for a detailed description of how they work.

A.4 Packages

A Maple *package* is a collection of commands to do specific tasks, such as plotting, solving differential equations, and linear algebra. A basic set of commands is always present whenever you start Maple, but to take advantage of the additional commands available in a package, the package has to be loaded using the `with` command. Here is the main package used in this book:

```
with(plots)
```

> *animate, animate3d, animatecurve, changecoords, complexplot, complexplot3d, conformal, contourplot, contourplot3d, coordplot, coordplot3d, cylinderplot, densityplot, display, display3d, fieldplot, fieldplot3d, gradplot, gradplot3d, implicitplot, implicitplot3d, inequal, listcontplot, listcontplot3d, listdensityplot, listplot, listplot3d, loglogplot, logplot, matrixplot, odeplot, pareto, pointplot, pointplot3d, polarplot, polygonplot, polygonplot3d, polyhedra-supported, polyhedraplot, replot, rootlocus, semilogplot, setoptions, setoptions3d, spacecurve, sparsematrixplot, sphereplot, surfdata, textplot, textplot3d, tubeplot*

The particular commands available in `plots` package are shown in the output above. To load all the commands in the package and have them available for use in a session use the `with` command:

```
with(plots);
```

Note: It is possible to load only a few commands from a package. For example,

```
with(plots,display,animate);
```

allows you to use the commands `display` and `animate` in a session without loading the whole `plots` package.

Bibliography

[Asp]   W. Aspray, Ed., *Computing Before Computers*, Iowa State University Press, Ames, 1990.

[Aug]   S. Augarten, *Bit by Bit, an Illustrated History of Computers*, Ticknor and Fields, New York, 1984.

[Ber]   H.C. Berg, *Random Walks in Biology*, Princeton University Press, Princeton, NJ, 1983.

[BG]   T.J. Bergin and R.G. Gibson, Eds., *History of Programming Languages II*, Addison-Wesley, Reading, MA, 1996.

[BM]   N.R. Blachman and M.J. Mossinghoff, *Maple V Quick Reference*, Brooks/Cole, Pacific Grove, CA, 1994.

[Bo]   K.P. Bogart, *Introductory Combinatorics*, Pitman, Boston, 1983.

[BW]   D.F. Brailsford and A. N. Walker, *Introductory Algol Programming*, Wiley, New York, 1979.

[Br]   R.A. Brualdi, *Introductory Combinatorics*, North-Holland, New York, 1977.

[Dug]   J. Dugundji, *Topology*, Allyn and Bacon, Boston, MA, 1966.

[FK]   M. Friedman and A. Kandel, *Fundamentals of Computer Numerical Analysis*, CRC Press, Boca Raton, 1994.

[GKW]   R.J. Gaylord, S.N. Kamin, and P.R. Wellin, *An Introduction to Programming with Mathematica*, Springer-TELOS, Santa Clara, CA, 1996.

[Hec]   A. Heck, *Introduction to Maple*, Springer-Verlag, New York, 1993.

[HHR] K.M. Heal, M.L. Hansen, and K.M. Rickard, *Maple V: Learning Guide*, Springer-Verlag, New York, 1996.

[Ho] C.A.R. Hoare, Quicksort, *Computer Journal*, **5** (1962).

[Hof] K. Hoffman, *Analysis in Euclidean Space*, Prentice-Hall, Englewood Cliffs, NJ, 1975.

[HZ] L.R. Hitt and X. Zhang, Dynamic Geometry of Polygons, *Elemente der Mathematik*, **56** (2001) 21-37.

[Lor] R. Lorentz, *Recursive Algorithms*, Ablex Publishing Corporation, Norwood, NJ, 1994.

[MH] J.E. Marsden and M.J. Hoffman, *Elementary Classical Analysis*, 2nd ed., W.H. Freeman, New York, 1993.

[MM] J. Mikusiński and P. Mikusiński, *An Introduction to Analysis, from Number to Integral*, Wiley, New York, 1993.

[Mon1] M.B. Monagan, K.O. Geddes, K.M. Heal, G. Labahn, and S. Vorkoetter, *Maple V: Programming Guide*, Springer-Verlag, New York, 1996.

[Mon2] M.B. Monagan, K.O. Geddes, K.M. Heal, G. Labahn, S. Vorkoetter, and J. McCarron, *Maple 6: Programming Guide*, Waterloo Maple Inc, Waterloo, 2000.

[Mon3] M.B. Monagan, K.O. Geddes, K.M. Heal, G. Labahn, S. Vorkoetter, J. McCarron, and P. DeMarco, *Maple 7: Programming Guide*, Waterloo Maple Inc., Waterloo, 2001.

[Pat] C. Wayne Patty, *Foundations of Topology*, PWS-Kent, Boston, 1993.

[Red] D. Redfern, *Maple Handbook: Maple V Release 3*, Springer-Verlag, New York, 1994.

[Ros1] S.M. Ross, *A First Course in Probability*, 4th ed., Macmillan, New York, 1994.

[Ros2] S. M. Ross, *Probability Models*, 7th ed., Harcourt/Academic, Burlington, MA, 2000.

[Rud] W. Rudin, *Principles of Mathematical Analysis*, McGraw-Hill, New York, 1976.

[Sch] M. Schramm, *Introduction to Real Analysis*, Prentice Hall, Upper Saddle River, NJ, 1996.

[Wex] R.L. Wexelblat, *History of Programming Languages* , Academic, New York, 1981.

[YG] D.M. Young and R.T. Gregory, *A Survey of Numerical Mathematics*, Vol. I, Addison-Wesley, Reading, 1972.

[Zac] J.L. Zachary, *Introduction to Scientific Programming: Computational Problem Solving using Maple and C*, Springer-Verlag, New York, 1996.

Index